MODERN

PHYSICS

AND

QUANTUM

MECHANICS

ELMER E. ANDERSON, Ph.D.

Chairman, Department of Physics,
Clarkson College of Technology
Potsdam, New York

W. B. SAUNDERS COMPANY · PHILADELPHIA · LONDON · TORONTO
1971

W. B. Saunders Company: West Washington Square
                          Philadelphia, Pa.   19105

                          12 Dyott Street
                          London, WC1A   1DB

                          1835 Yonge Street
                          Toronto 7, Ontario

Modern Physics and Quantum Mechanics   SBN 0-7216-1220-2

Print No.:   9   8   7   6   5   4   3   2   1

PREFACE

This book grew out of my experience in teaching a one-year course in quantum physics given to our seniors in the physics curriculum. It is designed to bridge the gap between the descriptive course at the sophomore level and a graduate course in quantum mechanics in which formal operator methods are used freely. I have written this text because I have been unable to find a book which suits our needs, although there are a number of excellent texts which are satisfactory for certain parts of the course.

Since many diverse areas of modern physics have developed simultaneously, a strictly chronological presentation would not necessarily be the most logical to follow. Therefore, I have chosen to develop the ideas that I consider necessary for an understanding of quantum physics without regard to historical order and with no attempt at complete coverage of all of twentieth century physics. The use of the word "Modern" in the title is not intended to suggest the same list of topics as is included in the sophomore course bearing that name. It is assumed, in fact, that the student has previously taken such a course at the sophomore level, as well as previous or concurrent courses in classical mechanics, electromagnetic waves, and thermodynamics. Much of the material of the first four chapters should not be new to the student but is included for the sake of logical completeness in the development of quantum mechanics.

Four great conceptual edifices form the principal themes for the first four chapters. These are: (1) the special theory of relativity and some examples of its role in contemporary physics; (2) the quantum theory of radiation and the concept of the photon as a wave packet which can display both particle and wave properties; (3) the Bohr theory of the atom and the quantization of angular momentum; and (4) the representation of a particle by a wave packet and the development of a wave mechanics for the description of a particle's dynamical states. In Chapter 5, Schrödinger's wave mechanics is applied to one-dimensional problems. In that chapter the linear harmonic oscillator is solved by both the series and the operator methods in order to familiarize the student with both of these approaches. This is followed in Chapter 6 by a summary of the postulates of quantum mechanics and the development of matrix mechanics. The angular momentum operators and their eigenfunctions are introduced in Chapter 7 where they are applied to the hydrogen atom. The complexities resulting from spin and its interactions with orbital angular momenta are treated in Chapter 8. The phenomenon of exchange degeneracy resulting from the symmetry properties of identical particles is also discussed in this chapter. Chapter 9 is an important chapter. Here some approximate methods are introduced which enable the reader to solve large classes of real problems by means of the solutions obtained for the ideal systems in earlier chapters. In Chapter 10, approximate methods are applied to some specific atomic problems, and Chapter 11 is an introduction to quantum mechanical scattering theory.

The book contains over 250 problems in addition to the numerical examples which are worked at appropriate points in the text. The importance of solving a large number of problems in order to achieve a mastery of the material cannot be overemphasized.

It is a pleasure to acknowledge the help of the following people and to thank them for their indispensable roles in the preparation of this book: my students over the past few years for their enthusiasm and inspiration; Agatha Hollister and Amelia Anderson for typing and preparing the manuscript; Rita Arajs for making all of the drawings; my colleagues, Professors Fred Otter and Sigurds Arajs, as well as Professors Henri Amar of Temple University, Paul H. Cutler of the Pennsylvania State University, Thomas B. W. Kirk of Harvard University, John Reading of Northeastern University, John G. Teasdale of San Diego State College, and an anonymous reviewer, for their most valuable critical comments; to John J. Hanley and others of the staff of the W. B. Saunders Co. for their assistance and encouragement; last of all, to my mother and to my family—Amy, Kenneth, Mark, Scott, Ruth, and Carl—to whom I dedicate this book.

Potsdam, New York   E. E. A.

PROLOGUE

Sometimes the over-dramatization of the concepts of modern physics tends to give the erroneous impression that success is achieved in the physical sciences in proportion to one's rejection of what was previously regarded as "intuitively obvious." The truth is, however, that the great natural philosophers in the modern era have not rejected the obvious but have forced us to redefine the obvious after taking a closer look at the evidence. The transition from 19th to 20th century physics is replete with illustrations of redefinitions of such concepts as space, time, causality, measurement, waves, particles, and so on. Admittedly, there have been some bold postulates made by modern physicists, but these generally arose out of the compelling urge either to incorporate new experimental facts into known theory or to make existing theories consistent with each other. It is hoped that the student of modern physics will learn to appreciate its conceptual structure as well as its application to specific physical problems.

CONTENTS

Chapter 7

Chapter 8

Chapter 8

Chapter 9

CHAPTER I

THE SPECIAL THEORY
OF RELATIVITY

The special theory of relativity is much more than a conceptual revolution of interest only to philosophers of science. The consequences of the postulates of the special theory of relativity to dynamical systems and to the interactions between matter and energy are so far-reaching that scarcely any area of contemporary physics is free of them, and the relativistic equations now play a significant role in the research activities of many scientists. Therefore, an understanding of the special theory of relativity is extremely important for the student of quantum physics, whether or not he ever encounters the formalism of relativistic quantum mechanics.

I. CLASSICAL PRINCIPLE OF RELATIVITY: GALILEAN TRANSFORMATION

The theory of relativity is concerned with the way in which observers who are in a state of relative motion describe physical phenomena. The idea of an absolute state of rest or motion has long been abandoned, since an observer "at rest" in an earth-bound laboratory is sharing the motion of the earth about its axis, the earth about the sun, the solar system through the Milky Way, and so on. It is also common knowledge that one can perform *simple* experiments with bouncing balls, oscillating springs, or swinging pendula in a laboratory fixed on the earth or in a smoothly running truck moving at constant velocity and obtain identical results in both sets of experiments.* We describe this fact by saying that the laws of motion are *covariant*, that is they retain the same form when expressed in the coordinates of either frame of reference.

Consider two frames having relative translational velocity V in their common x-directions. It is evident that the coordinates are related by

$$x' = x - Vt$$
$$y' = y$$
$$z' = z$$
$$t' = t,$$

---

* This is true only for experiments that are insensitive to the rotation of the earth.

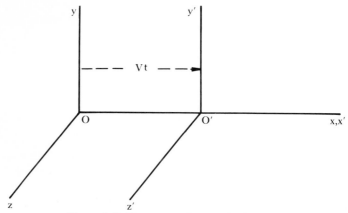

Figure 1-1 Two inertial frames of reference.

provided that $t = t' = 0$ when the origins coincide. Differentiating with respect to time we find,

$$u'_x = u_x - V$$
$$u'_y = u_y$$
$$u'_z = u_z,$$

where $u_x = dx/dt$, and so forth. This again checks with our every day experience with velocities. Differentiating once more we have,

$$a'_x = a_x$$
$$a'_y = a_y$$
$$a'_z = a_z,$$

where $a_x = du_x/dt$, and so forth. Then,

$$\vec{F}' = m\vec{a}' = m\vec{a} = \vec{F}. \tag{1.1}$$

Equation 1.1 says that Newton's laws have the same form in both coordinate frames, that is, they are covariant. Note that the coordinates and velocities are different in the two frames but each observer always knows how to obtain the other's values. Frames in which Newton's laws are covariant are called *inertial frames*. Inertial frames are *equivalent* in the sense that there is no mechanical experiment which can distinguish whether either frame is at rest or in uniform motion; hence, *there is no preferred frame*. This is known as the *Galilean* or *classical principle of relativity* and the coordinate transformation given above is called a *Galilean transformation*. Strictly speaking, the earth is not an inertial frame because of its rotation and its orbital motion, but it can often be treated as an inertial frame without serious error.

As a consequence of the principle of relativity, observers in different inertial frames would all discover the same set of mechanical laws, namely, Newton's laws of motion. By way of illustration, consider the following thought experiment. An observer, riding on a flat car which is moving at constant velocity past an observer on the ground, fires a ball vertically upward from a small cannon and notes its maximum height as well as the total time of flight for it to return to the muzzle of the cannon. The observer on the ground also measures the maximum height and the time of flight. Although the two observers will not agree on the shape of the trajectory of the ball, they *will* agree on the height, the time, the calculated muzzle velocity, and the value of g, the gravitational acceleration.

2. ELECTROMAGNETIC THEORY AND THE GALILEAN TRANSFORMATION

Long after the Galilean principle of relativity was well established, Maxwell formulated his electromagnetic field equations, out of which arose a finite and constant velocity for light in vacuum. In spite of the great success of Maxwell's equations in describing the behavior of electromagnetic phenomena in space and time, it was quite disturbing to find that they are *not* covariant under a Galilean transformation. Thus, the classical transformation which permits a covariant description of mechanical forces, does *not* hold for electromagnetic forces.

PROBLEM 1-1

Show that the equation $\vec{\nabla} \cdot \vec{D} = \rho$ is not covariant under a Galilean transformation. That is, assume the covariance of the Lorentz force, $\vec{F} = q(\vec{\mathscr{E}} + \frac{1}{c}\ \vec{v} \times \vec{\mathscr{B}})$, and show that this leads to a contradiction. Hint: require $\rho' = \rho$; otherwise, a simple measurement of charge would distinguish one frame from another.

Another serious difficulty arose from electromagnetic theory, in connection with a medium which was postulated to sustain the wave motion. This ethereal substance was simply called "the ether," and was endowed with infinite elasticity and inertia but zero mass. These contradictory properties were required in order to sustain the transverse vibrations of light waves and yet prevent any longitudinal vibrations. Furthermore, the ether had to pervade all of space without inhibiting the movements of celestial bodies, since it was known that these bodies move in a non-viscous medium.

It was hoped that the ether might provide the necessary reference frame for measuring absolute motion. However, these hopes were destined to be unfulfilled, as the following summary will show:

a. As early as 1728 it was known that the observed position of a star varies slightly throughout the year.[1,2] This phenomenon, which is called the *aberration of starlight*, is due to the fact that the observed direction of a light ray depends upon the relative velocity of the light source and the observer. Since the earth moves in its orbit at about 30 km/sec, a telescope should be tilted towards a star at an angle $\theta \sim 30/300,000 = 1 \times 10^{-4}$ radian or 20 seconds of arc. Six

---

[1] J. Bradley, *Phil. Trans.* **35**, 637 (1728).
[2] A. B. Stewart, "The Discovery of Stellar Aberration," *Scientific American*, p. 100 (March 1964).

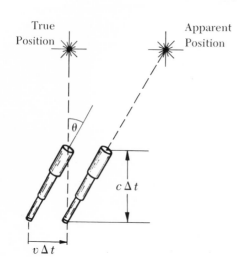

Figure 1-2 The angular position of a star is shifted by the amount θ because the telescope moves with the earth during the time the light travels down the tube. In the course of a year, the telescope sweeps out a cone of abberation of angular diameter 2θ.

months later, the telescope should be tilted the same amount in the opposite direction. See Figure 1-2. An important conclusion can be drawn from the fact that aberration of the correct amount is experimentally observed, namely, that the *earth moves through the ether*. The ether, then, was assumed fixed with respect to the stars.

b. A number of experiments were devised to measure the earth's motion through the ether, that is, to detect an ether wind with respect to the earth. The most famous of these was the Michelson-Morley experiment[3] which will be described below. This experiment was repeated many times but always with a null result, that is, no ether wind was detected. Another famous but unsuccessful attempt to detect an ether wind was the experiment of Trouton and Noble.[4] They sought to detect a torque on a charged capacitor due to its motion through the ether.

c. All attempts to explain the absence of an ether wind while retaining the ether concept were also refuted. It could not be argued that the earth carries the ether along with it since this would contradict the observed aberration of starlight. Some proposed that the velocity of light adds vectorially to the velocity of the source. This theory (historically called the emission theory) would easily account for the null result of the Michelson-Morley experiment, but it contradicts a known fact about wave motion, namely, that the velocity of a wave depends only upon the properties of the medium. Perhaps the most convincing evidence for rejecting the emission theory comes from the study of the periods of binary stars as proposed by De Sitter.[5] The relative velocity (to an earth observer) of a distant star performing circular motion will vary sinusoidally with time. Therefore, the measured Doppler shift* of the light emitted by such a star would show the same sinusoidal

---

[3] A. A. Michelson, *Amer. Journ. of Science* (3), **22**, 20 (1881); A. A. Michelson and E. W. Morley, *ibid.* **34**, 333 (1887). These results were confirmed by R. J. Kennedy, *Proc. Nat. Acad.* **12**, 621 (1926) and K. K. Illingworth, *Phys. Rev.* **30**, 692 (1927).

[4] F. T. Trouton and H. R. Noble, *Phil. Trans. Roy. Soc.*, **A202**, 165 (1903); *Proc. Roy. Soc.* (London), **72**, 132 (1903).

[5] W. De Sitter, *Proc. Amsterdam Acad.* **16**, 395 (1913).

* The mathematical details of the Doppler shift, which will be discussed in section 11 of this chapter, are not required for the present argument.

variation if the speed of light were constant. On the other hand, if the emission theory were true, the speed of the emitted light would be greater as the star approached and less as the star receded. In this event the measured Doppler shifts would no longer show a sinusoidal variation but the curve would be distorted as in Figure 1-3(a). An actual plot of the Doppler shifts measured for the components of the binary star Castor C is shown in Figure 1-3(b).[6] The absence of distortion in the curves is strong evidence for the validity of the constancy of the speed of light. Many such binaries have been studied and there is no indication of distortion other than that due to known eccentricities in the orbits. More recent experiments which refute the emission theory of light are measurements of the velocity of light in the form of gamma rays emitted from positron annihilation[7] and from the decay of $\pi°$ mesons.[8]

A third attempt to explain the null result of the Michelson-Morley experiment was the Lorentz-FitzGerald contraction theory. It was postulated that a length is contracted by a factor $(1 - v^2/c^2)^{\frac{1}{2}}$, where v is the component of the velocity parallel to the length and c is the velocity of light. This contraction is just the right amount to explain the null Michelson-Morley result, although it was not derived from first principles. The Lorentz-FitzGerald

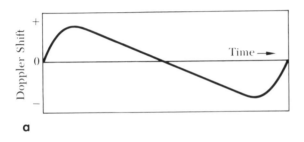

a

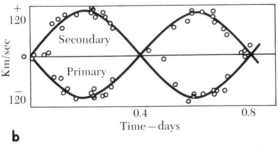

b

Figure 1-3 The measured Doppler shifts due to the orbital velocity of a star performing circular motion would appear as in (a) if the speed of light were to increase as the star approached and to decrease as the star receded from the observer. Shown in (b) are the relative velocities of the components of the binary star Castor C as determined from the Doppler shifts of their spectra. The absence of distortion is strong evidence for the constancy of the speed of light. (The data are from A. H. Joy and R. F. Sanford, *Astrophysical Journal* **64**, 250 (1926).)

[6] A. H. Joy and R. F. Sanford, *Astrophys. J.* **64**, 250 (1926).
[7] D. Sadeh, *Phys. Rev. Letters* **10**, 271 (1963).
[8] T. Alväger, F. J. M. Farley, J. Kjellman and I. Wallin, *Phys. Letters* **12**, 260 (1964).

contraction was extremely difficult to refute,[9] and its physical significance did
not become clear until after the work of Einstein.

3. THE MICHELSON-MORLEY EXPERIMENT

As mentioned in the previous section, if the ether is assumed to be at rest
in the frame of the "fixed" stars, then the earth's motion through the ether
should result in an ether wind for an earth observer. Michelson[10] proposed
to detect this wind by looking at the interference pattern produced by two
coherent beams of light, where the optical paths of the two beams are respectively parallel to and perpendicular to the earth's motion through the
ether. If the two beams are initially in phase and if their optical paths differ
by zero or an integral number of wavelengths, they will still be in phase when
they are brought together again. In general, the optical paths of the two
beams will differ by an arbitrary amount, and either destructive or constructive
interference can occur. The arrangement used by Michelson is shown in
Figure 1-4. One of the mirrors of the interferometer is usually tilted slightly so
that a pattern of alternately bright and dark lines, called fringes, is viewed
in the eyepiece. Each dark fringe corresponds to a path difference of an odd
number of half wavelengths in the two arms of the instrument. Slight variations
in either path produced while observing through the eyepiece will result in
an apparent motion of the fringe pattern. Shifts corresponding to a fraction
of a wavelength can be easily detected.

Suppose the interferometer is set up so that one light path, say ℓ_2, is
colinear with the earth's motion through the ether with velocity v. Then,
assuming a Galilean transformation, the time for one round trip along that
path is

$$t_2 = \frac{\ell_2}{c - v} + \frac{\ell_2}{c + v} = \frac{2c\ell_2}{c^2 - v^2} = \frac{2\ell_2}{c}(1 - \beta^2)^{-1} \sim \frac{2\ell_2}{c}(1 + \beta^2),$$

where $\beta = v/c$. Since the velocity of the earth in its orbit is about 30 kilometers
per second, $\beta \sim 10^{-4}$ and terms higher than β^2 are neglected in the expansion.
For the path that is perpendicular to the relative motion the situation is
analogous to the problem of rowing a boat across a river having a current v.
If the rower's speed is c, his path is actually of length

$$ct = \sqrt{\ell_1^2 + v^2 t^2},$$

where t is the time for one transit. Then, his velocity for a direct crossing is
$\ell_1/t = \sqrt{c^2 - v^2}$, and the time for a round trip transverse to the current is

$$t_1 = \frac{2\ell_1}{\sqrt{c^2 - v^2}} = \frac{2\ell_1}{c}(1 - \beta^2)^{-\frac{1}{2}} \sim \frac{2\ell_1}{c}(1 + \tfrac{1}{2}\beta^2).$$

---

[9] R. J. Kennedy and E. M. Thorndike, *Phys. Rev.* **42**, 400 (1932).
[10] See footnote 3, and R. S. Shankland, *Amer. J. Phys.* **32**, 16 (1964).

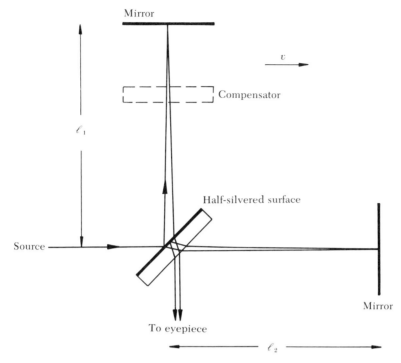

Figure I-4 Michelson interferometer with arm ℓ_2 parallel to the earth's motion.

The phase difference between the two beams as they enter the eyepiece is $\Delta\phi_1 = (2\pi/\lambda) \cdot \text{(path difference)} = (2\pi/\lambda) \cdot c(t_2 - t_1) = (2\pi/\lambda) \cdot [2(\ell_2 - \ell_1) + \beta^2(2\ell_2 - \ell_1)]$. By rotating the interferometer through 90° the roles of the two beams are interchanged and the phase difference for this case is

$$\Delta\phi_2 = \frac{2\pi}{\lambda} \cdot [2(\ell_1 - \ell_2) + \beta^2(2\ell_1 - \ell_2)].$$

Thus, the effect of rotating the instrument through 90° is to enable the observer to detect a total phase difference of $\Delta\phi_1 + \Delta\phi_2$. This corresponds to a fringe shift of

$$\frac{1}{2\pi}(\Delta\phi_1 + \Delta\phi_2) = \frac{\beta^2}{\lambda}(\ell_1 + \ell_2). \tag{1.2}$$

For $\ell_1 \sim \ell_2 = 10$ meters and $\lambda \sim 5000$ Å, Equation 1.2 predicts a shift of about 0.4 fringe. Such a shift could be easily detected, as Michelson's instrument could resolve 0.01 of a fringe at that wavelength.

The Michelson-Morley experiment was repeated many times, at all hours of the day and in all portions of the earth's orbit, yet no fringe shift has ever been observed. The null result requires us either to accept the non-physical Lorentz-FitzGerald contraction or to reject the ether theory, since the ether cannot be fixed with respect to the earth as well as the distant stars.

Show that the Lorentz-FitzGerald contraction can account for the null result of the Michelson-Morley experiment.

4. THE POSTULATES OF SPECIAL RELATIVITY

The failure of such efforts as the Michelson-Morley and the Trouton-Noble experiments to discover a preferred frame for the electromagnetic equations of Maxwell suggested that the latter must conform to a principle of relativity. However, the fact that the Galilean principle of relativity— which was known to be valid for classical mechanics—failed for Maxwell's equations was a considerable source of frustration to physicists at the turn of the century. After a critical examination of the concepts of space, time, and simultaneity, Einstein[11] discarded the Galilean principle of relativity and postulated in its stead a principle of relativity for all physical laws. His postulates may be stated as follows:

(1) There is no preferred or absolute inertial system. That is, all inertial frames are equivalent for the description of all physical laws.

(2) The speed of light in vacuum is the same for all observers who are in uniform, rectilinear, relative motion and is independent of the motion of the source. Its free space value is the universal constant c given by Maxwell's equations.

Since the first postulate says that Maxwell's equations as well as Newton's laws are covariant, a new set of relativistic transformation equations must be obtained to replace those of Section 1. The second postulate follows from the first, since, if an observer were to measure, for example, the value $c - v$ for the speed of light, he would then have a means of determining the relative speed v, which would violate the principle of relativity.

Maxwell's equations enable one to express the universal constant c in terms of the fundamental constants of electromagnetic theory. Thus, in m.k.s. units, $c = (\epsilon_0 \mu_0)^{-\frac{1}{2}}$, and in c.g.s. units, c is given by the ratio of the electrostatic unit of charge (esu) to the absolute unit (emu). The value of c is very nearly 3×10^8 meters per second. Frequent attempts have been made to determine c to as many significant figures as possible. Some of the methods employed have involved direct measurements of distance and transmission time; others have been concerned with precise measurements of λ and v in order to calculate c from their product, and still others have utilized interferometry. There is a large amount of literature dealing with this experimental problem and a number of good summaries exist which are replete with references.[12,13,14] At the present time the accepted value for the speed of light is

$$c = 299{,}792.5 \pm 0.4 \text{ km/sec.}$$

[11] A. Einstein, *Ann. Physik* **17,** 891 (1905).
[12] J. F. Mulligan, *Am. J. Phys.* 20, 165 (1952).
[13] J. F. Mulligan and D. F. McDonald, *Am. J. Phys.* **25,** 180 (1957).
[14] J. H. Sanders, *The Velocity of Light*, New York: Pergamon Press (1965).

This value was adopted by the International Union of Geodesy and Geophysics in 1957.[14]

Before deriving the new relativistic transformation, let us look briefly at the concept of time. According to classical physics, $t = t'$, and it was tacitly assumed that clocks in different inertial frames could be easily synchronized. The concept of simultaneity was not questioned. Einstein, however, realized that the process of synchronization requires the transmission of signals at a finite velocity which can be at most the velocity of light in free space. Furthermore, simultaneity is a meaningful concept only in a frame in which light sources, clocks and observers are all at rest. By way of illustration, suppose observer O has measured two equal distances x_1 and x_2 from his reference position and installs identical flash lamps S_1 and S_2 as shown below:

He can fire these flash lamps simultaneously by means of matched cables or radio operated relays, and he will assert that the flashes reach him simultaneously. However, to a second observer O', who is in a state of relative motion parallel to the line joining S_1 and S_2, the flashes will not appear to be simultaneous because of the finite distance through which he moves during the transit time of the light waves. Thus, if the flashes occur just as O' passes O, then O' will claim that S_2 flashed earlier than S_1.

Since simultaneity cannot be defined independently of spatial coordinates for observers who are in relative motion, their clocks are not synchronized and their measured times are different. As a consequence of this, taken together with the constancy of the observed speed of light in either frame, we conclude that the statement $t' = t$ is no longer valid.

5. THE LORENTZ TRANSFORMATION

Although the Galilean transformation is not the correct expression of the principle of relativity, the new transformation should reduce to the Galilean transformation for small relative velocities, since the latter gives correct results for Newton's laws in our macroscopic world. Also, the transformation must be linear in order to make the intervals between events independent of the choice of origin in the space-time coordinate system. The simplest transformation that will satisfy both of these requirements will have the form

$$\left.\begin{aligned} x' &= C(x - vt) \\ y' &= y \\ z' &= z, \end{aligned}\right\} \tag{1.3}$$

where $C \to 1$ as $v \to 0$, and where v is parallel to the common direction of the x and x' axes. Since the transformed time must, in general, depend upon both time and space coordinates, the simplest form to try is,

$$t' = At + Bx, \tag{1.4}$$

where $A \to 1$ and $B \to 0$ as $v \to 0$.

Consider two observers in different inertial frames having relative velocity v along their common x-directions. At the instant that their origins coincide a light is flashed by the unprimed observer and a spherical wave emanates from his origin, the equation of which is

$$c^2t^2 = x^2 + y^2 + z^2.$$

However, by Einstein's postulates, the primed observer also must see a spherical wave emanating at speed c from *his* origin as its center. If he were not to see a spherical wave but, say, an ellipsoidal wavefront, he would have a means of determining which frame were moving with velocity v. The equation of the wave front in the primed system is,

$$c^2t'^2 = x'^2 + y'^2 + z'^2$$

Then we have,

$$x^2 + y^2 + z^2 - c^2t^2 = x'^2 + y'^2 + z'^2 - c^2t'^2 \tag{1.5}$$

If the new transformation satisfies Equation 1.5, it will be consistent with Einstein's postulates.

PROBLEM 1-3

By means of Equations 1.3, 1.4, and 1.5, determine the Lorentz transformation,

$$\left. \begin{array}{l} x' = \gamma(x - \beta ct) \\ y' = y \\ z' = z \\ ct' = \gamma(ct - \beta x), \end{array} \right\} \tag{1.6}$$

where

$$\beta = v/c \quad \text{and} \quad \gamma = (1 - \beta^2)^{-\frac{1}{2}}.$$

PROBLEM 1-4

Show that the wave equation,

$$\nabla^2 \phi(x, y, z, t) = \frac{1}{c^2} \frac{\partial^2}{\partial t^2} \phi(x, y, z, t),$$

is covariant under a Lorentz transformation.

6. RELATIVISTIC KINEMATICS

We define an event as a physical occurrence that is localized in space and time. That is, to an observer who is at rest in the frame of the event, the time interval of the event can be measured by means of a single clock and the position of the event can be determined by means of a meter stick. We call the time interval in the rest frame the *proper time interval*, τ_0.

If the event consists of the measurement of the length of an object in its rest frame, the coordinates of the two ends of the object can be measured simultaneously by an observer in that frame. We call the length of the object measured in this fashion its *proper length*, L_0. To an observer in any inertial frame other than that of the event, a measurement of either a length or a time interval will require the use of *both* a meter stick and at least one clock. We will now look in detail at the way lengths and time intervals transform under a Lorentz transformation.

a. Preservation of lengths perpendicular to the direction of relative motion. Consider the following thought experiment. The lengths of two meter sticks are known to be exactly the same in a frame where they are both at rest. If they are now placed in relative motion in a direction perpendicular to their lengths, as in Figure 1-5, what do observers O and O' conclude about their lengths? O can install synchronized flash cameras at each end of his meter stick which he can trigger just as O' passes him. O' will agree that the flash pictures were simultaneous since he is at all times equidistant from the two flash lamps. Since both observers agree that the flashes were simultaneous, they must agree with the result shown on the photographs. That is, the result is an *absolute* one, not a relative one. Now the experiment can be repeated with the flash cameras in the primed frame and once again both observers will agree that the flashes were simultaneous. The same result must be obtained in both sets of experiments. If, for example, the stick in the unprimed frame were found to be shorter in the first set of pictures, then the stick in the unprimed frame must be found to be shorter in the second set of pictures. However, such a result would contradict the principle of relativity since in that case motion to the right would stretch a stick and motion to the left would shrink it. Therefore, we conclude that the lengths of the two sticks must remain unchanged. This result is consistent with the symmetrical conditions of the experiments.

b. Contraction of lengths parallel to the direction of relative motion. Let the two meter sticks of Figure 1-5 now be placed in relative motion parallel to their common x-axes and the photographs repeated as before. The first set

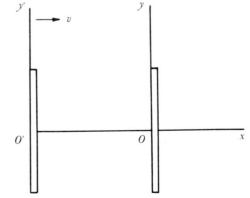

Figure I-5 Length measurements perpendicular to the line of motion.

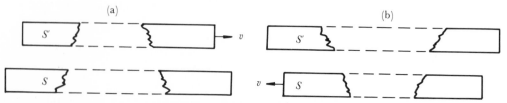

Figure 1-6 Length measurements parallel to the line of motion. In (a) the photographs are simultaneous to observer O in S; in (b) they are simultaneous to observer O' in S'.

of photographs, taken by O, would look like Figure 1-6(a), and the second, taken by O', would look like Figure 1-6(b). In the first, O would say that the "moving" stick was contracted. O' would explain the photographs by claiming that the flashes fired by O were *not* simultaneous. In fact, he would say that the leading end of his meter stick was photographed *before* it reached the coincidence position and the lagging end of his meter stick was photographed *after* it had passed the position of coincidence. Thus, both observers accept the photographs but since they do not agree that the flashes were simultaneous they have different explanations for them. In the second experiment the roles are reversed and now observer O' says that the meter stick in S is contracted by virtue of its "motion." We cannot argue, as we did in the previous thought experiment, that the sticks should appear equal in length since *there is no agreement on simultaneity* in the present case. We must conclude that the stick that is moving with respect to the observer is contracted.

This contraction may be demonstrated analytically as follows. Suppose observer O' has a meter stick situated so that the end points are at x_1' and x_2' for any time t' in his frame. By the Lorentz transformation, an observer in S will obtain the result

$$x_1' = \gamma(x_1 - \beta c t_1)$$
$$x_2' = \gamma(x_2 - \beta c t_2),$$

from which we obtain

$$x_2' - x_1' = \gamma(x_2 - x_1) - \beta\gamma c(t_2 - t_1).$$

But $x_2' - x_1' = L_0$, the proper length, since the stick is at rest in S'. Furthermore, if x_2 and x_1 are measured simultaneously in S, then $x_2 - x_1 = L$, the length of the stick as observed in S. Thus, for $t_1 = t_2$,

$$L_0 = \gamma L \quad \text{or} \quad L = \sqrt{1 - \beta^2}\, L_0. \tag{1.7}$$

The same result can be obtained by using the inverse transformation, but the algebra requires a few more steps. To illustrate this we write,

$$x_1 = \gamma(x_1' + \beta c t_1')$$
$$x_2 = \gamma(x_2' + \beta c t_2')$$
$$x_2 - x_1 = \gamma(x_2' - x_1') + \beta\gamma c(t_2' - t_1').$$

But the time transformations must be used here since the simultaneous measurement of x_1 and x_2 in frame S will not appear simultaneous in frame S'. Then,

$$ct_1' = \gamma(ct_1 - \beta x_1)$$
$$ct_2' = \gamma(ct_2 - \beta x_2),$$

from which

$$c(t_2' - t_1') = \gamma c(t_2 - t_1) - \beta\gamma(x_2 - x_1).$$

Putting $t_1 = t_2$ and eliminating $(t_2' - t_1')$ from the above equations,

$$x_2 - x_1 = \gamma(x_2' - x_1') - \beta^2\gamma^2(x_2 - x_1),$$

which reduces immediately to Equation 1.7.

This result is identical with the Lorentz-FitzGerald contraction, but it now has a theoretical basis in the postulates of Einstein. Note that the proper length is the *maximum* length that can be measured by any observer.

c. Time dilation. Suppose an observer in S measures a proper time interval τ_0 by a single clock and that the coordinates of the events marking the start and the finish of the interval are (x_0, t_1) and (x_0, t_2). An observer in the inertial frame S' will observe the coordinates of the events as follows:

$$ct_1' = \gamma(ct_1 - \beta x_0)$$
$$ct_2' = \gamma(ct_2 - \beta x_0)$$
$$c(t_2' - t_1') = \gamma c(t_2 - t_1) - \beta\gamma(x_0 - x_0).$$

Letting $t_2' - t_1' = \tau$ and $t_2 - t_1 = \tau_0$,

$$\tau = \gamma\tau_0 \quad \text{or} \quad \tau_0 = \sqrt{1 - \beta^2}\,\tau. \tag{1.8}$$

As in the case of Equation 1.7, the inverse transformations can be used to derive Equation 1.8, but additional algebraic steps would be required to eliminate x_1' and x_2'. Note that the proper time interval is the *minimum* time interval that any observer can measure between two events. This phenomenon is called *time dilation* and provides the basis for the statement that "moving clocks run slow." If the two events are two "ticks" of a clock, Equation 1.8 says that the minimum interval between ticks will occur for the clock in the rest frame of the observer, that is, the proper clock will run fastest.

It is possible to perform thought experiments in order to compare measurements of time intervals by clocks which are in relative motion. In the course of such experiments it is assumed that any number of clocks in the same rest frame as an observer can be synchronized to any desired degree of precision. It is also assumed that clocks in different inertial frames can be photographed or compared visually from either frame at the instant that the clocks pass the same point in space. Consider the following example. Let observer O' send a light flash a distance ℓ to a mirror at rest in his frame, S', at the instant that he arrives at the origin of the unprimed frame where observer O is stationed. O' sees the light reflected directly back to his position and he records the proper time interval for the round trip by means of a single clock. The proper time interval is $\tau_0 = 2\ell/c$. In the unprimed frame the light beam is reflected

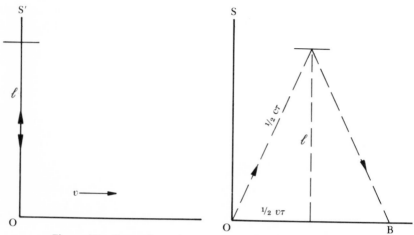

Figure 1-7 Comparison of time intervals measured in two frames.

from a point to the right of observer O, as shown in Figure 1-7, since the mirror is moving to the right with velocity v. The time at which the beam returns is measured by observer B on a clock which had previously been synchronized with the one used by observer O. If the elapsed time in the unprimed frame is τ, then from the figure we have,

$$(\tfrac{1}{2}c\tau)^2 = \ell^2 + (\tfrac{1}{2}v\tau)^2$$

$$\tau^2 = \left(\frac{2\ell}{c}\right)^2 + (\beta\tau)^2$$

$$(1 - \beta^2)\tau^2 = \tau_0^2$$

$$\tau = \gamma\tau_0,$$

as in Equation 1.8. Observer O thus concludes that the clock used by O' runs slow. Had O' photographed the clocks at O and at B as he passed those points, his photos would confirm the time interval reported by O. However, O' would claim that the clocks used by O and B were *not* synchronized but that the clock used by B was fast by the amount $\beta^2\tau$. Therefore, according to O', the time interval in the unprimed frame should be

$$\tau - \beta^2\tau = (1 - \beta^2)\tau = (1 - \beta^2)\gamma\tau_0 = \sqrt{1 - \beta^2}\,\tau_0,$$

and O' would also conclude that "moving clocks run slow."

Before concluding this section, mention should be made of the visual appearance of objects traveling at relativistic speeds, since misconceptions are prevalent in the literature. A visual image formed at the retina or a camera film is produced by photons that have *arrived* simultaneously, but they did not necessarily *leave* the object simultaneously. Thus, photographs of high speed objects must be corrected for the transit time of the photons to the various points of the image in order to determine the effects of the Lorentz contraction.[15]

[15] See V. T. Weisskopf, *Physics Today* **13**, 24 (1960); also, G. D. Scott and M. R. Viner, *Amer. J. Phys.* **33**, 534 (1965).

PROBLEM 1-5

A meter stick lies at rest in the xy-plane at an angle of $45°$ with the x-axis. What are the length and orientation of the stick to an observer moving at relative velocity βc in the x-direction?

PROBLEM 1-6

If the galaxy has a diameter of 10^5 light-years, what will the diameter appear to be to a cosmic ray particle traveling at the relative speed $\beta = 0.99$? How long will the trip take as measured by a clock riding with the particle?

PROBLEM 1-7

What synchronization error will an observer moving at relative speed βc detect in two stationary clocks separated by a distance L along the path of the motion?

7. RELATIVISTIC DYNAMICS

By means of the Lorentz transformation given in Equations 1.6 we can readily obtain expressions for the velocity components which each observer will measure in his own frame. Thus, first taking the differentials,

$$dx' = \gamma(dx - \beta c\, dt)$$
$$dy' = dy$$
$$dz' = dz$$
$$dt' = \gamma\left(dt - \frac{\beta}{c}\, dx\right),$$

we define,

$$u'_x = \frac{dx'}{dt'} = \frac{u_x - \beta c}{1 - \frac{u_x \beta}{c}}$$

$$u'_y = \frac{dy'}{dt'} = \frac{u_y}{\gamma\left(1 - \frac{u_x \beta}{c}\right)} \qquad (1.9)$$

$$u'_z = \frac{dz'}{dt'} = \frac{u_z}{\gamma\left(1 - \frac{u_x \beta}{c}\right)}.$$

Note that β, the relative velocity of the frames, is in the common x-direction in Equations 1.9. The numerators of these expressions are just the velocity transformations as given by the Galilean principle of relativity; these are the transformations that are normally used in the everyday world. The denominators represent the correction due to the special theory of relativity; they all approach unity as β gets small. These equations can be used to obtain the relative velocity of one body with respect to the other when two bodies are moving with respect to the laboratory frame. For example, suppose two electrons are fired in opposite directions, each with speed V with respect to the laboratory. Let the S frame be the rest frame of the electron moving in the negative x-direction, and the S' frame be the rest frame of the laboratory. Then, in the S frame the laboratory is moving to the right with velocity $\beta c = V$; in the S' frame, $u'_x = V$ is the velocity of the other electron. Since we seek u_x, we must use the inverse transformation,

$$u_x = \frac{u'_x + \beta c}{1 + \frac{u'_x \beta}{c}} = \frac{V + V}{1 + \frac{V^2}{c^2}}. \tag{1.10}$$

If V is large, such as $0.9c$, then $u_x = (1.80/1.81)c$, which is very close to, but still less than, the velocity of light in free space. An observer in the laboratory will assert that the electrons are separating at a relative velocity of $1.80c$. This is not in conflict with the theory of relativity, since neither electron is moving at a speed greater than c in any frame of reference. If the particles in this example were photons fired by two lasers, Equation 1.10 tells us that relative to each photon, the other would have a velocity of

$$u_x = \frac{2c}{2} = c,$$

which is the result required by Einstein's second postulate.

PROBLEM 1-8

In the rest frame of an electron, another electron approaches at a speed v. What is the relative velocity of an observer who measures equal and opposite velocities for the electrons?

PROBLEM 1-9

A cosmic ray mu meson moves vertically through the atmosphere at a relative speed of $\beta = 0.99$. Its mean lifetime in its own rest frame is 2.22 microseconds.
(a)   What is the mean lifetime measured by an observer on earth?
(b)   What mean path length will be "seen" by the meson and by the earth observer?

PROBLEM 1-10

A stream of electrons has velocity components $V_x = 0.9c$, $V_y = 0.3c$, $V_z = 0$, with respect to the rest frame of the electron gun from which the electrons were ejected. Find the velocity components and the total velocity of the electrons with respect to an observer approaching the electron gun at relative velocity $\beta = 0.8$ along the x-axis.

PROBLEM 1-11

Consider three inertial frames such that S and S' have relative speed $\beta_1 c$ along their common x-directions and S' and S'' have relative speed $\beta_2 c$ along their common x-directions. Find the expression for transforming velocities from S to S''. Express βc, the relative speed of S'' with respect to S, in terms of $\beta_1 c$ and $\beta_2 c$.

PROBLEM 1-12

(a) Derive the result expressed by Equation 1.7 by allowing one observer to measure how long it takes for a fixed length in a moving frame to pass him.

(b) Show that Equation 1.8 can be obtained if an observer in S measures the coordinates of two events which occur at the same point x_0' in S'.

PROBLEM 1-13

A radioactive atom is moving with respect to the laboratory at a speed of $0.3c$ in the x-direction. If it emits an electron having a speed of $0.8c$ in the rest frame of the atom, find the velocity of the electron with respect to a laboratory observer when:

(a) the electron is ejected in the x-direction, (b) it is ejected in the negative x-direction, and (c) it is ejected in the y-direction.

(Ans.: $0.89c$; $0.659c$; $0.819c$ at $68.5°$.)

PROBLEM 1-14

An observer in S places a laser so that the beam makes an angle θ with his x-axis. What angle does an observer in S' see if the relative velocity between S and S' is βc in their common x-directions? (This is the relativistic equation for the aberration of starlight.) What does the S' observer see when $\cos \theta \leq \beta$?

The velocities defined by Equations 1.9 were obtained by differentiating with respect to the local time in each observer's frame. It is convenient to define the *proper velocity* by differentiating with respect to the proper time, that is,

$$u_x^0 = \frac{dx}{d\tau_0} = \frac{dx}{dt}\frac{dt}{d\tau_0} = \frac{u_x}{\sqrt{1 - \dfrac{u^2}{c^2}}} \qquad (1.11)$$

where $u^2 = u_x^2 + u_y^2 + u_z^2$ is the square of the local velocity.

The other components follow in like fashion. The advantage of using the proper velocity will be evident when we discuss four-vectors in the next section.

Now consider the following thought experiment. Two identical volley balls, having rest mass m_0 when measured in the same rest frame, are given to two observers in different inertial frames with relative velocity βc in their common x-directions. Each observer is instructed to throw his volley ball so that it will make a head-on collision with the other volley ball at the instant that the line joining the two observers is perpendicular to their common x-axes (see Figure 1.8). Each observer claims that his own volley ball moves parallel to his own y-axis both before and after the collision and that it undergoes a momentum change of $2m_0u_y$. Moreover, each claims that the other volley ball has an x-component of velocity before and after the collision. Since Newton's laws are covariant and momentum is conserved, either observer can write,

$$2m_0u_y = (2m_0u_y)' = 2m_0'\sqrt{1 - \beta^2}\,u_y, \quad \text{provided} \quad u_y \ll c.$$

Then, letting $m_0' = m$, we may write

$$m = \gamma m_0, \qquad (1.12)$$

which is the expression for the relativistic change of mass. That is, a body of rest mass m_0 has its mass increased by the factor γ when it moves at a

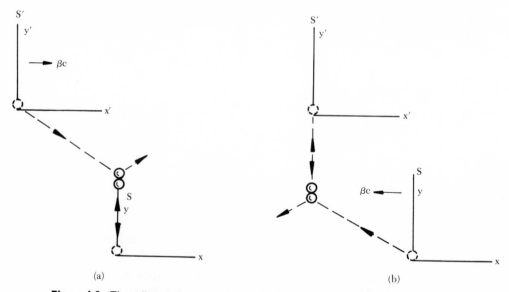

(a)   (b)

Figure I-8 The collision of two identical volley balls: (a) as observed in S, (b) as observed in S'. The y-axes of S and S' are colinear at the instant of impact.

velocity v relative to an observer. After squaring Equation 1.12, it can be written as

$$m^2c^2 - m^2v^2 = m_0^2c^2.$$

Differentiating with respect to both m and v, we obtain

$$2mc^2\, dm - 2mv^2\, dm - 2m^2v\, dv = 0,$$

or

$$c^2\, dm = v^2\, dm + mv\, dv. \tag{1.13}$$

Also,

$$\vec{F} = \frac{d\vec{p}}{dt} = \frac{d}{dt}(m\vec{v}) = m\frac{d\vec{v}}{dt} + \vec{v}\frac{dm}{dt},$$

and the change of kinetic energy is

$$dT = \vec{F}\cdot d\vec{s} = m\frac{d\vec{s}}{dt}\cdot d\vec{v} + \vec{v}\cdot\frac{d\vec{s}}{dt}\,dm = mv\, dv + v^2\, dm, \tag{1.14}$$

where the last expression holds for $d\vec{v}$ parallel to $\vec{v}$. From (1.13) and (1.14) we have

$$dT = c^2\, dm,$$

which can be integrated as follows:

$$\int_{v=0}^{v} dT = \int_{m_0}^{m} c^2\, dm.$$

Then,

$$T = (m - m_0)c^2 = (\gamma - 1)m_0c^2, \tag{1.15}$$

which says that the kinetic energy of a body is equal to the product of its mass change times the square of the velocity of light. This result can be compared with the classical expression for the kinetic energy by writing

$$T = \frac{m_0c^2}{\sqrt{1 - \beta^2}} - m_0c^2 = m_0c^2[(1 - \beta^2)^{-\frac{1}{2}} - 1]$$
$$= m_0c^2(1 + \tfrac{1}{2}\beta^2 + \cdots - 1) \quad \text{(for small } \beta)$$
$$= m_0c^2(\tfrac{1}{2}\beta^2 + \cdots)$$
$$\sim \tfrac{1}{2}m_0v^2.$$

Thus, if β is small enough to permit discarding all terms higher than β^2, Equation 1.15 reduces to the classical result.

Still more information can be gleaned from Equation 1.15 by writing it in the form,

$$E = mc^2 = T + m_0c^2, \tag{1.16}$$

where we now call E the total energy and m_0c^2 the rest mass energy. Equations 1.15 and 1.16 express the equivalence of mass and energy and provide the basis for restating the conservation laws of classical mechanics. Neither mass nor the quantity $(T + V)$ need be conserved separately in relativity*; instead, the quantity mc^2 is conserved.

The ordinary momentum is

$$\vec{p} = m\vec{v} = \gamma m_0 \vec{v},$$

and its magnitude may be written as

$$p = mv = \frac{mc^2 v}{c^2} = \frac{Ev}{c^2} = \frac{\beta E}{c}. \tag{1.17}$$

Using Equations 1.16 and 1.17, we find that

$$E^2(1 - \beta^2) = (m_0 c^2)^2$$

$$E^2\left[1 - \left(\frac{pc}{E}\right)^2\right] = (m_0 c^2)^2$$

or

$$E^2 = (pc)^2 + (m_0 c^2)^2. \tag{1.18}$$

Further,

$$(T + m_0 c^2)^2 = (pc)^2 + (m_0 c^2)^2$$

and

$$(pc)^2 = T^2 + 2Tm_0 c^2. \tag{1.19}$$

Equations 1.18 and 1.19 provide useful relationships between the momentum of a body and its energy.

It is worth pointing out that the kinetic energy of a body, according to Equation 1.15, will increase rapidly as the velocity increases. In the limit as $v \to c$, γ would become infinite and hence the kinetic energy would also be infinite. We conclude that c must represent a limiting velocity that cannot be achieved by any particle that has a rest mass. Conversely, we also conclude that photons and neutrinos must have zero rest mass since they are known to travel with speed c. Another way of stating this is to say that only those entities for which a rest frame exists can have a rest mass. Since an entity traveling at the speed of light has *no* rest frame (its speed will be c to all observers), it can have no rest mass. A zero rest mass particle *can*, however, have a momentum and kinetic energy. From Equation 1.18 we see that the momentum and kinetic energy are $p = E/c$ and $T = pc = E$, respectively.

PROBLEM 1-15 ▬▬▬▬▬▬▬▬▬▬▬▬▬▬▬▬▬▬▬

Find the velocity and the momentum of an electron whose kinetic energy equals its rest mass energy.

PROBLEM 1-16 ▬▬▬▬▬▬▬▬▬▬▬▬▬▬▬▬▬▬▬

Through what potential difference must an electron "fall" in order to achieve a speed of $0.95c$? What is the mass of the

---

* A potential energy V may be added to each side of Equation 1.15, and then Equation 1.16 becomes $E = mc^2 = T + V + m_0 c^2$.

electron in the L-frame? What is its momentum? How could you verify experimentally that it actually has that mass and speed?

PROBLEM 1-17

Derive an expression for the radius of the orbit of a particle of mass m_0 and charge q moving with kinetic energy T at a relativistic velocity in a transverse magnetic field of intensity B. Express your answer in terms of q, B, c, T, and m_0.

8. FOUR-VECTOR NOTATION

In ordinary three-space the components of a vector $\vec{A}$ in two coordinate systems S and S' with a common origin are related by:

$$
\begin{aligned}
A_1' &= A_1 \cos (x_1', x_1) + A_2 \cos (x_1', x_2) + A_3 \cos (x_1', x_3) \\
A_2' &= A_1 \cos (x_2', x_1) + A_2 \cos (x_2', x_2) + A_3 \cos (x_2', x_3) \\
A_3' &= A_1 \cos (x_3', x_1) + A_2 \cos (x_3', x_2) + A_3 \cos (x_3', x_3).
\end{aligned}
\tag{1.20}
$$

These equations may be summarized by $A_i' = \sum_{k=1}^{3} c_{ik}A_k = c_{ik}A_k$, where the summation over the repeated index is understood in the last expression. One can readily show that for a simple rotation through an angle θ in two dimensions this reduces to the form,

$$
\begin{aligned}
x' &= x \cos \theta + y \sin \theta \\
y' &= -x \sin \theta + y \cos \theta,
\end{aligned}
\tag{1.21}
$$

which can be written in matrix notation as follows:

$$
\begin{pmatrix} x' \\ y' \end{pmatrix} = \begin{pmatrix} \cos \theta & \sin \theta \\ -\sin \theta & \cos \theta \end{pmatrix} \begin{pmatrix} x \\ y \end{pmatrix} = \begin{pmatrix} x \cos \theta + y \sin \theta \\ -x \sin \theta + y \cos \theta \end{pmatrix}.
$$

In general

$$
\begin{pmatrix} x' \\ y' \\ z' \end{pmatrix} = \begin{pmatrix} c_{11} & c_{12} & c_{13} \\ c_{21} & c_{22} & c_{23} \\ c_{31} & c_{32} & c_{33} \end{pmatrix} \begin{pmatrix} x \\ y \\ z \end{pmatrix}.
$$

The c_{ij} are not all independent since they are direction cosines and must satisfy the relations

$$
c_{ij}c_{ik} = \delta_{jk} = \begin{cases} 0, & \text{for } j \neq k \\ 1, & \text{for } j = k \end{cases}
$$

and

$$
\det c_{ij} = 1.
\tag{1.22}
$$

Show that the relations in Equations 1.22 must hold for a Cartesian coordinate system.

In addition to their transformation properties, another important property of vectors is the invariance of lengths. That is,

$$(\Delta s)^2 = (\Delta x)^2 + (\Delta y)^2 + (\Delta z)^2 = (\Delta x')^2 + (\Delta y')^2 + (\Delta z')^2 = \text{constant}.$$

We have seen that in special relativity we require the Lorentz invariance of the quantity

$$(\Delta s)^2 = (\Delta x)^2 + (\Delta y)^2 + (\Delta z)^2 - c^2(\Delta t)^2.$$

Therefore, if we can treat these quantities as the four components of a vector in four-space and the Lorentz transformations as a rotation of coordinates in four-space, then the desired invariance properties are automatically satisfied.

a. The position four-vector. The position four-vector is easily constructed by adding to the three-vector components the fourth component, ict. Thus,

$$\mathbf{X}_\mu = \begin{pmatrix} X_1 \\ X_2 \\ X_3 \\ X_4 \end{pmatrix} = \begin{pmatrix} x \\ y \\ z \\ ict \end{pmatrix}. \tag{1.23}$$

When the index is a Greek letter, a four dimensional space is implied; a Roman letter will represent a three dimensional vector. With this convention

$$\Delta \mathbf{X}_\mu \, \Delta \mathbf{X}_\mu = (\Delta s)^2, \quad \text{as given above.}$$

The matrix of the Lorentz transformation is given by

$$\mathbf{\Gamma} = \begin{pmatrix} \gamma & 0 & 0 & i\beta\gamma \\ 0 & 1 & 0 & 0 \\ 0 & 0 & 1 & 0 \\ -i\beta\gamma & 0 & 0 & \gamma \end{pmatrix} \tag{1.24}$$

in terms of which the transformation becomes,

$$\mathbf{X}'_\mu = \mathbf{\Gamma}_{\mu\nu}\mathbf{X}_\nu \quad \text{or} \quad \mathbf{X}' = \mathbf{\Gamma}\mathbf{X}, \tag{1.25}$$

in either tensor or matrix notation. We will show that Equation 1.25 is equivalent to Equation 1.6 by writing the full matrices, that is:

$$\begin{pmatrix} X'_1 \\ X'_2 \\ X'_3 \\ X'_4 \end{pmatrix} = \begin{pmatrix} \gamma & 0 & 0 & i\beta\gamma \\ 0 & 1 & 0 & 0 \\ 0 & 0 & 1 & 0 \\ -i\beta\gamma & 0 & 0 & \gamma \end{pmatrix} \begin{pmatrix} X_1 \\ X_2 \\ X_3 \\ X_4 \end{pmatrix}.$$

By means of matrix multiplication we obtain the four equations:

$$X_1' = \gamma(X_1 + i\beta X_4)$$
$$X_2' = X_2$$
$$X_3' = X_3$$
$$X_4' = \gamma(X_4 - i\beta X_1)$$

Substituting the components from Equation 1.23, we obtain

$$x' = \gamma(x - \beta ct)$$
$$y' = y$$
$$z' = z$$
$$ct' = \gamma(ct - \beta x),$$

which was to be shown.

PROBLEM 1-19

Show that the Lorentz transformation is equivalent to a rotation in four-space, that is, that the transformation is similar to Equation 1.21 with the circular functions replaced by hyperbolic functions. Hint: Use the $x_1 - x_4$ plane as in the accompanying figure and derive the expression, $\theta =$ arc tan $(i\beta)$.

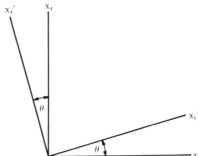

PROBLEM 1-20

Show that the Lorentz transformation from the primed to the unprimed system may be expressed by a matrix equation of the form

$$\mathbf{X} = \mathbf{\Gamma}^{-1}\mathbf{X}',$$

where $\mathbf{\Gamma}^{-1}$ is the matrix inverse of $\mathbf{\Gamma}$ given by

$$\mathbf{\Gamma}^{-1} = \begin{pmatrix} \gamma & 0 & 0 & -i\beta\gamma \\ 0 & 1 & 0 & 0 \\ 0 & 0 & 1 & 0 \\ i\beta\gamma & 0 & 0 & \gamma \end{pmatrix}.$$

b. The velocity four-vector. The components of the four-velocity are obtained by differentiating the position four-vector components with respect to the proper time. This was done for one component in Equation 1.11 and was defined as the proper velocity. Thus, the first three components of the velocity four-vector are the components of the proper velocity, that is to say, the local velocity multiplied by

$$\gamma' = \frac{1}{\sqrt{1 - \dfrac{u^2}{c^2}}},$$

where u^2 is the square of the local velocity. The fourth component is

$$U_4 = \frac{d}{d\tau_0} X_4 = \frac{d}{dt}(ict) \cdot \frac{dt}{d\tau_0} = \frac{ic}{\sqrt{1 - \dfrac{u^2}{c^2}}} = \gamma'ic.$$

Then the four-velocity is

$$\mathbf{U}_\mu = \begin{pmatrix} U_1 \\ U_2 \\ U_3 \\ U_4 \end{pmatrix} = \begin{pmatrix} \gamma'u_x \\ \gamma'u_y \\ \gamma'u_z \\ \gamma'ic \end{pmatrix}, \tag{1.26}$$

and its transformation is given by Equation 1.25 as $\mathbf{U}'_\mu = \mathbf{\Gamma}_{\mu\nu}\mathbf{U}_\nu$.

c. The momentum four-vector. The four-momentum is formed simply by multiplying the four-velocity by the rest mass. Then,

$$\mathbf{P}_\mu = \begin{pmatrix} \gamma'm_0u_x \\ \gamma'm_0u_y \\ \gamma'm_0u_z \\ \gamma'm_0ic \end{pmatrix} = \begin{pmatrix} mu_x \\ mu_y \\ mu_z \\ imc \end{pmatrix} = \begin{pmatrix} p_x \\ p_y \\ p_z \\ i\dfrac{E}{c} \end{pmatrix}. \tag{1.27}$$

The first three components reduce to the three-momentum components and the fourth is the total energy multiplied by i/c.

PROBLEM 1-21

Find the scalar invariants associated with the four-velocity and the four-momentum.

d. The force four-vector.* The components of the four-force are obtained by differentiating the four-momentum with respect to the proper time. For

---

* This is frequently called the *Minkowski force* after the man who first formulated the four-vector notation.

example,

$$F_1 = \gamma' \frac{d}{dt} P_1 = \gamma' \frac{dp_x}{dt} = \gamma' F_x,$$

and

$$F_4 = \gamma' \frac{d}{dt} P_4 = \gamma' \frac{i}{c} \frac{dE}{dt} = \gamma' \frac{i}{c} \vec{F} \cdot \vec{u}.$$

Thus, the first three components of the four-force are the three-vector force components multiplied by γ'.

$$\mathbf{F}_\mu = \begin{pmatrix} \gamma' F_x \\ \gamma' F_y \\ \gamma' F_z \\ \gamma' \frac{i}{c} \vec{F} \cdot \vec{u} \end{pmatrix}. \tag{1.28}$$

e. The current density four-vector. If ρ_0 is the charge density in a frame where the charges are at rest, then a current density four-vector may be defined as the product of ρ_0 and the four-velocity. That is,

$$\mathbf{J}_\mu = \rho_0 \mathbf{U}_\mu = \begin{pmatrix} \gamma' \rho_0 u_x \\ \gamma' \rho_0 u_y \\ \gamma' \rho_0 u_z \\ \gamma' \rho_0 ic \end{pmatrix} = \begin{pmatrix} \rho u_x \\ \rho u_y \\ \rho u_z \\ \rho ic \end{pmatrix} = \begin{pmatrix} j_x \\ j_y \\ j_z \\ \rho ic \end{pmatrix}. \tag{1.29}$$

Here we have defined $\rho = \gamma' \rho_0$, illustrating the increase in charge density due to the change in volume element. Then the product of local charge density and local velocity gives the ordinary three-space current density, (j_x, j_y, j_z).

f. The electromagnetic potential four-vector. Since a charge distribution can be at rest in only one frame, it will be observed as a current accompanied by a magnetic field by all other observers. Therefore, we would not expect the components of either electric or magnetic fields to become components of the four-vector. Using the vector and scalar potentials defined by

$$\vec{\mathcal{B}} = \vec{\nabla} \times \vec{A} \quad \text{and} \quad \vec{\mathcal{E}} = -\vec{\nabla}\phi - \frac{\partial \vec{A}}{\partial t},$$

we define the potential four-vector as

$$\mathbf{A}_\mu = \begin{pmatrix} A_x \\ A_y \\ A_z \\ \frac{i\phi}{c} \end{pmatrix}, \tag{1.30}$$

where the A_i are the components of the three-space vector potential.

g. The electromagnetic field four-tensor. It is often desirable to work with the components of the electric and magnetic fields themselves rather than with charges, currents, or potentials. However, in order to express the electromagnetic fields in four-space we require a second-rank tensor rather than a vector. The construction of this tensor is generally done in a course in electromagnetic theory so it will merely be given here[16]:

$$
\mathbf{T}_{\mu v} = \begin{pmatrix} 0 & \mathscr{B}_z & -\mathscr{B}_y & -\dfrac{i}{c}\mathscr{E}_x \\[12pt] -\mathscr{B}_z & 0 & \mathscr{B}_x & -\dfrac{i}{c}\mathscr{E}_y \\[12pt] \mathscr{B}_y & -\mathscr{B}_x & 0 & -\dfrac{i}{c}\mathscr{E}_z \\[12pt] \dfrac{i}{c}\mathscr{E}_x & \dfrac{i}{c}\mathscr{E}_y & \dfrac{i}{c}\mathscr{E}_z & 0 \end{pmatrix} .
\tag{1.31}
$$

PROBLEM 1-22

By means of the transformation $\mathbf{T}'_{\mu v} = \mathbf{\Gamma}_{\mu v \rho \sigma}\mathbf{T}_{\rho \sigma}$, find the Lorentz transformation equations for the six components of the electric and magnetic fields. Show that these may be expressed by the following four equations:

$$
\mathscr{B}'_{\parallel} = \mathscr{B}_{\parallel} \qquad\qquad \mathscr{E}'_{\parallel} = \mathscr{E}_{\parallel}
$$

$$
\mathscr{B}'_{\perp} = \gamma\left(\mathscr{B}_{\perp} - \frac{\vec{v} \times \vec{\mathscr{E}}}{c^2}\right) \qquad \mathscr{E}'_{\perp} = \gamma(\mathscr{E}_{\perp} + \vec{v} \times \vec{\mathscr{B}}), \tag{1.32}
$$

where "parallel" and "perpendicular" refer to the line of relative motion.

PROBLEM 1-23

A meter stick is at rest and at equilibrium under the action of the forces $\vec{F}$ as shown.
(a) What angle, θ', will the meter stick appear to make

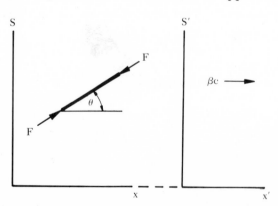

[16] See, for example, W. K. H. Panofsky and M. Phillips, *Classical Electricity and Magnetism.* Addison-Wesley Publishing Co., Inc., Reading, Mass. 1955.

with the x-axis in a frame moving in the positive x-direction at relativistic speed βc?

(b)   What angle will the forces make with the meter stick in the moving frame?

(c)   Calculate the torque in the moving frame.

h. The proper time as a four-scalar.   Using the symbol $(\Delta r)^2$ for an interval in three-space, we may express the length of an interval in four-space as

$$\Delta X_\mu \, \Delta X_\mu = (\Delta s)^2 = (\Delta r)^2 - c^2(\Delta t)^2 \equiv \text{invariant.}$$

Now, when two events are observed by the same clock, $\Delta r = 0$ and $\Delta t = \Delta \tau_0$, the proper time interval. Therefore, in any inertial frame the square of the interval is

$$(\Delta s)^2 = -c^2 \, (\Delta \tau_0)^2.$$

Since this is a scalar quantity in all frames, $\Delta \tau_0$ is called a four-scalar. We can show that the proper time is the minimum time interval by rearranging the above expression as follows,

$$(\Delta t)^2 = (\Delta \tau_0)^2 + \frac{(\Delta r)^2}{c^2}.$$

When $(\Delta \tau_0)^2 > 0$, $\Delta \tau_0$ is *real* and is called a *time-like interval*. In this case one can always find an inertial frame traveling at a velocity less than c such that the two events occur at the same point in space but at different times. Two events separated by a time-like interval can never be regarded as simultaneous by any observer. When $(\Delta \tau_0)^2 < 0$, $\Delta \tau_0$ is *imaginary* and is called a *space-like interval*. In such a case one can always find an inertial frame traveling at a velocity less than c such that the two events occur simultaneously but at different points in space. No observer can eliminate the space interval between the events regardless of his state of relative motion. The imaginary proper time interval corresponds to the transit time for a light pulse between the two spatial points.

At this point it is worthwhile to summarize the principal features of the four-vector formulation. First, all four-vectors are Lorentz covariant, that is, they transform like Equation 1.25. Second, the "length" of a four-vector is a scalar that is invariant in any inertial frame, as, indeed, is any scalar product of two four-vectors. Third, the first three components of a four-vector form a vector in three-space, but not all three–vector components are components of a four-vector (for example, the three–velocity). Fourth, unlike a three-vector, a four-vector *can* have zero magnitude without all of its components being zero.

9.   RELATIVISTIC COLLISIONS

In the majority of experiments involving relativistic velocities, the observer is at rest in the laboratory and the events under study are taking place between particles which, for the most part, are moving at high velocities with respect to the laboratory. The physical nature of these events, however, depends upon the amount of energy available to do work in the zero-momentum frame of the particles. (The zero-momentum frame is the center-of-mass frame of classical mechanics.) If two particles approach each other and exchange momentum and energy we call the event a *scattering* event. If the particles are not the same in *number* or in *kind after* the collision we say that a *reaction* has occurred. In either case, it is often convenient to do the theoretical analysis in the zero-momentum frame and then transform the results to the laboratory frame in order to check against experiments.

a. The zero-momentum frame (C-frame). We will designate the total energy, total kinetic energy and total momentum in the zero-momentum frame by E^*, T^*, and $\vec{P}^*$, respectively. Of course, $\vec{P}^* = 0$, by definition. Consider two identical particles of rest mass m_0, one particle having momentum $\vec{P}_a$ and kinetic energy T_a and the other being at rest in the laboratory (*L*-frame). If we call β^* the relative velocity of the *C*-frame, the Lorentz transformation may be written,

$$
\begin{pmatrix} P_1 \\ P_2 \\ P_3 \\ P_4 \end{pmatrix}^* = \begin{pmatrix} \gamma^* & 0 & 0 & i\beta^*\gamma^* \\ 0 & 1 & 0 & 0 \\ 0 & 0 & 1 & 0 \\ -i\beta^*\gamma^* & 0 & 0 & \gamma^* \end{pmatrix} \begin{pmatrix} P_1 \\ P_2 \\ P_3 \\ P_4 \end{pmatrix},
$$

where $P_1 = P_a$, $P_2 = P_3 = 0$, and $P_4 = iE/c = (i/c)(E_a + E_b) = (i/c)(T_a + 2m_0c^2)$.

$$
P_1^* = \gamma^*(P_1 + i\beta^* P_4)
$$
$$
P_2^* = P_2
$$
$$
P_3^* = P_3
$$
$$
P_4^* = \gamma^*(P_4 - i\beta^* P_1).
$$

Substituting for P_1 and P_4 it follows that:

$$
P^* = \gamma^*\left(P_a - \frac{\beta^* E}{c}\right) = 0
$$

and

$$
E^* = \gamma^*(E - \beta^* c P_a).
$$

From the invariance of the length of the momentum four-vector we have,

$$
E^{*2} = E^2 - P_a^2 c^2.
$$

From the first of these three equations we see that the velocity of the *C*-frame is given by

$$
\beta^* = \frac{P_a c}{E} = \sqrt{1 - \left(\frac{m_0 c^2}{E}\right)^2}.
$$

Putting $P_a c = \beta^* E$ into the second or third equation above,

$$E^* = \sqrt{1 - \beta^{*2}}\, E. \tag{1.34}$$

From the third equation we can easily obtain an expression for the kinetic energy of each particle in the C-frame in terms of the laboratory kinetic energy of the incident particle, T_a. Thus,

$$
\begin{aligned}
E^{*2} &= (E_a + m_0 c^2)^2 - P_a^2 c^2 \\
&= (E_a^2 - P_a^2 c^2) + 2 E_a m_0 c^2 + (m_0 c^2)^2 \\
&= (m_0 c^2)^2 + 2 E_a m_0 c^2 + (m_0 c^2)^2 \\
&= 2 m_0 c^2 (E_a + m_0 c^2) \\
&= 2 m_0 c^2 (T_a + 2 m_0 c^2). \tag{1.35}
\end{aligned}
$$

Since the energy is equally divided between the two particles in the C-frame, the energy of either particle is

$$E_a^* = T_a^* + m_0 c^2 = \tfrac{1}{2} E^*,$$

and the kinetic energy of each particle is

$$T_a^* = \tfrac{1}{2} E^* - m_0 c^2.$$

EXAMPLE

In the rest frame of an electron, a positron is observed coming directly toward the electron with momentum given by $pc = 4$ MeV. Find the relative velocity of the C-frame, the total energy in the C-frame, and the kinetic energy of each particle in the C-frame.

Each particle has a rest mass energy of $m_0 c^2 = 0.51$ MeV. Then the energies of the positron and the electron in the L-frame are:

$$
\begin{aligned}
E_a &= \sqrt{(pc)^2 + (m_0 c^2)^2} = \sqrt{4^2 + (0.51)^2}\ \text{MeV} = 4.03\ \text{MeV} \\
E_b &= m_0 c^2 = 0.51\ \text{MeV}
\end{aligned}
$$

and

$$E = E_a + E_b = 4.54\ \text{MeV}.$$

The relative velocity of the C-frame is given by Equation 1.33:

$$\beta^* = \frac{4.00}{4.54} = 0.88,$$

Then the total energy in the C-frame is

$$E^* = \sqrt{1 - \beta^{*2}}\, E = 2.14\ \text{MeV},$$

and

$$T_a^* = T_b^* = 1.07 - 0.51 = 0.56\ \text{MeV}.$$

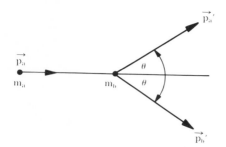

Figure 1-9 Symmetric collision in the L-frame.

b. Elastic collision of two particles of equal mass. Consider two uncharged particles of rest mass m_0 such that particle a is moving toward particle b, which is at rest in the L-frame. In order to keep the algebra simple let us treat the symmetric collision, after which the momenta of the two masses make equal angles with the direction of the incident particle in the L-frame. Primes will be used to denote values *after* the collision. Using the conservation of momentum, $p_a' \sin \theta = p_b' \sin \theta$, or $p_a' = p_b'$. Then, $p_a = p_a' \cos \theta + p_b' \cos \theta = 2p_a' \cos \theta$. It follows that

$$\cos \theta = \frac{p_a}{2p_a'} = \frac{\sqrt{E_a^2 - (m_0 c^2)^2}}{2\sqrt{E_a'^2 - (m_0 c^2)^2}} \tag{1.36}$$

From conservation of energy we have, $E_a + m_0 c^2 = E_a' + E_b' = 2E_a'$, since the particles share equally the momentum and the energy. Then Equation 1.36 becomes

$$\cos \theta = \sqrt{\frac{E_a^2 - (m_0 c^2)^2}{(E_a + m_0 c^2)^2 - (2m_0 c^2)^2}} = \sqrt{\frac{E_a + m_0 c^2}{E_a + 3m_0 c^2}}. \tag{1.37}$$

In the classical limit the kinetic energy of the incident particle is small compared to its rest mass, and $E_a \sim m_0 c^2$. Then $\cos \theta = \sqrt{2}/2$, and the angle between the paths of the particles is 90°. This result holds even for asymmetric scattering in the classical limit as long as the masses are equal. As the incident energy increases so that $E_a > m_0 c^2$, the scattering angle gets smaller and both particles tend to go off in the forward direction.

c. Inelastic collisions. If kinetic energy is not conserved during a collision we call that collision inelastic, and the kinetic energy that disappears is converted to mass or potential energy. By way of illustration consider the following example: A particle of rest mass m_0 moving with velocity $\beta_1 c$ collides inelastically with and sticks to a stationary particle of rest mass M_0. Find the velocity of the composite particle. The momentum in the L-frame is

$$p = \gamma_1 m_0 \beta_1 c,$$

and the total energy in the L-frame is

$$E = \gamma_1 m_0 c^2 + M_0 c^2.$$

Then, from Equation 1.33, if we call $\beta*$ the relative velocity of the C-frame,

$$\beta* = \frac{pc}{E} = \frac{\gamma_1 m_0 \beta_1 c^2}{\gamma_1 m_0 c^2 + M_0 c^2} = \frac{\gamma_1 m_0 \beta_1}{\gamma_1 m_0 + M_0},$$

and $\beta*c$ is the velocity of the composite particle in the L-frame. All of the kinetic energy that existed in the C-frame *before* the collision has been converted to mass, since there is no kinetic energy in the C-frame after the collision.

PROBLEM 1-24

What is the rest mass of the composite particle in the above example after the collision?

PROBLEM 1-25

A 1 GeV proton is moving toward an alpha particle which is at rest in the L-frame.
(a) What is the velocity of the proton in the L-frame?
(b) What is the velocity of the C-frame (zero momentum frame) relative to the L-frame?
(c) What is the total kinetic energy in the C-frame *before* the collision?
(Ans.: 0.875c; 0.298c; 740 MeV.)

d. The Q equation for inelastic scattering. The change of kinetic energy in the C-frame is often designated by the symbol Q, that is,

$$Q = T'* - T*,$$

where the prime denotes a value after the collision. Since the change in kinetic energy is numerically equal to the change in rest mass energy in the C-frame we can write,

$$Q = T'* - T* = \left(\sum_i m_i\right)c^2 - \left(\sum_i m_i\right)'c^2. \tag{1.38}$$

For $Q = 0$, the collision is elastic; for $Q > 0$, the rest mass decreases, the kinetic energy in the C-frame increases, and the reaction is exoergic; for $Q < 0$, the rest mass increases, the kinetic energy decreases and the reaction is endoergic. Equation 1.38 is most useful when Q is expressed in terms of the energy of the incident particle and its angular deflection by the collision. Let us designate the incident particle as particle a with rest mass m and kinetic energy T_a in the L-frame. The target particle b has a mass M and is initially at rest in the L-frame. Then from conservation of momentum we have

$$\vec{p}_a = \vec{p}'_a + \vec{p}'_b$$

$$p'^2_b = p^2_a + p'^2_a - 2p_a p'_a \cos \theta,$$

where θ is the angle through which particle a is scattered in the L-frame. Conservation of energy in the L-frame requires that

$$E_a + E_b = E'_a + E'_b$$

$$T_a + (M + m)c^2 = T'_a + m'c^2 + T'_b + M'c^2.$$

Then, Equation 1.38 becomes:

$$Q = (M + m - M' - m')c^2 = T'_a + T'_b - T_a. \qquad (1.39)$$

Since this equation gets quite complicated for relativistic velocities it will be derived here for the classical case instead. It has been shown[17] that the correct relativistic expression can be obtained from the classical result by substituting for each rest mass m_i the value $m_i + (T_i/2c^2)$, where T_i is the value in the L-frame. Then, using the classical expression for the kinetic energy, Equation 1.39 becomes:

$$Q = \frac{p'^2_a}{2m'} + \frac{p'^2_b}{2M'} - \frac{p^2_a}{2m} = \frac{p'^2_a}{2m'} - \frac{p^2_a}{2m} + \frac{1}{2M'}(p^2_a + p'^2_a - 2p_a p'_a \cos \theta)$$

$$= \frac{1}{2}\left(\frac{1}{m'} + \frac{1}{M'}\right)p'^2_a + \frac{1}{2}\left(\frac{1}{M'} - \frac{1}{m}\right)p^2_a - \frac{p_a p'_a}{M'}\cos \theta$$

or,

$$Q = T'_a\left(1 + \frac{m'}{M'}\right) - T_a\left(1 - \frac{m}{M'}\right) - \frac{2\sqrt{mm' T_a T'_a}}{M'}\cos \theta. \qquad (1.40)$$

Equation 1.40 is frequently used in nuclear physics to obtain the energy released by a nuclear reaction or the threshold energy to induce a reaction, both of which are given by the value of Q. Note that T_a, T'_a, and θ are all measured in the L-frame.

10. THE CREATION AND ANNIHILATION OF PARTICLES

An important consequence of the equivalence of mass and energy is that under certain circumstances particles can be created or destroyed, the mass difference being accounted for by the disappearance or appearance of energy. Whether or not such a transformation occurs in a given physical situation is governed by the requirements of conservation laws and the availability of

---

[17] A. B. Brown, C. W. Snyder, W. A. Fowler, and C. C. Lauritsen, *Phys. Rev.* **82**, 159 (1951).

competing reactions or processes. The conservation laws with which we shall be concerned here include the classical requirements on total linear momentum and total angular momentum, as well as the relativistic statement of the conservation of mass + energy. The classical conservation laws for mass and energy are no longer valid separately, but the quantity conserved is the total *relativistic energy* which includes the rest masses of all particles in the system under consideration. In addition, we shall accept and make use of the conservation laws for total charge and total spin* for a system of particles.

These laws are very simple to apply. For example, if a system has, initially, a total charge of zero and a total spin of zero, it must remain in a state of no net charge and no net spin. Thus, if a charged particle is created or destroyed, it must be accompanied by another particle of equal but opposite charge. The same rule applies to spin, where we can use the terminology "positive or negative" or "up or down" to designate the orientation of a hypothetical vector which represents the spin.

In the case of particle creation there must be sufficient kinetic energy in the C-frame to provide for the masses of the new particles. Consequently, in the L-frame the experimentalist must have a means of obtaining projectiles of very high energy. In the case of heavy particle production it is sometimes more practical to use two colliding beams of high energy projectiles in order to optimize the energy available in the C-frame. For instance, if two particles of equal mass and velocity could be directed toward each other in a head-on collision, the C-frame as a whole would have no kinetic energy, and the available energy in the C-frame would be a maximum.

By way of illustration, consider the production of a proton–antiproton pair by bombarding protons with protons. Since the proton has a charge and a spin angular momentum, the conservation laws require that its antiparticle (having opposite charge and spin) must be created simultaneously. Multiplying the mass of two protons by c^2 gives a rest energy of 1876 MeV. Therefore, the threshold energy in the C-frame for this reaction is 1876 MeV. The question we must answer is, what must be the energy of the bombarding proton in the L-frame in order to initiate the reaction?

The reaction may be represented by the expression,

$$p + p \rightarrow p + p + p + \bar{p}.$$

Here the bar over the particle symbol indicates an anti-particle. Note that before the collision there are two particles, but after the collision there are four particles of equal rest mass. In what follows we will assume that the target proton was initially at rest in the L-frame. After the collision, the minimum total energy in the C-frame is

$$E^* = 4M_0c^2,$$

assuming that all four particles are at rest in this frame. Before the collision, the total energy in the C-frame was

$$2M_0c^2 + T^*.$$

---

* Spin will be discussed in detail in section 9 of Chapter 3 and in Chapter 8.

Therefore, in the C-frame the threshold value of T^* is $2M_0c^2$. Then

$$T^* = 2M_0c^2 = 2(M - M_0)c^2 = 2(\gamma^* - 1)M_0c^2$$

or

$$\gamma^* = 2.$$

From Equation 1.34, the total energy in the L-frame must be

$$E = \gamma^* E^* = 2E^* = 8M_0c^2.$$

Since $E = T + 2M_0c^2$, we find that the threshold kinetic energy in the L-frame for this reaction is $T = 6M_0c^2$.

PROBLEM 1-26

A positive pion can be produced through the reaction $p + p \rightarrow p + n + \pi^+$ by bombarding protons at rest with high–energy protons. Find the minimum kinetic energy for the incident protons (in the L-frame) to initiate this reaction. Take the π^+ rest mass energy to be about 140 MeV.

The previous examples have illustrated the creation of particles from kinetic energy, but it is also possible to convert electromagnetic radiation quanta† to rest mass energy. A common example of this is the production of a positron-electron pair by the annihilation of a gamma ray quantum, a process frequently referred to as "pair-production." Since the electron and positron each require 0.51 MeV of rest mass energy, it is evident that the threshold photon energy for this process is slightly over 1 MeV. Pair production occurs only in the presence of a nucleus or some other particle which can be given some of the initial momentum of the photon.

PROBLEM 1-27

Show that the conservation laws cannot be satisfied for the conversion of an isolated photon into a positron-electron pair. Let the energy of the photon be $h\nu$ and take its momentum to be $h\nu/c$.

The inverse process, the annihilation of a particle–antiparticle pair, also occurs with the emission of two or more energetic photons. At least two photons are always required to conserve linear momentum, but we will see that the

---

† The quantum theory of radiation will be developed in Chapter 2. However, the concept of the photon as a quantum of energy of amount $h\nu$ will be used freely here, since it should be familiar to the reader from an earlier course.

number of photons emitted is determined by considering the spin conservation law as well as the dynamical conservation laws.

Since electrons and positrons each have a spin of $\frac{1}{2}$, there are two possibilities for the initial spin state. If the particle spins are parallel, the total spin of the system is unity; if they are antiparallel, the system has zero spin. Photons, on the other hand, each have a spin of 1, so the production of two photons would necessarily result in either a spin of 0 or a spin of 2. In order to obtain a state having a total spin of 1 and satisfying the linear momentum conservation law, we must have at least three photons! Therefore, for electron-positron annihilation two photons will be emitted in the antiparallel spin case and three photons will generally be emitted in the case of parallel spins.

PROBLEM 1-28

Assuming that electrons and positrons having equal energies collide head-on, what laboratory kinetic energy per particle is required to produce the reaction,

$$e^+ + e^- \rightarrow \pi^+ + \pi^-?$$

Use 139.6 MeV for the rest mass energies of the pions. (Ans.: 139.1 MeV.)

II. THE RELATIVISTIC DOPPLER SHIFT

The four-vector formalism already derived may be used to obtain the Doppler frequency shift in a straightforward manner. We need only incorporate the Planck expression for the energy of a photon, $E = h\nu$, and the relativistic expression for the momentum, $p = h\nu/c$, where ν is the frequency. Consider a monochromatic source at rest in the unprimed frame as shown in Figure 1-10. An observer moving to the right at velocity βc will detect a frequency which is

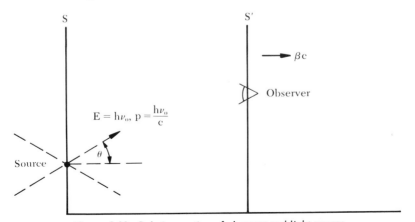

Figure I-10   Relative motion of observer and light source.

lower than the proper frequency. For the case of the moving observer we have the transformation equation,

$$\mathbf{P}'_\mu = \mathbf{\Gamma P}_\mu.$$

The equation for the fourth component becomes

$$(P_4)' = \gamma(P_4 - i\beta P_1),$$

or,

$$\frac{ih\nu'}{c} = \gamma\left(\frac{ih\nu_0}{c} - i\beta \frac{h\nu_0}{c}\cos\theta\right)$$

and

$$\nu' = \frac{\nu_0(1 - \beta\cos\theta)}{\sqrt{1 - \beta^2}}.$$

As the observer in the S' frame moves toward the source from the left, he sees light emitted at the angle $\theta = \pi$ and he measures the frequency,

$$\nu' = \frac{\nu_0\sqrt{1 + \beta}}{\sqrt{1 - \beta}}. \qquad \text{(Approaching observer)}$$

As he passes the source he sees light emitted at the angle $\theta = \pi/2$ and he measures,

$$\nu' = \frac{\nu_0}{\sqrt{1 - \beta^2}}. \qquad \text{(Transverse Doppler effect)}$$

As he recedes to the right, the observer in S' measures light emitted at the angle $\theta = 0$ and he obtains,

$$\nu' = \frac{\nu_0\sqrt{1 - \beta}}{\sqrt{1 + \beta}}. \qquad \text{(Receding observer)}$$

Now suppose we consider the inverse transformation in which the observer is regarded as stationary in the S frame and the source is moving. In matrix form this is (see Problem 1-20),

$$\mathbf{P}_\mu = \mathbf{\Gamma}^{-1}\mathbf{P}'_\mu.$$

The equation for the fourth component is

$$P_4 = \gamma(P'_4 + i\beta P'_1)$$

or,

$$\nu = \frac{\nu_0(1 + \beta\cos\theta')}{\sqrt{1 - \beta^2}}.$$

Again we have the three cases:

(a) Approaching source: $\theta' = 0$, and $\nu = \dfrac{\nu_0\sqrt{1 + \beta}}{\sqrt{1 - \beta}}$

(b) Receding source: $\theta' = \pi$, and $\nu = \dfrac{\nu_0 \sqrt{1 - \beta}}{\sqrt{1 + \beta}}$

(c) Transverse motion: $\theta' = \dfrac{\pi}{2}$, and $\nu = \dfrac{\nu_0}{\sqrt{1 - \beta^2}}$.

The above results are consistent with the principle of relativity in that it is not possible to distinguish between motion of the source or motion of the observer. A relative velocity of separation decreases the observed frequency and a relative velocity of approach increases the observed frequency. Note that the transverse effect is simply an example of the phenomenon of time dilation, that is, that "moving" clocks run slowly. These results may be summarized by the vector equation,

$$\nu' = \frac{\nu_0(1 - \beta \hat{n} \cdot \hat{v})}{\sqrt{1 - \beta^2}},$$

where β is always positive, $\hat{n}$ is the unit wave vector, and $\hat{v}$ is the unit vector in the direction of the velocity of the primed frame.

PROBLEM I-29

The light emitted by hydrogen atoms which are traveling at a velocity of $0.1c$ is observed in the laboratory from the forward and backward directions by means of a mirror.[18] Calculate the wavelength separation between the forward and backward beams for the spectral line of wavelength $\lambda_0 = 4861$ Å.
(Ans.: 977 Å.)

PROBLEM I-30

Assume that the stars of a binary pair revolve about their center of mass with a linear velocity of 6000 km/sec. If one of the stars emits a spectral line of wavelength 6000 Å, what will an observer measure for this line when the emitting star is receding, approaching, and moving transverse to the line of sight if (a) the observer is in the rest frame of the center of mass, and (b) the observer is moving with speed $0.6c$ with respect to the center of mass frame of the binary?
(Ans.: (a) 6120 Å, 5880 Å, 5999 Å; (b) 12,240 Å, 3060 Å, 5094 Å.)

---

[18] This is the experiment of H. E. Ives and G. R. Stilwell, *J. Opt. Soc. Am.* **28,** 215 (1938) and **31,** 369 (1941), but with an exaggerated atomic velocity.

PROBLEM 1-31

In its own rest frame, a π° meson decays into two photons of equal energy and equal but opposite momenta. If a π° has a velocity v_0 in the laboratory, what are the highest and lowest photon frequencies that can be produced when it decays?

12. GEOMETRIC REPRESENTATIONS

It is possible to make two-dimensional plots of space-time events if we confine our attention to only two components of $\mathbf{X}_\mu$, say $X_1 = x$ and $X_4 = ct$. Such a diagram is shown in Figure 1-11. The line given by $x = ct$ is the "world line" of a light wave, since it represents the progress of events moving in space-time at velocity c. The world line of a particle can trace out a curve of any shape provided the angle between the time axis and the tangent to the curve $\leq 45^\circ$. This restriction is obvious, since

$$\tan \theta = \frac{1}{\dfrac{d}{dx}(ct)} = \frac{1}{c}\frac{dx}{dt} = \frac{v_x}{c} \leq 1.$$

Now consider a frame S' traveling at velocity βc relative to S in the x-direction. The line $x' = 0$, which corresponds to $ct = x/\beta$ (from Equation 1.6), becomes the time axis in the primed frame. The line $ct' = 0$, which corresponds to $ct = \beta x$, is the space axis in the primed frame. Hence, the angle α between

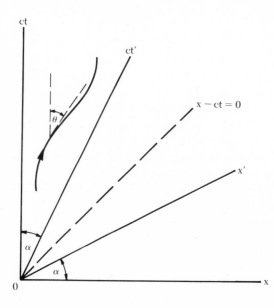

Figure 1-11 Space-time diagram. The curved line represents the world line of a particle. The angle through which the primed axes are rotated is given by $\tan \alpha = \beta$.

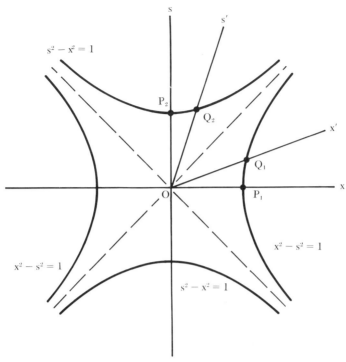

Figure I-12 Calibration curves for the coordinate axes. The segments OP_1, OP_2, OQ_1 and OQ_2 each represent unit interval along their respective axes.

the two space axes and between the two time axes is given by

$$\tan \alpha = \beta,$$

and these axes can be drawn once β is known. Now it is necessary to obtain the calibrations for the two sets of axes. For convenience define the new variables $s = ct$ and $s' = ct'$. Then the Lorentz transformations are:

$$
\begin{aligned}
x' &= \gamma(x - \beta s) & x &= \gamma(x' + \beta s') \\
s' &= \gamma(s - \beta x) & s &= \gamma(s' + \beta x')
\end{aligned}
\tag{1.41}
$$

Now draw the hyperbolae $s^2 - x^2 = \pm 1$ as shown in Figure 1-12. Event Q_1 is the space-time point determined by the intersection of the line $s = \beta x$ and the right branch of the hyperbola, $x^2 - s^2 = 1$. Then the coordinates of Q_1 are $x = \gamma$ and $s = \gamma\beta$ in the unprimed frame, but in the primed frame they are $x' = 1$ and $s' = 0$. The latter are obtained from the relations in Equation 1.41. Therefore, interval $\overline{OQ_1}$ corresponds to unit length along the spatial axis in the primed frame. Similarly, $\overline{OQ_2}$ corresponds to unit time along the time axis in the primed frame. Likewise, the vertices of the hyperbolae give unit calibrations of the unprimed axes.

The concept of simultaneity is readily visualized from space-time diagrams, as a glance at Figure 1-13 will show. Events P_1 and P_2 are simultaneous in the primed frame, but occur at times s_1/c and s_2/c in the unprimed frame.

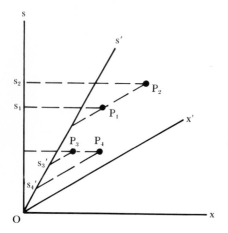

Figure 1-13 Concept of simultaneity. P_1 and P_2 are simultaneous in the primed frame while P_3 and P_4 are simultaneous in the unprimed frame.

Conversely, events P_3 and P_4 are simultaneous in the unprimed frame but not in the primed frame. The Lorentz contraction and time dilation are shown in Figure 1-14, where L_0 and τ_0 represent the proper lengths and proper times in each case.

In discussions of cosmology, the form of space-time diagram that is frequently used is the Minkowski or world-time diagram such as that shown in Figure 1-15. Note the similarity between this figure and Figure 1-12, with the calibration curves removed.

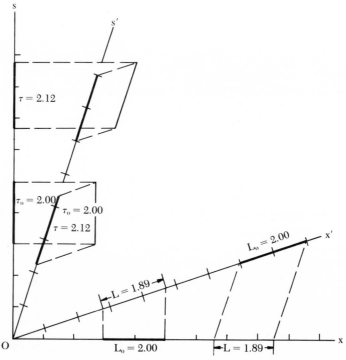

Figure 1-14 Graphical illustration of time dilation and the Lorentz contraction. The scale used here is for $\gamma = 1.06$.

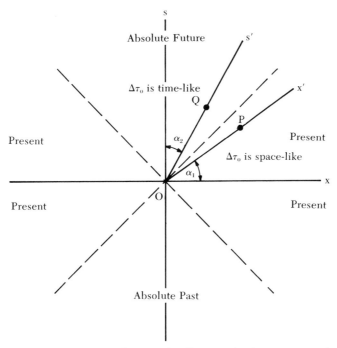

Figure I-15 World-time diagram. An event P will appear simultaneous to an observer at O by transforming to a frame x'. An event Q will appear to be localized at O by transforming to a frame s'.

This section has been merely a brief introduction to some of the basic ideas of the geometrical approach to the solution of problems in special relativity. To some the geometrical approach is fresh and appealing; others prefer the analytical methods stressed in this book. Those readers who wish to pursue the intricacies of the geometrical method are referred to the recent book by Shadowitz.[19]

SUMMARY

The two postulates of the special theory of relativity are that (1) all inertial frames are equivalent for the description of all physical laws, and (2) that the speed of light in vacuum is the same for all observers. Since the Lorentz transformation follows directly from these postulates, any physical law which retains the same mathematical form under a Lorentz transformation automatically satisfies the principle of relativity. Such a law is said to be *Lorentz covariant*. Two important kinematical effects which derive from the theory of relativity are the contraction of lengths parallel to the relative motion and the dilation of time. An important dynamical effect is the increase of mass with velocity, which leads to the statement of the equivalence of mass and

---

[19] Albert Shadowitz, *Special Relativity*. W. B. Saunders Company, Philadelphia, 1968.

energy, $E = mc^2$. The conservation of total energy, which includes rest mass energy, now replaces the separate conservation theorems for mass and energy in classical physics. The concept of a four-vector is developed from the treatment of time as a fourth dimension on an equal footing with the three dimensions of Euclidean geometry. The invariance of the "length" of a four-vector is shown to be a useful property when solving physical problems. Applications are made to particle collisions, particle creation and annihilation, and the Doppler shift. The use of the space-time diagram to solve problems graphically is briefly introduced.

**SUGGESTED
REFERENCES**

P. G. Bergmann, *Introduction to the Theory of Relativity*. Prentice-Hall, Englewood Cliffs, N.J., 1962.

A. Einstein, *The Meaning of Relativity*. Princeton University Press, Princeton, N.J., 1946.

A. P. French, *Special Relativity*. W. W. Norton, New York, 1968.

J. D. Jackson, *Classical Electrodynamics*. John Wiley and Sons, New York, 1962, Chap. 11, 12.

R. B. Leighton, *Principles of Modern Physics*. McGraw-Hill, New York, 1959, Chap. 1.

C. Møller, *The Theory of Relativity*. Oxford University Press, New York, 1952.

W. K. H. Panofsky and M. Phillips, *Classical Electricity and Magnetism*. Addison-Wesley, Reading, Mass., 1955, Chap. 14–17.

R. Resnick, *Introduction to Special Relativity*. John Wiley and Sons, New York, 1968.

W. G. V. Rosser, *An Introduction to the Theory of Relativity*. Butterworths, London, 1964.

J. H. Smith, *Introduction to Special Relativity*. W. A. Benjamin, New York, 1965.

F. W. Van Name, Jr., *Modern Physics*. Prentice-Hall, Englewood Cliffs, N.J., 1962, Chap. 3.

F. K. Richtmyer, E. H. Kennard, and J. N. Cooper, *Introduction to Modern Physics*. McGraw-Hill, New York, 1969, 6th edition.

A. Shadowitz, *Special Relativity*. W. B. Saunders, Philadelphia, 1968.

CHAPTER 2

THE BEGINNINGS OF THE QUANTUM THEORY

In this chapter we will review some of the significant events which led to the realization that electromagnetic radiation is quantized, and further, that these quanta—called photons—display both wave and particle characteristics. What at first seems like a conflict between contradictory natures will be resolved by adopting the point of view that a photon is a wave packet which can participate in either wave-like or particle-like interactions.

I. BLACKBODY RADIATION

The quantum theory of radiation had its origin in the search for an explanation of the spectral distribution of the radiant energy emitted by a heated body. By the term *spectral distribution*, we refer to the relative amount of energy associated with each wavelength interval of the emitted radiation. It has been known for a long time that the color of a heated object changes to a dull red at about 1100°K and that the color of the visible light emitted shifts toward the blue end of the spectrum as the temperature rises further. Experimental curves of the distribution of energy with wavelength for a given equilibrium temperature show the same general characteristics, regardless of the material of the body. Hence, it is natural to define an ideal *blackbody*, which is a perfect absorber (and emitter) of radiation. Since it reflects no light at all, it must appear perfectly black unless it is *emitting* light in the visible region of the spectrum. If a pulse of visible light, made up of a narrow band of frequencies, were incident upon such a blackbody, all of the light would be absorbed and would then be reradiated in all directions at greatly reduced intensity and with a different spectral distribution. It turns out that a blackbody which is in thermal equilibrium with its surroundings has a constant spectral distribution of radiated energy which is characteristic of all blackbodies maintained at that same temperature. Moreover, the spectral distribution curve for a real object can be predicted by multiplying the blackbody curve by the absorptivity of the real body.

A study of the radiation from an ideal blackbody can be approximated experimentally by observing the light emerging from a small hole in an isothermal enclosure, such as a hollow block of carbon. If the hole is sufficiently

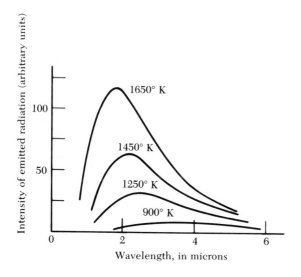

Figure 2-1 Spectral distribution of blackbody radiation for several different temperatures.

small, the energy radiated through it will have a negligible effect upon the equilibrium state in the cavity. For such a cavity, a typical plot of the spectral distribution of energy for several absolute temperatures is shown in Figure 2-1. The curves shown were obtained by Lummer and Pringsheim[1] in 1900, although their general characteristics were known earlier. It should be noted that (1) the short wavelength cutoff advances toward the origin as the temperature increases, (2) raising the temperature increases the energy of all spectral components, and (3) the peak of the curve shifts to shorter wavelengths as the temperature increases. The shift of the peak of the curve was found to obey the following empirical relationship, commonly called *Wien's displacement law*,[2]

$$\lambda_p T = \text{constant}, \tag{2.1}$$

where the symbol λ_p refers to the value of the wavelength corresponding to the peak of the curve. A thermodynamic expression also exists which relates the total power radiated per unit area of a blackbody to its absolute temperature. This is known as the Stefan-Boltzmann law and is expressed mathematically as

$$E(T) = \sigma T^4, \tag{2.2}$$

where $\sigma = 5.6699 \times 10^{-8}$ watts m^{-2} deg^{-4}. Thus the total energy radiated in a given time by a heated object is proportional to the fourth power of its absolute temperature. The monochromatic emissive power, $E(\lambda, T)$, is the power radiated per unit area at a given wavelength. This is related to the total power radiated per unit area by the integral,

$$E(T) = \int_0^\infty E(\lambda, T) \, d\lambda.$$

[1] O. Lummer and E. Pringsheim, *Verhandlungen der Deutschen physikalischen Gesellschaft* **1**, 23, 215 (1899); **2**, 163 (1900).
[2] W. Wien, *Ann. Physik* **58**, 662 (1896).

Wien proposed an empirical form for the monochromatic emissive power, $E(\lambda, T)$, by merely constructing a mathematical function to fit the experimental blackbody curves. Although he did not completely define the functional form, but wrote only $E(\lambda, T) = \lambda^{-5} f(\lambda, T)$, the following expression is commonly known as Wien's law:

$$E(\lambda, T) = \frac{ae^{-b/\lambda T}}{\lambda^5}.$$

$\qquad\qquad\qquad\qquad\qquad\qquad\qquad\qquad\qquad\qquad\qquad\qquad\qquad\qquad$ (2.3)

The quantities a and b are not derived but are simply curve-fitting parameters. Although Wien's radiation law is consistent with both the Stefan-Boltzmann law and the displacement law, it is quite unsatisfactory as a theory since it is not derivable from a physical model. That is, there was no attempt to relate the emitted radiation to physical processes taking place within the radiating body.

PROBLEM 2-1

The peak of the radiation curve for a certain blackbody occurs at a wavelength of 10,000 Å. If the temperature is raised so that the total radiated energy is increased 16-fold, at what wavelength will the new intensity maximum be found?
(Ans.: 5000 Å.)

PROBLEM 2-2

From Wien's law as given in Equation 2.3, derive the displacement law and the Stefan-Boltzmann law. Express the constants in Equations 2.1 and 2.2 in terms of a and b.
$\left(\text{Ans.:} \quad \sigma \doteq \dfrac{6a}{b^4}; \quad \lambda_p T = \dfrac{b}{5}.\right)$

2. THE RAYLEIGH-JEANS THEORY

The simplest model of a radiating body is to regard it as a collection of a large number (on the order of 10^{23}) of linear oscillators performing simple harmonic motion. Since the particles undergoing the oscillations are, in general, charged particles, they will radiate electromagnetic waves. In the case of a cavity as discussed above, at thermal equilibrium the electromagnetic energy density inside the cavity will equal the energy density of the atomic oscillators situated in the cavity walls. When the walls are raised to a higher temperature, the following events take place: more energy is put into existing oscillator modes by increasing their amplitudes, new modes corresponding to stiffer spring constants (higher frequencies) are excited, and the radiation density in the cavity is increased until a new equilibrium point is reached.

According to the classical theory of the equipartition of energy, an average energy of $\frac{1}{2}kT$ is associated with each position coordinate or momentum that appears as a quadratic term in the Hamiltonian. Since the Hamiltonian for a linear oscillator may be written in the form

$$H = T + V = \frac{p_x^2}{2m} + \tfrac{1}{2}kx^2,$$

it has two such variables, p_x and x (called degrees of freedom), and hence the average total energy is simply kT. All that is necessary, then, to obtain the spectral energy density in the cavity is to find $n(\nu)$, the number of oscillators per unit volume at each frequency ν, and to multiply this number by kT. This number is known as *Jeans' number* and is calculated in Appendix A of this chapter. Its value is*

$$n(\nu) = \frac{8\pi\nu^2}{c^3} \quad \text{or} \quad n(\lambda) = \frac{8\pi}{\lambda^4}. \tag{2.4}$$

We then obtain the important Rayleigh-Jeans law,[3] namely,

$$I(\lambda, T) = \frac{8\pi kT}{\lambda^4}, \tag{2.5}$$

where $I(\lambda, T)$ is the energy per unit volume at wavelength λ at the equilibrium temperature $T°$K. The relationship between $I(\lambda, T)$ and $E(\lambda, T)$ is given in Problem 2-5.

A difficulty with Equation 2.5 immediately appears when we consider very small wavelengths. Although the Rayleigh-Jeans law describes the experimental curve quite well in the long wavelength region, it diverges (in the mathematical sense of having no limit) as the wavelengths approach zero. This failure was such a crushing blow to classical physics that it is historically referred to as "the ultraviolet catastrophe." We can gain some appreciation of the difficulty physicists faced at that time if we realize that the Rayleigh-Jeans theory utilized the important equipartition theorem of classical physics, and further, that it had no adjustable parameters. Its failure implied that there was something fundamentally wrong with either the equipartition theorem or the theory of electromagnetic radiation.

3. PLANCK'S QUANTUM THEORY OF RADIATION

Being well aware of the shortcomings of both the Rayleigh-Jeans and the Wien radiation laws, Planck examined the mathematical statistics of these

---

* Conversion between the parameters ν and λ is easily done if one remembers that $n(\nu)\,d\nu = n(\lambda)\,d\lambda$. Then

$$n(\nu) = n(\lambda) \cdot \left|\frac{d\lambda}{d\nu}\right| = n(\lambda) \cdot \frac{c}{\nu^2}$$

[3] Lord Rayleigh, *Phil. Mag.* **49,** 539 (1900) and J. H. Jeans, *Phil. Mag.* **10,** 91 (1905).

theories to ascertain what changes, if any, might result in a reasonable description of the experimental radiation curve. As a result of this work, he was led to certain conclusions about the nature of the electromagnetic oscillators which are in equilibrium with the energy density within the blackbody cavity. These postulates, which have become the foundation of the quantum theory of radiation, are as follows:[4]

(1) The amount of energy emitted or absorbed by an oscillator is proportional to its frequency. Calling the constant of proportionality h, we then write for the change in oscillator energy,

$$\Delta \epsilon = h\nu.$$

(2) An oscillator cannot have an arbitrary energy but must occupy one of a discrete set of energy states given by

$$\epsilon_n = nh\nu,$$

where n is an integer or zero. It was assumed that the ground state corresponded to the zero energy state. The value Planck gave for the constant of proportionality, h, was 6.55×10^{-27} erg-sec. He obtained this value by fitting curves such as those shown in Figure 2-1 with his radiation law (which will be derived below [see Equation 2.8]). However, the best value[5] for h is now believed to be

$$h = 6.626196 \times 10^{-27} \text{ erg-sec.}$$

Planck's constant is a universal constant which plays an important role in all quantum phenomena.

The previous picture of a continuum of oscillator states is now replaced by a discrete set of "quantized" states. Furthermore, the amount of energy emitted or absorbed is also quantized, since each quantum must correspond to the energy difference between two states of a given oscillator. Each quantum of electromagnetic energy is called a photon. The absorption of a photon of frequency ν will raise the energy of an oscillator of frequency ν by an amount given by $h\nu$; it will have no effect on an oscillator of frequency $\nu' \neq \nu$. Emission of a photon occurs when the oscillator energy drops to the next lower energy; the frequency of the emitted light will correspond to the oscillator frequency.

In what state is an oscillator most likely to be found? If nothing excites it, it is most likely to be found in its lowest energy state or ground state. Hence, at absolute zero one would expect to find all oscillators in the zero energy state according to the above model. This would be true in classical mechanics and can be assumed here without affecting our answer. But we will see later that quantum mechanics predicts a so-called "zero-point motion" at absolute zero instead of the complete cessation of all vibration. At higher temperatures thermal agitation excites some oscillators to higher states so that some sort of distribution of oscillators over all possible states will exist for each temperature.

[4] M. Planck, *Ann. Physik* **4**, 553 (1901).
[5] The best value has been obtained from measurements of e/h using the effect in superconductors predicted by B. D. Josephson in *Physics Letters* **1**, 251 (1962). See B. N. Taylor, W. H. Parker, and D. N. Langenberg, *Reviews of Modern Physics* **41**, 375 (1969).

The required distribution is the Maxwell–Boltzmann function derived in Appendix B of this chapter, that is,

$$N(n) = N_0 e^{-\epsilon_n/kT}. \tag{2.6}$$

The higher energy states are thus less likely to be populated, and as the energy increases indefinitely the number of such oscillators becomes vanishingly small.

It can now be seen that this is going to eliminate the problem of the ultraviolet catastrophe, because the latter arose as a result of the assumption that oscillators of all energies were excited with equal probability; hence, all energies contributed equally to the emitted radiation. From the quantum hypothesis we see that a high energy oscillator can contribute more radiated energy than a low energy oscillator only *if* it is excited. But its probability of being excited is so low that the energy which appears at the high-frequency end of the spectrum is much less than the classical Rayleigh-Jeans theory predicted.

Mathematically, this can be shown by a calculation of the average energy per oscillator, $\bar{\epsilon}$, based on Planck's quantum hypothesis.

$$\bar{\epsilon} = \frac{\sum\limits_{n=0}^{\infty} N(n)\epsilon_n}{\sum\limits_{n=0}^{\infty} N(n)}, \text{ where the sums are taken over the energy levels, } \epsilon_0, \epsilon_1, \epsilon_2 \cdots$$

$$\bar{\epsilon} = \frac{\sum\limits_{n=0}^{\infty} N_0 e^{-nh\nu/kT} nh\nu}{\sum\limits_{n=0}^{\infty} N_0 e^{-nh\nu/kT}} = \frac{0 + h\nu e^{-h\nu/kT} + 2h\nu e^{-2h\nu/kT} + 3h\nu e^{-3h\nu/kT} + \cdots}{1 + e^{-h\nu/kT} + e^{-2h\nu/kT} + e^{-3h\nu/kT} + \cdots}$$

$$= h\nu x \left(\frac{1 + 2x + 3x^2 + 4x^3 + \cdots}{1 + x + x^2 + x^3 + \cdots} \right), \text{ where } x = e^{-h\nu/kT}$$

$$= h\nu x \cdot \frac{(1-x)^{-2}}{(1-x)^{-1}} = \frac{h\nu x}{1-x} = \frac{h\nu}{e^{h\nu/kT} - 1}. \tag{2.7}$$

PROBLEM 2-3

By treating the summation above as an integration over an energy continuum, show that the average energy per oscillator becomes the classical value, $\bar{\epsilon} = kT$.

PROBLEM 2-4

Show that $\bar{\epsilon}$ may be written as $\bar{\epsilon} = -d/d\alpha \ln \sum\limits_{n=0}^{\infty} x^n = -d/d\alpha \ln (1-x)^{-1}$, where $\alpha = 1/kT$ and $x = e^{-\alpha h\nu}$. Perform the differentiation and obtain Equation 2.7.

Using Jeans' number, Equation 2.4, and the average energy per oscillator obtained in Equation 2.7, we write the Planck radiation law as

$$I(\lambda, T) = n(\lambda) \cdot \bar{\epsilon} = \frac{8\pi hc}{\lambda^5} (e^{hc/\lambda kT} - 1)^{-1}$$

or,

$$I(\nu, T) = n(\nu) \cdot \bar{\epsilon} = \frac{8\pi h\nu^3}{c^3} (e^{h\nu/kT} - 1)^{-1}$$

(2.8)

Planck's law in this form is in units of energy per unit volume per interval of wavelength (or frequency). To express it in terms of radiated power per unit area per interval of wavelength (or frequency), one need merely multiply by the factor $c/4$.

PROBLEM 2-5

Assume that $I(\lambda, T)$ results from N beams, each of energy density $U(\lambda, T)$ and speed c. Derive the result that $E(\lambda, T) = \frac{c}{4} I(\lambda, T)$, where $E(\lambda, T)$ is the radiated power per unit area per wavelength interval.

PROBLEM 2-6

Using Planck's law as given in Equation 2.8,
(a) Show that it reduces to the Rayleigh-Jeans law in the limit as λ gets large.
(b) Show that it reduces to Wien's law in the short wavelength limit. Evaluate a and b.
(c) Evaluate the Stefan-Boltzmann constant.
(d) Evaluate the constant in the Wien displacement law.

(Ans.: (b) $a = 2\pi hc^2 = 3.73 \times 10^{-16}$ watt-m²;

$$b = \frac{hc}{k} = 1.43 \times 10^{-2} \text{ m-°K}$$

(c) $\sigma = \frac{2\pi^5 k^4}{15c^2 h^3} = 5.67 \times 10^{-8}$ watt/m²-°K⁴

(d) $\lambda_p T = \frac{hc}{4.965k} = 2.898 \times 10^{-3}$ m-°K)

As Problem 2-6 illustrates, the Planck theory of radiation incorporates all that is valid from the older theories as special cases. It thus serves as an excellent example of a conceptual advance which opened exciting new frontiers while still preserving much of the older physics.

PROBLEM 2-7

Einstein[6] considered the thermal vibrations of a solid to be equivalent to a large number of 3-dimensional oscillators, all having the same frequency ν. If the average energy per 1-dimensional oscillator is E_{av}, then the total energy per mole is $U = 3N_0 E_{av}$.

(a) Using the value for E_{av} given by the equipartition theorem, show that the specific heat is given by the Dulong-Petit value,

$$C_V = \frac{\partial U}{\partial T} = 3R,$$

where $R = N_0 k$, the molar gas constant.

(b) Now use Planck's value for E_{av} and obtain C_V. Show that this reduces to $3R$ at high temperatures.

PROBLEM 2-8

Show that the Stefan-Boltzmann law can be derived from thermodynamics if one starts with the classical expression

$$P = \tfrac{1}{3}u;$$

where P is the pressure and u the energy density of isotropic radiation. Assume that the ideal gas law holds in the cavity.

PROBLEM 2-9

Assume that the sun radiates as a blackbody. If $\lambda_p = 5000$ Å for the solar spectrum, what is the surface temperature of the sun?

4. EINSTEIN'S TRANSITION PROBABILITIES

Einstein[7] made use of Planck's quantum hypothesis to study the probability of radiative transitions for a system in equilibrium with electromagnetic radiation. The radiation field may be thought of as a photon gas of energy density $I(\nu)$. Let the atomic system consist of N atoms, with N_1 atoms occupying energy level E_1 and N_2 atoms excited to energy E_2; that is, $E_2 > E_1$. Photons

---

[6] A. Einstein, *Ann. Physik* **22**, 180 (1907).

[7] A. Einstein, *Verhandl. deut. physik. Ges.* **18**, 318 (1916); also *Physik. Z.* **18**, 121 (1917).

with energy $h\nu = E_2 - E_1$ can be absorbed by atoms in the E_1 state, thus raising them to the E_2 state; or, they can induce atoms in the E_2 state to emit photons of energy $h\nu$ and drop to the E_1 state. The photons radiated by induced or *stimulated emission* are in phase with the radiation field and are thus in phase with each other; that is, they are *coherent*. Einstein also assumed that *spontaneous* transitions can occur from E_2 to E_1, but not in the other direction. The probabilities for these three types of transitions may be summarized as follows:

$B_{12}I(\nu)$ is the probability of absorption of a photon,

$B_{21}I(\nu)$ is the probability of stimulated emission of a photon, and

A_{21} is the probability of spontaneous emission of a photon,

where A and B are constants. Then the transition rate for emission is

$$[A_{21} + B_{21}I(\nu)] \cdot N_2,$$

and the transition rate for absorption is

$$B_{12}I(\nu) \cdot N_1.$$

At equilibrium these two rates must be equal, so we have

$$\frac{N_1}{N_2} = \frac{A_{21} + B_{21}I(\nu)}{B_{12}I(\nu)}. \qquad (2.9)$$

Now the equilibrium populations, N_1 and N_2, are given by the Boltzmann distribution function (Equation 2.6) as:

$$\frac{N_1}{N_2} = \exp\left(-\frac{E_1}{kT}\right)\exp\left(\frac{E_2}{kT}\right) = \exp\left(\frac{h\nu}{kT}\right),$$

where T is the temperature corresponding to the equilibrium state. Then,

$$\exp\left(\frac{h\nu}{kT}\right) \cdot B_{12}I(\nu) = A_{21} + B_{21}I(\nu)$$

and

$$I(\nu) = \frac{A_{21}}{B_{12}\exp\left(\dfrac{h\nu}{kT}\right) - B_{21}}.$$

For extremely large energy densities it is reasonable to assume that $N_1 \sim N_2$ and that $B_{21}I(\nu) \gg A_{21}$. Then from Equation 2.9 we get the result that

$$B_{12} = B_{21},$$

and

$$I(\nu) = \frac{\dfrac{A_{21}}{B_{21}}}{\exp\left(\dfrac{h\nu}{kT}\right) - 1}. \tag{2.10}$$

It is evident that this is the Planck radiation law (Equation 2.8) and that the ratio of the Einstein coefficients is

$$\frac{A_{21}}{B_{21}} = \frac{8\pi h\nu^3}{c^3}.$$

The B coefficient is readily obtained* quantum mechanically by regarding the atom as a dipole oscillator driven by the time-dependent electric field intensity of the radiation field. Contributions from other charge configurations and from interactions with the magnetic field may also be considered, but the most pronounced effect of the radiation field is simply to give each atom an induced electric dipole moment. Hence, the radiation resulting from the interactions between atoms and electromagnetic radiation is normally called *electric dipole radiation*. The calculation of the relative intensities of the spectral lines arising from electric dipole transitions is one of the triumphs of modern physics.

The extension of Einstein's analysis to non-equilibrium cases has become the basis for the amplification of signals in laser and maser devices. Note that if the population of the upper level is increased or the population of the lower level is decreased such that $N_2/N_1 > 1$, more photons of energy $h\nu$ will be emitted than absorbed. Furthermore, the emitted photons are in phase with the incident beam, so that amplification can result. In order to achieve the non-equilibrium state, an additional process is required or at least one other energy level must be involved. For example, suppose a state $E_3 > E_2$ exists such that its lifetime is short and its most probable spontaneous decay involves a transition to state E_2 rather than a return to E_1 (or any other lower state). Suppose further that E_2 is a metastable state.† Then an external signal whose frequency corresponds to $(E_3 - E_1)/h$ can be utilized to "pump" atoms from the E_1 state to the E_3 state. This has the effect of depopulating state E_1 and, if

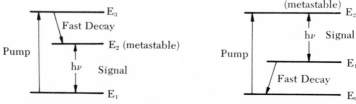

Figure 2-2 Three-level lasers.

* This is done in section 6 of Chapter 9.
† A state is said to be metastable if the probability for a *spontaneous* transition out of that state is very small or zero. The physical basis for metastable states will become clear later when quantum mechanical transitions are discussed.

the $E_3 \rightarrow E_2$ decay is fast, over-populating state E_2. Alternatively, suppose a state E_0 exists such that the $E_1 \rightarrow E_0$ transition is fast and E_2 is a metastable state. Then pumping can occur between the E_0 and E_2 levels. There are also four-level systems which combine the processes shown in Figure 2-2. In all cases of amplification of a continuous signal by means of stimulated emission, the pump must run continuously to maintain the non-equilibrium populations required for successful operation.

PROBLEM 2-10

Using the Boltzmann distribution, Equation 2.6, find the temperature that corresponds to equal populations for two levels. What temperature is required in order for the population of the upper level to be greater than that of the lower level? How does this statistical use of the term "temperature" compare with its use in thermodynamics?

PROBLEM 2-11

Which of the systems shown in Figure 2-2 would operate more efficiently if the laser material were cooled? Show this analytically.

PROBLEM 2-12

A certain semiconductor has electron energy levels just 0.1 eV below its conduction band. In order to control the number of electrons in the conduction band, an experimenter plans to cool the material and excite electrons to the conduction band by means of a weak source of photons of the correct energy. He decides to use a carbon rod inside the cryostat as a source of radiation.

(a)   If the sample is kept at 4°K, what is the probability that an electron will be thermally excited to the conduction band?

(b)   At what temperature must the carbon rod be kept so that the peak of the blackbody radiation curve corresponds to the energy gap of 0.1 eV?

Much of the significance of laser (or maser) action lies in the fact that it can provide an intense source of *coherent* radiation. In fact, an incoherent source acting as a pump can be converted to a coherent source at the pump frequency if an appropriate pair of energy levels can be found in a physical system.

Although the term "laser" can be used for all such devices employing photons, it is common practice to regard a device operating in the microwave region as a maser and one operating at optical frequencies as a laser. However, this distinction cannot be held too rigidly, since it is possible to pump at microwave frequencies and to amplify at optical frequencies, or vice versa. Since there are regions of the electromagnetic spectrum wherein it is extremely difficult either to obtain a monochromatic source or to amplify an existing signal, it would be attractive to devise lasers which could operate at these frequencies but which could be pumped at frequencies that are readily attainable (such as "white" light). This goal provides impetus for a great deal of research on the energy levels of atomic, molecular and many-body systems.[8]

5. THE PARTICLE NATURE OF PHOTONS

The measurement of the charge of the electron by Millikan[9] in 1909 established the fact that charge, as well as energy, is quantized. That is, any accumulation of charge must consist of an integral multiple of the electronic charge. Once the value of the electronic charge was known, it was possible to determine the electron's mass from the value of e/m obtained by Thomson[10] and others. As a result of a great deal of research with cathode ray tubes, a number of interesting demonstrations were devised to verify the particle nature of the electron. The electron's mass is $m_0 = 9.109 \times 10^{-28}$ gram and its charge is $e = 4.803 \times 10^{-10}$ statcoulomb $= 1.602 \times 10^{-19}$ coulomb.

It is easy to accept the particle nature of the electron because we can define its mass, we can accelerate it, and we can make it behave as we think a particle ought to behave. A photon, on the other hand, has no rest mass and cannot be accelerated; however, we have seen that it has momentum associated with it in both classical electromagnetic theory and in the special theory of relativity (see Equation 1.18). We will now discuss two very important experimental events which can be best explained by assuming that the electron interacts with a single photon as if the photon were a localized particle rather than a wave front. Thus, the quantum or particle nature of light dominates its wave nature in these experiments. The first is the well-known *photoelectric effect* and the second is the *Compton effect*.

6. THE PHOTOELECTRIC EFFECT

Light incident upon a metal surface can, under some conditions, eject electrons from the surface. These electrons are called photoelectrons, not that they differ from other electrons, but merely to identify their source. The following facts must be explained by a satisfactory theory of the photoelectric

---

[8] B. Lengyel, *Introduction to Laser Physics.* John Wiley and Sons, Inc. New York, 1966. See also, *Lasers and Light, Readings from Scientific American.* W. H. Freeman and Co., San Francisco, 1969.

[9] R. A. Millikan, *Phil. Mag.* **19,** 209 (1910); *Phys. Rev.* **32,** 349 (1911).

[10] J. J. Thomson, *Phil. Mag.* **44,** 293 (1897).

effect:

(1) If the incident light has no frequency component above a certain threshold frequency, no photoelectrons will be emitted.

(2) The threshold frequency appears to depend upon the properties of the metal.

(3) Provided that the threshold frequency is exceeded, photoelectrons will be emitted *instantly*,[11] regardless of how low the intensity of the light source is.

(4) The photoelectric current (number of electrons emitted per second) increases with the intensity of the light but is independent of the frequency.

(5) The maximum kinetic energy of the photoelectrons is independent of the intensity of the light and depends only upon its frequency.

Attempts to explain the above phenomena by means of the classical radiation theory met with defeat. Classically, an electron would be ejected as soon as it could accumulate enough energy to overcome the force which binds it to the metal. This energy is called the *work function* of the metal. Since an electromagnetic wave has an energy density associated with it, one would expect that the metal should accumulate enough energy, ultimately, to eject an electron, regardless of the frequency or the intensity of the source. However, the classical picture was completely at a loss to explain all of the above experimental facts except number four. From the classical point of view it was reasonable to expect the photoemission to increase when the energy density of the incident radiation is increased.

The answer, of course, is the one for which Einstein[12] was awarded the Nobel prize in 1921, and which is neatly expressed by the following equation:

$$h\nu = W + T. \tag{2.11}$$

This merely states the conservation of energy; $h\nu$ is the energy of the incident photon, W is the work function of the metal and T is the kinetic energy of the photoelectron. The threshold frequency is immediately given by the condition

$$\nu_c = \frac{W}{h},$$

which states that all of the energy of the photon of frequency ν_c is required to remove the electron and no energy is left over to provide its kinetic energy. Thus, all of the above experimental results are accounted for on the basis of the exchange of energy between a single photon and a single electron, if we but add the postulate that the number of photoelectrons should increase with the number of photons having $\nu \geq \nu_c$.

The maximum kinetic energy of the photoelectrons is determined experimentally by applying a stopping voltage, or reversed potential, to the photocell as is shown in Figure 2-3. T_{max} is given by the zero-current intercept in

---

[11] Lawrence and Beams reported that the delay must be less than 10^{-9} sec. *Phys. Rev.* **29,** 903 (1927).

[12] A. Einstein, *Ann. Physik* **17,** 132 (1905).

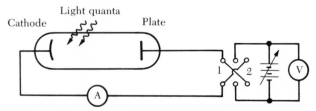

Figure 2-3 Apparatus for determining the photocurrent (switch in position I) and the maximum kinetic energy of the electrons produced by photoemission (switch in position 2).

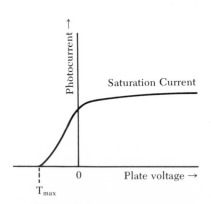

Figure 2-4 Plot of photocurrent vs. plate voltage for a given photon frequency. T_{max} is determined from the value of the reversed potential that cuts off the photocurrent.

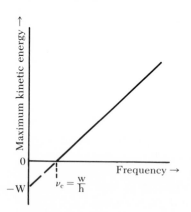

Figure 2-5 Maximum kinetic energy of photoelectrons vs. photon frequency.

Figure 2-4. Figure 2-5 shows a plot of T_{max} *vs.* ν which permits the experimental determination of both h and W in Einstein's equation. The first experimental verification of this equation was made by Millikan in 1916.[13]

PROBLEM 2-13

If a retarding potential of 5 volts just stops the fastest photoelectrons emitted from cesium, what is the wavelength of the most energetic incident photons? Use 1.8 eV for the work function of cesium.
(Ans.: 1840 Å.)

PROBLEM 2-14

If the photoelectric threshold wavelength of sodium is 5420 Å, calculate the maximum velocity of photoelectrons ejected by photons of wavelength 4000 Å.
(Ans.: 5.36 × 10⁵ m/sec.)

---

[13] R. A. Millikan, *Phys. Rev.* **7,** 18,355 (1916).

PROBLEM 2-15

(a) Light of wavelength 4000 Å is incident upon lithium. If the work function for lithium is 2.13 eV, find the kinetic energy of the fastest photoelectrons.

(b) What would be the maximum wavelength of photons capable of ejecting photoelectrons from lithium at a velocity of $0.95\,c$?

(Ans.: 0.97 eV; 0.01 Å.)

PROBLEM 2-16

If the photocurrent of a photocell is cut off by a retarding potential of 0.92 volts for monochromatic radiation of 2500 Å, what is the work function of the material? (Ans.: 4.08 eV.)

With the advent of lasers capable of emitting coherent radiation at high power levels, two-photon photoemission has now been observed in sodium metal.[14] Theory predicts that the double-quantum photocurrent should be proportional to the square of the power of the incident radiation as opposed to the nearly linear relationship that holds for the single-quantum photoeffect.[15]

7. THE COMPTON EFFECT

It was generally known among the early workers with monochromatic x-rays that a scattered beam always contained a longer wavelength component in addition to the incident wavelength. Compton[16] made a systematic study of the scattering of x-rays from carbon and obtained the spectrum shown in Figure 2-6. The surprising thing is that the wavelength shift is independent of the wavelength of the source and the scattering material, although it varies with the scattering angle. However, Compton succeeded in explaining this peculiar effect by treating the x-ray photon as a particle which undergoes a collision with a rest mass particle such as an electron or an atom as a whole. This is shown schematically in Figure 2-7.

The mathematical analysis of the collision requires only the conservation of energy and momentum. From momentum conservation we have

$$p_0 = p' \cos \theta + p \cos \phi$$

and

$$0 = p' \sin \theta - p \sin \phi,$$

---

[14] M. C. Teich, J. M. Schroeer, and G. J. Wolga, *Phys. Rev. Letters* **13,** 611 (1964).
[15] R. L. Smith, *Phys. Rev.* **128,** 2225 (1962).
[16] A. H. Compton, *Phys. Rev.* **22, 409** (1923).

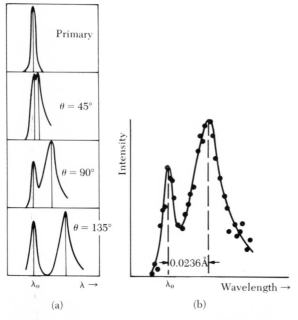

Figure 2-6 Scattering of molybdenum K$_\alpha$ x-radiation from graphite. (a) Variation of the shifted line with scattering angle. The peak at λ_0 is due to the incident x-ray beam. (b) Compton's data for the wavelength shift for 90° scattering. (From A. H. Compton, *Phys. Rev.* **22,** 409 (1923), used with permission.)

from which we can eliminate ϕ by isolating the term in ϕ on the left side of each equation, squaring, and adding. Then,

$$p^2 = p_0^2 + p'^2 - 2p_0 p' \cos \theta. \tag{2.12}$$

The conservation of energy requires that

$$E_0 - E' = T,$$

or

$$p_0 - p' = \frac{T}{c}. \tag{2.13}$$

Squaring this,

$$p_0^2 + p'^2 - 2p_0 p' = \left(\frac{T}{c}\right)^2. \tag{2.14}$$

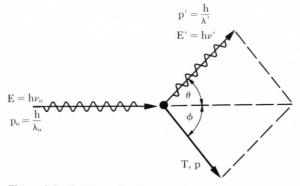

Figure 2-7 Collision of a photon with a free, stationary mass.

Subtracting Equation 2.14 from Equation 2.12,

$$p^2 - \frac{T^2}{c^2} = 2p_0 p'(1 - \cos\theta).\tag{2.15}$$

Since the velocity of the particle after the collision could well be relativistic, we must use the relativistic momentum, Equation 1.19,

$$p^2 = \frac{1}{c^2}(T^2 + 2m_0 c^2 T).$$

Then

$$p^2 - \frac{T^2}{c^2} = 2m_0 T = 2m_0 c(p_0 - p'),\tag{2.16}$$

where Equation 2.13 has been used to obtain the last expression. Using Equation 2.15 and Equation 2.16,

$$m_0 c(p_0 - p') = p_0 p'(1 - \cos\theta)$$

$$\frac{1}{p'} - \frac{1}{p_0} = \frac{1}{m_0 c}(1 - \cos\theta)$$

or,

$$\lambda' - \lambda = \frac{h}{m_0 c}(1 - \cos\theta).\tag{2.17}$$

Using the mass of the electron for m_0, the quantity $h/m_0 c = 0.0243$ angstrom and is now called the Compton wavelength. Notice that for 90° scattering, Equation 2.17 predicts a new x-ray line just 0.0243 angstrom longer than the primary line. Figure 2-6 shows this line as the large peak measured by Compton. The presence of the primary peak at 90° might at first seem surprising. Compton explained this by considering the scattering of a photon from the atom as a whole. Thus, if one uses the mass of a whole carbon atom instead of the electronic mass in Equation 2.17, the wavelength shift will be reduced by a factor of 20,000 and amounts to roughly one millionth of an angstrom. Therefore, the line scattered by an atom is for all practical purposes unshifted.

Compton's work provided rather convincing evidence that a photon can undergo particle-like collisions with both atoms and unbound electrons. Later studies of the recoil electrons and their energies added further confirmation to the predictions of the theory.[17]

PROBLEM 2-17

Show that a photon cannot transfer all of its energy to a free electron. What is the maximum recoil kinetic energy that can be given to an electron by a photon of energy E_0? (See Problem 1-27.)

---

[17] C. T. R. Wilson, *Proc. Royal Soc. (London)* **104**, 1 (1923); W. Bothe, *Z. Physik* **20**, 237 (1923); A. A. Bless, *Phys. Rev.* **29**, 918 (1927).

PROBLEM 2-18

What is the recoil kinetic energy of a free electron, initially at rest, after a Compton scattering event in which a 1 MeV gamma photon is scattered at 90°? Find the recoil angle of the electron.

(Ans. 0.662 MeV; 18.7°.)

PROBLEM 2-19

A 2 MeV gamma photon is scattered through an angle of 180° by an electron. What is the recoil kinetic energy of the electron?

(Ans. 1.77 MeV.)

PROBLEM 2-20

Photons of energy 0.1 MeV undergo Compton scattering. Find the energy of a photon scattered at 60°, the recoil angle of the electron, and the recoil kinetic energy of the electron.

(Ans. 0.91 MeV; 0.009 MeV; 55.4°.)

PROBLEM 2-21

Gamma rays of energy 1.02 MeV are scattered from electrons which are initially at rest. Find the angle for symmetric scattering at this energy (that is, the angles θ and ϕ are equal). What is the energy of the scattered photon for this case?

(Ans. 41.5°; 0.68 MeV.)

8. THE DUAL NATURE OF THE PHOTON

The particle nature of light, as illustrated by the photoelectric effect and the Compton effect, is no longer viewed as irreconcilable with the overwhelming evidence for its wave-like behavior. Such phenomena as interference and diffraction are peculiar to a wave description wherein the region of interaction is extended over a large portion of the wavefront, in contrast with the localized interactions of particles. Regarding the photon as a *wave packet* consisting of a superposition of many waves imparts to it some of the properties of both waves and material particles. Thus, the photon exhibits its wave nature when it interacts with an object such as a grating, where the details of the instantaneous phase of each of the constituent waves are important. On the other hand, it manifests its particle nature when energy and momentum of the packet as a whole are transferred to another particle. In the latter case the details of the phases of the constituent waves are unimportant. That both the wave and particle aspects of photons are required for the description of light is known historically as *Bohr's principle of complementarity*. That is, the wave and particle aspects complement each other. In order to acquire a better understanding of the nature of a wave packet, let us first consider the superposition of two plane waves.

In classical physics we frequently represent a plane wave traveling in the positive x-direction by either the real or the imaginary part of one of the following equivalent expressions

$$\Psi = Ae^{i(kx-\omega t)} = Ae^{i2\pi(x/\lambda-t/T)} = Ae^{i\omega(kx/\omega-t)}. \qquad (2.18)$$

Here A is the wave amplitude, $k = 2\pi/\lambda$ is the propagation constant, $\omega = 2\pi\nu$ is the angular frequency, and T is the period of the harmonic oscillation. The velocity of propagation of the wave front is the phase velocity, $u = \omega/k = \lambda\nu$.

Suppose that two waves having slightly different frequencies and wavelengths are propagating together through a medium. For simplicity let us take their amplitudes and initial phases to be equal. Then we may represent these two waves by the expressions,

$$\Psi_1 = A \sin (kx - \omega t)$$

and

$$\Psi_2 = A \sin [(k + dk)x - (\omega + d\omega)t],$$

where dk and $d\omega$ are infinitesimal quantities. Making use of the trigonometric identity,

$$\sin \alpha + \sin \beta = 2 \cos \tfrac{1}{2}(\alpha - \beta) \cdot \sin \tfrac{1}{2}(\alpha + \beta),$$

we may express the displacement resulting from the superposition of these two waves by,

$$\Psi = \Psi_1 + \Psi_2 = 2A \cos \frac{d\omega}{2} \left(\frac{x}{\dfrac{d\omega}{dk}} - t \right) \cdot \sin \omega' \left(\frac{x}{\dfrac{\omega'}{k'}} - t \right).$$

Note that $k' = k + dk/2 \sim k$, and $\omega' = \omega + d\omega/2 \sim \omega$, since $dk \ll k$ and $d\omega \ll \omega$. Then the factor containing the sine is essentially the same function as Ψ_1 and may be thought of as a "carrier wave" of phase velocity $u = \omega'/k' \sim \omega/k$. The cosine factor has the effect of modulating the amplitude of the carrier wave, and the modulation envelope moves at the so-called group velocity given by $v = d\omega/dk$. The transmission of energy (that is, a signal) must occur at the group velocity and not with the phase velocity. The reason for this will be clearer after a discussion of the Fourier integral theorem in Chapter 4, but it may be noted here that the transmission of a signal always involves modulation of one kind or another. An infinitely long wave train of a single frequency can never be used to transmit information at its phase velocity. Signaling always involves chopping (keying), amplitude modulation, frequency

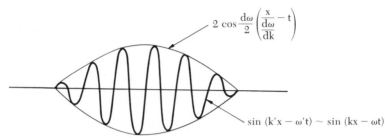

Figure 2-8 One half-wavelength of the modulation envelope formed from the superposition of two plane waves of nearly equal frequencies and equal amplitude. This pattern is repeated continuously throughout space.

modulation or phase modulation. These processes all result in the superposition of many plane waves of different frequencies grouped around some center frequency, that is, the formation of a *wave packet*. Since the envelope of the packet must have a finite spatial extent, the sum of the amplitudes of all of its plane wave components must be zero everywhere except where the packet is localized. The energy carried by the packet at its group velocity is thus analogous to the kinetic energy of a free particle of the same velocity. The packet shown in Figure 2-8, which is made up of only two components, is not zero elsewhere, but its envelope is repeated throughout space. In principle, however, the amplitudes elsewhere can be cancelled if enough component frequencies are used. How this is done analytically will be shown in Chapter 4, where the Fourier integral theorem will be used to construct localized packets to represent material particles.

If all of the wavelengths of a packet travel through a medium at the same phase velocity, there is said to be no *dispersion*. A physical consequence of this case is that the packet retains its shape as it propagates, regardless of the number of frequencies which comprise it. When dispersion occurs, the packet changes shape as it propagates. The dependence of frequency upon the propagation constant is called the *dispersion relation* and may be written,

$$\omega = u(k) \cdot k.$$

If the phase velocity u is a constant for all wavelengths, there is no dispersion and the group velocity and the phase velocity are equal. That is,

$$v = \frac{d\omega}{dk} = u.$$

9. THE HEISENBERG UNCERTAINTY PRINCIPLE

In the previous section it was argued that a localized packet of electromagnetic energy must be composed of a mixture of a large number of plane waves of differing frequencies. Paradoxically, the more nearly monochromatic the packet is, the broader it becomes. Thus, if the frequency spread is nearly zero, then the packet is so broad (its wave train is so long) that its emission time is extremely large. Light radiated by atoms during electronic transitions has a finite wave train which corresponds roughly to the lifetime of the state. In the time domain these bursts, which are of the order of 10^{-8} seconds, result in a spread of frequencies rather than in a single emitted frequency, and produce what is called the *natural linewidth* of the spectral line. The longer the lifetime, the fewer extraneous frequencies in the spectral line.

Denoting the uncertainty in the frequency (the frequency spread) by $\Delta\omega$ and the uncertainty in emission time by Δt, we can write,

$$\Delta\omega \sim \frac{2\pi}{\Delta t}.^{*}$$

On the other hand, if the degree of monochromaticity is described in terms of the spread in k-values (where $k = \omega/c$), the smaller the value of Δk, the greater the extent of the packet in coordinate space. That is, the uncertainty in the propagation constant and the uncertainty in position are related by

$$\Delta k_x \sim \frac{2\pi}{\Delta x}.^{*}$$

Making use of the Planck expression for the energy of the photon and the relativistic value for the photon momentum, the above relations suggest that

$$\Delta E \cdot \Delta t \sim h$$

and   (2.19)

$$\Delta p_x \cdot \Delta x \sim h.$$

This is an intrinsic limitation to the precision with which certain pairs of physical variables can be simultaneously measured under ideal circumstances in which there are no experimental or instrumental errors. Although these limitations are of great importance in quantum systems, they are completely unnoticeable in macroscopic systems because of the smallness of the Planck constant h. Thus, for a momentum which is known to within 10^{-7} g-cm/sec, the uncertainty in x is $\sim 10^{-20}$ cm, which is certainly not detectable. The relations in Equation 2.19 are two examples of the well-known *uncertainty principle* of Heisenberg, enunciated in 1927.[18] The same constraints apply to any generalized coordinate and the generalized momentum associated with that coordinate. Such variables are known as *canonically conjugate* variables in classical mechanics. Thus, if q is a generalized coordinate, then p_q is the generalized momentum conjugate to q. Other examples of pairs of conjugate variables are (y, p_y), (z, p_z), (θ, p_θ), and so on. In formal language, the uncertainty principle may be stated as follows: The product of the uncertainties in the measurement of two canonically conjugate variables must be greater

---

* These relationships will be derived in section 8 of Chapter 4.
[18] W. Heisenberg, Z. *Physik* **43**, 172 (1927).

than a quantity of the order of $\hbar$. That is,

$$\Delta q \cdot \Delta p_q \geq \hbar. \tag{2.20}$$

The uncertainty principle will be discussed quantitatively in sections 8 and 11 of Chapter 4 in connection with the behavior of wave packets and the concept of expectation value. Numerous examples of how it can be applied to physical problems will occur throughout the subsequent chapters of this book. The reader who is interested in pursuing this topic further at this point is referred to Heisenberg's own discussion of the subject.[19]

SUMMARY

The concept of the quantization of electromagnetic energy was the key to Planck's explanation of the spectral distribution of the energy radiated from a thermally excited blackbody. As a consequence of the Planck theory we now regard the process of absorption or emission of electromagnetic energy as the exchange of a photon—which is a quantum of energy and momentum*— between matter and the electromagnetic field. Although the photon can interfere with itself to produce diffraction effects, it behaves like a zero rest mass particle in such processes as the photoelectric effect and Compton scattering. These classically contradictory properties are reconciled by regarding the photon as a wave packet, that is, a superposition of a large number of waves representing a continuum of frequencies whose center frequency is given by $\nu = E/h$, that is, the packet energy divided by Planck's constant. The packet momentum is $p = h/\lambda$, which is required for consistency with the special theory of relativity. The particle and wave aspects of a photon are thus regarded as complementary properties which are manifestations of the particular way in which the photon interacts with a specific detector. Thus the photon exhibits its wave nature when it interacts with an object such as a grating, where the details of the instantaneous phase of each of the component waves are important. On the other hand, it displays its particle nature when it transfers energy and momentum to a material particle.

Also discussed in this chapter are the Einstein probabilities for spontaneous and induced transitions. These are shown to lead to a radiation energy density that is consistent with the Planck law. They are then used to illustrate the essential ideas behind the operation of a laser.

---

[19] See Chapter II of W. Heisenberg, *The Physical Principles of the Quantum Theory*. Dover Publications, Inc., New York.

* The photon also has unit spin, which is ignored in the present discussion.

CHAPTER 2   APPENDICES A to C

APPENDIX A.   CALCULATION OF JEANS' NUMBER

Consider the one-dimensional wave equation,

$$\frac{\partial^2 u}{\partial x^2} = \frac{1}{c^2}\frac{\partial^2 u}{\partial t^2},$$

where u is the displacement and c is the velocity of the wave. Since the variables are independent, we assume a product solution of the form $u = T(t) \cdot \phi(x)$. Substituting this into the wave equation we obtain the separated equation,

$$c^2\frac{\phi''(x)}{\phi(x)} = \frac{\ddot{T}(t)}{T(t)} = -\omega^2, \tag{A1}$$

where $-\omega^2$ is simply the separation constant. Solutions of the two equations (A1) are

$$T = C \sin \omega t + D \cos \omega t$$

and

$$\phi = \gamma \sin kx + \delta \cos kx, \tag{A2}$$

where $k = \omega/c = 2\pi/\lambda$.

An infinite medium can have any values of ω and λ, but a finite medium such as a string fastened at each end has only certain allowed values. These allowed values are called *eigenvalues*, and the modes are known as standing waves. If we impose the constraint that $u = 0$ for $x = 0$ and $x = L$, then we find that $\delta = 0$ and $k = n\pi/L$, where n is an integer. The general solution to (A1) becomes

$$u(x, t) = \gamma \sin \frac{n\pi x}{L}(C \sin \omega t + D \cos \omega t),$$

which may be rewritten in the form,

$$u(x, t) = F_n \sin (k_n x) \cos (\omega_n t) + G_n \sin (k_n x) \sin (\omega_n t). \tag{A3}$$

Although we will leave (A3) in its present form, one can show that it can be written as the superposition of waves traveling to the right and to the left. Also, the most general solution is one including a summation over all possible ω_n. If we consider one frequency, ω_n, we note that

$$\omega_n = k_n c = \frac{n\pi c}{L} = 2\pi\nu_n,$$

or,

$$n = \frac{2L}{c} \nu_n.$$

If the frequency modes are closely spaced we can express the density of modes as

$$dn = \frac{2L}{c} d\nu,$$

where dn is the number of modes in the frequency interval between ν and $\nu + d\nu$.

An analogous procedure can be used for three dimensions, where we now write the wave equation as

$$\nabla^2 u = \frac{1}{c^2} \frac{\partial^2 u}{\partial t^2},$$

and constrain $u(x, y, z, t)$ to be zero for $x = 0, L$; $y = 0, L$; $z = 0, L$. Since all of the variables are independent, the solution is a product function, the first term of which is

$$F_n \sin\left(\frac{n_x \pi x}{L}\right) \cdot \sin\left(\frac{n_y \pi y}{L}\right) \cdot \sin\left(\frac{n_z \pi z}{L}\right) \cdot \cos \omega_n t, \tag{A4}$$

provided that $\omega_n^2 = (\pi c/L)^2 (n_x^2 + n_y^2 + n_z^2) = (\pi c n/L)^2$. That is, n is now defined in terms of the three integers n_x, n_y, and n_z. Rearranging the last expression,

$$n^2 = \left(\frac{L\omega_n}{\pi c}\right)^2 = \left(\frac{2L}{c}\right)^2 \nu_n^2. \tag{A5}$$

Each set of integers (n_x, n_y, n_z) determines a point in lattice space and each such point occupies unit volume of the lattice. (Each unit cube contains eight points, but each lattice point is contained in eight such cubes.) Since there is a one-to-one correspondence between the volume of the space, the number of lattice points contained in that volume, and the number of allowed modes of a given frequency, we can count modes by merely calculating the volume in the lattice space. Therefore, the number of modes of frequency ν_n in the first octant of a sphere of radius n is

$$N = \frac{1}{8} \cdot \frac{4\pi}{3} n^3 = \frac{\pi}{6} \left(\frac{2L}{c}\right)^3 \nu_n^3. \tag{A6}$$

Then the number of modes in the frequency interval between ν_n and $\nu_n + d\nu$ is

$$dN = \frac{4\pi}{c^3} L^3 \nu_n^2 \, d\nu,$$

and the number of modes per unit volume in the same frequency interval is

$$\frac{dN}{L^3} = \frac{4\pi}{c^3} v_n^2 \, dv.$$

In the case of transverse waves there are actually two times this number of modes, one for each sense of polarization. Finally, we obtain Jeans' number, which is

$$n(v) = \frac{8\pi}{c^3} v^2. \tag{A7}$$

APPENDIX B.   THE MAXWELL-BOLTZMANN DISTRIBUTION FUNCTION

Suppose that a given volume is divided into z cells and that n distinguishable but otherwise identical particles are to be distributed among the z cells. If each particle has the same *a priori* probability of occupying any unit volume of the space, then intuition tells us that in the final distribution the number of particles occupying each cell should be proportional to the volume of the cell. Thus, a *uniform* distribution is the most probable distribution, and if the cell volumes are the same each cell would be expected to contain n/z particles. We will prove this conclusion formally before deriving the Maxwell-Boltzmann distribution function.

Let g_i be the fractional volume and n_i the number of particles occupying the ith cell such that

$$\sum_{i=1}^{z} g_i = 1 \quad \text{and} \quad \sum_{i=1}^{z} n_i = n.$$

Then the probability of a given distribution with n_1 particles in g_1, n_2 in g_2, and so on, is

$$W = \frac{n!}{n_1! n_2! \cdots n_z!} (g_1)^{n_1} (g_2)^{n_2} \cdots (g_z)^{n_z}. \tag{B1}$$

The first factor is the expression from combinatorial algebra for the number of configurations corresponding to a given distribution (that is, the number of ways it can be obtained) when neither the *order* in which particles enter the cells nor the *designation* of specific particles is important. The remaining factors give the *a priori* probability of obtaining a single configuration. For example, consider the case of distributing 9 particles over 3 equal cells such that $n_1 = 2$, $n_2 = 3$, and $n_3 = 4$. The 2 particles in the first cell can be put there in 2! equivalent ways, $n_2 = 3$ can be obtained in 3! ways, and $n_3 = 4$ can be obtained in 4! ways. Thus, the number of different configurations corresponding to the distribution (2, 3, 4) is given by

$$\frac{9!}{2! 3! 4!}.$$

But the *a priori* probability of putting any particle in a cell is $\frac{1}{3}$; for two particles in the same cell it is $(\frac{1}{3})^2$; and so on for n_i particles. Therefore, the probability of the distribution is

$$W = \frac{9!}{2!3!4!} \, (\tfrac{1}{3})^2 (\tfrac{1}{3})^3 (\tfrac{1}{3})^4,$$

which agrees with (B1). Before maximizing (B1) it is convenient to take its logarithm as follows:

$$\ln(W) = \ln(n!) - \ln(n_1!) - \ln(n_2!) - \cdots - \ln(n_z!)$$
$$+ n_1 \ln(g_1) + n_2 \ln(g_2) + \cdots + n_z \ln(g_z).$$

This can be simplified by means of Stirling's formula,

$$\ln(n!) \sim n \ln(n) - n,$$

where the approximation is valid for large n. Then,

$$\ln(W) = n \ln(n) + n_1 \ln\left(\frac{g_1}{n_1}\right) + n_2 \ln\left(\frac{g_2}{n_2}\right) + \cdots + n_z \ln\left(\frac{g_z}{n_z}\right). \quad (B2)$$

In order to find the most probable distribution we must maximize (B2). For this we will use the method of Lagrange's multipliers, which is as follows. Suppose we wish to find the extremum, given by (x_0, y_0, z_0), of a function $f(x, y, z) = 0$, subject to the constraints $\phi_1(x, y, z) = 0$ and $\phi_2(x, y, z) = 0$. We introduce two new parameters, λ_1 and λ_2, and write,

$$F = f + \lambda_1 \phi_1 + \lambda_2 \phi_2 = 0.$$

The conditions for the extremum are given by the five equations

$$\frac{\partial F}{\partial x} = 0, \quad \frac{\partial F}{\partial y} = 0, \quad \frac{\partial F}{\partial z} = 0, \quad \frac{\partial F}{\partial \lambda_1} = 0, \quad \frac{\partial F}{\partial \lambda_2} = 0,$$

which permit us to obtain the five parameters x_0, y_0, z_0, λ_1, and λ_2. The method of Lagrange's multipliers provides a direct and symmetric approach to all such problems involving large numbers of variables subject to any number of constraints. To apply the method to the present problem we note that $f = \ln(W)$, and since there is only one constraint we need only one multiplier, λ. The constraint is

$$\phi = n - \sum n_i = 0.$$

Then, for the ith equation we have,

$$\frac{\partial}{\partial n_i}[\ln(W)] + \lambda \frac{\partial \phi}{\partial n_i} = 0,$$

and we must solve the set of z equations obtained by letting i run from 1 to z. From the equation obtained by differentiating with respect to n_1 we obtain:

$$\ln\left(\frac{g_1}{n_1}\right) = \lambda + 1$$

or

$$g_1 = n_1 e^{\lambda+1} \tag{B3}$$

Similar expressions are obtained from all z equations, which, when added, yield

$$(g_1 + g_2 + \cdots + g_z) = (n_1 + n_2 + \cdots + n_z)e^{\lambda+1}$$

or

$$\frac{1}{n} = e^{\lambda+1}.$$

Then (B3) becomes

$$g_i = \frac{n_i}{n},$$

which says that the number of particles per cell is proportional to the size of the cell. That is, the most probable distribution is the uniform distribution.

Now we will consider the same problem with an additional constraint on the energy of the system. The cells may now be regarded as a discrete set of energy states, ϵ_i, such that the total energy is

$$\epsilon = \sum_1^z n_i \epsilon_i.$$

The ith Lagrange equation is

$$\frac{\partial}{\partial n_i}[\ln(W)] + \lambda_1 \frac{\partial \phi_1}{\partial n_i} + \lambda_2 \frac{\partial \phi_2}{\partial n_i} = 0,$$

where $\phi_1 = n - \sum n_i = 0$ and $\phi_2 = \epsilon - \sum n_i \epsilon_i = 0$. We then obtain the set of equations,

$$n_1 = g_1 e^{-(\lambda_1+1)} \cdot e^{-\lambda_2 \epsilon_1}$$

$$n_2 = g_2 e^{-(\lambda_1+1)} \cdot e^{-\lambda_2 \epsilon_2}$$

$$\cdots \cdots \cdots \cdots \cdots$$

Adding,

$$n = e^{-(\lambda_1+1)}(g_1 e^{-\lambda_2 \epsilon_1} + g_2 e^{-\lambda_2 \epsilon_2} + \cdots + g_z e^{-\lambda_2 \epsilon_z})$$

$$= e^{-(\lambda_1+1)}Z,$$

where we have defined the *partition function*, Z, by

$$Z = \sum g_i e^{-\lambda_2 \epsilon_i} = n e^{\lambda_1+1}. \tag{B4}$$

Then,

$$n_i = \frac{g_i n}{Z} e^{-\lambda_2 \epsilon_i}. \tag{B5}$$

To obtain the Maxwell-Boltzmann function we define $\lambda_2 = 1/kT$. Then,

$$n_i = n_0 e^{-\epsilon_i/kT}.$$

PROBLEM 2-22

Examine the function, $\int_0^a (\ddot{y})^2 \, dt$, for extrema subject to the constraint, $\int_0^a y \, dt = c$, where $y(0) = 0$ and $\dot{y}(0) = 0$. (Ans.: $\ddot{y} = -\lambda/2$.)

APPENDIX C.　THE MAXWELL VELOCITY DISTRIBUTION FUNCTION

A useful application of the Maxwell-Boltzmann function can be made to a monatomic gas to derive the Maxwell distribution law for velocities. Consider a volume element in velocity space,

$$d\tau = dv_x \, dv_y \, dv_z.$$

Since the cells are infinitesimal, we can replace the summation over g_i by an integration over $d\tau$. That is,

$$\sum g_i \rightarrow \int d\tau = \iiint_{-\infty}^{\infty} dv_x \, dv_y \, dv_z = \int_0^{\infty} 4\pi v^2 \, dv,$$

where v is the radius of a sphere in velocity space. The total number of molecules may be expressed using Equation B5 as

$$n = \sum_i n_i = \sum_i A g_i e^{-\lambda_2 \epsilon_i} = A \sum_i g_i \exp\left(-\frac{\lambda_2 m v_i^2}{2}\right)$$

$$n = 4\pi A \int_0^{\infty} v^2 \exp\left(-\frac{\lambda_2 m v_i^2}{2}\right) dv. \tag{C1}$$

Similarly, the total energy is

$$E = \sum_i n_i \epsilon_i = \sum_i A g_i \epsilon_i \exp\left(-\lambda_2 \epsilon_i\right)$$

$$E = 2\pi A m \int_0^{\infty} v^4 \exp\left(-\frac{\lambda_2 m v^2}{2}\right) dv. \tag{C2}$$

The integrals occurring above appear frequently enough in theoretical physics to justify the following listing:

$$I_0 = \int_0^\infty e^{-\alpha x^2} dx = \frac{1}{2}\sqrt{\frac{\pi}{\alpha}} \qquad \text{(Gauss' probability integral)}$$

$$I_1 = \int_0^\infty x e^{-\alpha x^2} dx = \frac{1}{2\alpha}$$

$$I_2 = \int_0^\infty x^2 e^{-\alpha x^2} dx = -\frac{dI_0}{d\alpha} = \frac{1}{4}\sqrt{\frac{\pi}{\alpha^3}}$$

$$I_3 = \int_0^\infty x^3 e^{-\alpha x^2} dx = -\frac{dI_1}{d\alpha} = \frac{1}{2\alpha^2}$$

$$I_4 = \int_0^\infty x^4 e^{-\alpha x^2} dx = \frac{d^2 I_0}{d\alpha^2} = \frac{3}{8}\sqrt{\frac{\pi}{\alpha^5}}$$

$$I_5 = \int_0^\infty x^5 e^{-\alpha x^2} dx = \frac{d^2 I_1}{d\alpha^2} = \frac{1}{\alpha^3}$$

. . . .

$$I_{2n} = (-1)^n \frac{d^n}{d\alpha^n} I_0$$

$$I_{2n+1} = (-1)^n \frac{d^n}{d\alpha^n} I_1.$$

Then,

$$n = A\left(\frac{2\pi}{\lambda_2 m}\right)^{\frac{3}{2}} \qquad\qquad \text{(C3)}$$

and

$$E = \frac{3}{2\lambda_2} A\left(\frac{2\pi}{\lambda_2 m}\right)^{\frac{3}{2}} = \frac{3n}{2\lambda_2}. \qquad\qquad \text{(C4)}$$

Using the result that the energy of an ideal gas is $\frac{3}{2}nkT$, we have

$$\frac{3n}{2\lambda_2} = \tfrac{3}{2}nkT,$$

or

$$\lambda_2 = \frac{1}{kT}$$

which is consistent with the definition of the previous section. Then,

$$A = n\left(\frac{m}{2\pi kT}\right)^{\frac{3}{2}}.$$

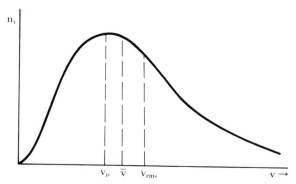

Figure 2-9 Maxwell velocity distribution function.

From (C1) we then find that the number of molecules having speeds between v and $v + dv$ is:

$$n_v = 4\pi n \left(\frac{m}{2\pi kT}\right)^{\frac{3}{2}} v^2 \exp\left(-\frac{mv^2}{2kT}\right), \tag{C5}$$

which is the Maxwell velocity distribution.

PROBLEM 2-23

Using the Maxwell velocity distribution and the integrals evaluated above, find the average molecular speed, the rms speed, and the most probable speed for the molecules of an ideal gas of temperature T. Show that these speeds are proportional to $1.12:1.22:1$, respectively.

SUGGESTED REFERENCES

A. Beiser, *Perspectives of Modern Physics*. McGraw-Hill Book Co., Inc., New York, 1969.

C. H. Blanchard, C. R. Burnett, R. G. Stoner, and R. L. Weber, *Introduction to Modern Physics, 2nd ed.* Prentice-Hall, Inc., Englewood Cliffs, New Jersey, 1969.

M. Born, *Atomic Physics, 6th ed.* Hafner Publishing Co., New York, 1959.

R. M. Eisberg, *Fundamentals of Modern Physics*. John Wiley and Sons, Inc., New York, 1961.

Richard P. Feynman, Robert B. Leighton, and Matthew Sands, *The Feynman Lectures on Physics*. Addison-Wesley Publishing Co., Inc., Reading, Mass., 1963.

R. B. Leighton, *Principles of Modern Physics*. McGraw-Hill Book Co., Inc., New York, 1959.

F. K. Richtmyer, E. H. Kennard, and J. N. Cooper, *Introduction to Modern Physics, 6th ed.* McGraw-Hill Book Co., New York, 1969.

Henry Semat, *Introduction to Atomic and Nuclear Physics, 4th ed.* Holt, Rinehart and Winston, New York, 1969.

Paul A. Tipler, *Foundations of Modern Physics*. Worth Publishers, Inc., New York, 1969.

F. W. Van Name, Jr., *Modern Physics, 2nd ed.* Prentice-Hall, Inc., Englewood Cliffs, N.J., 1962.

Hugh D. Young, *Fundamentals of Optics and Modern Physics*. McGraw-Hill Book Co., New York, 1968.

CHAPTER 3

THE CONCEPT OF THE NUCLEAR ATOM

This chapter addresses the question of the structure of the atom, its relationship to the more familiar elementary particles, the origin of electromagnetic spectra, and the source of the atomic magnetic moment. The Bohr model, when combined with the Planck theory, leads to a simple but unified quantum theory of matter and radiation.

I. THE ATOMIC MODELS OF THOMSON AND RUTHERFORD

The assumption that electrons are constituents of all atoms was a reasonable inference from such experiments as the photoelectric effect in metals, the ionization of gases in discharge tubes, x-ray bombardment, and so forth. A difficulty arose, however, in attempting to design a stable atom which could contain both electrons and the positive charges necessary to make the atom electrically neutral. It was supposed that these positive charges were many times heavier than electrons, and this was confirmed by the e/m measurements of positive ions by Thomson.[1] The hydrogen ion turned out to be 1836 times as heavy as the electron if one assumed the same magnitude of charge for each. Thomson proposed an atomic model in which electrons were embedded in a massive matrix of positive charge filling a volume of roughly one atomic diameter (which was known to be about 1 angstrom). This model has been called the "plum-pudding" model and the "jellium" model, the former term referring to the role of the electrons analogous to the raisins in a pudding, the latter deriving from the fact that the electrons were permitted to vibrate in order to account for the radiation spectrum discussed in the previous chapter.

After the identification of alpha particles by Rutherford and his coworkers,[2,3] a series of experiments was performed by Geiger and Marsden[4-6] in

---

[1] J. J. Thomson, *Phil. Mag.* **13,** 561 (1907).
[2] E. Rutherford and H. Geiger, *Proc. Roy. Soc. (London)* **81,** 141 (1908).
[3] E. Rutherford and T. Royds, *Phil. Mag.* **17,** 281 (1909).
[4] H. Geiger and E. Marsden, *Proc. Roy. Soc. (London)* **82,** 495 (1909).
[5] H. Geiger, *Proc. Roy. Soc. (London)* **83,** 492 (1910).
[6] H. Geiger and E. Marsden, *Phil. Mag.* **25,** 605 (1913).

which alpha particles of known energies were scattered from gold foils. The deflection of an alpha particle caused by a Thomson atom can be estimated by the following argument. Neglecting the binding energy of the electron to the atom, the maximum momentum that can be transferred to an electron of mass m_0 by an alpha particle of mass M and velocity v is $2m_0v$. Then the maximum angle of deflection of the alpha particle is approximately

$$\Delta\phi_{\max} = \frac{2m_0v}{Mv} = \frac{2m_0v}{4(1836)m_0v} \sim 10^{-4} \text{ radian,}$$

since the momentum given to the electron represents the change in momentum of the alpha particle. Similarly, one can estimate the deflection due to the continuous charge within the atom by calculating the total impulse given the alpha particle by the Coulomb force during its transit through the positive jellium. Thus,

$$\Delta\phi_{\max} = \frac{(\Delta F)_{\max} \cdot \Delta t}{mv} \sim \frac{\frac{2R}{v} \cdot (\Delta F)_{\max}}{mv} = \frac{2R}{mv^2}\int_0^{Ze} \frac{2e\,dq}{R^2} = \frac{4Ze^2}{MRv^2},$$

where R is the atomic radius ($\sim 10^{-8}$ cm) and $v \sim 2 \times 10^9$ cm/sec. Substituting numerical values results in a $\Delta\phi_{\max} \sim 10^{-4}$ or 10^{-5} radian. Since the probability of scattering from more than one electron within a single atom is quite small, we will assume that $\Delta\phi_{\max} \sim 10^{-4}$ radian per atom. Multiple scattering from a layer of such atoms is completely random and obeys the laws

$$\overline{\Delta\phi} = \sqrt{N} \cdot \Delta\phi_{\max} \tag{3.1}$$

and

$$N(\phi) = N(0)e^{-(\phi/\overline{\Delta\phi})^2}, \tag{3.2}$$

where N is the number of atoms contributing to the scattering and $\overline{\Delta\phi}$ is the average total deflection of the alpha particle. Geiger[7] found that the average deflection of alpha particles passing through a 0.5 micron gold foil is about 1°. This seems reasonable in the light of the above discussion, since a one micron film would be about 10^4 atoms thick and the average deflection for random scattering would be

$$\overline{\Delta\phi} = \sqrt{10^4} \cdot 10^{-4} \text{ rad} \sim 10^{-2} \text{ rad} \sim 1°.$$

However, for angles much larger than 1°, Equation 3.2 predicts an infinitesimal probability. In particular, the fraction of particles scattered through 90° would be $\sim e^{-90^2}$! The earlier work of Geiger and Marsden[8] showed that about one alpha particle in 10^4 was scattered 90° or more. Though this might seem like a small number, it is many orders of magnitude greater than any prediction based on the Thomson model of the atom.

[7] See footnote 5.
[8] See footnote 4.

In order to account for the unexpected large-angle scattering of alpha particles, Rutherford[9] proposed the model of the nuclear atom in which most of the atomic mass and all of the positive charge is concentrated in a nucleus of dimension much less than the atomic dimension. By considering only the Coulomb interaction between the incident alpha particle and the target nucleus, Rutherford calculated the scattering cross section and obtained results which agreed remarkably with the experimental results. His calculation will be shown in the next section after a brief discussion of classical scattering theory.

2. CLASSICAL SCATTERING CROSS SECTIONS

There are two principal methods of studying the forces between elementary particles, namely, the study of the bound states of the particle system and a statistical study of the scattering of a beam of particles of one type by a target containing the other particles. The study of bound states is more limited because we must be content with the bound systems provided by nature and we must adapt our detectors to the energies required by those states. In the case of the neutron-proton system (the deuteron) there is only one bound state, so that there is not a great deal that one can learn. However, in a scattering experiment a wide range of particle energies and scattering angles may be studied, and the resolving power of the scattering apparatus is potentially very great, because the de Broglie wavelength* of high-energy particles can become quite small in comparison with the range of the forces being studied. Furthermore, such quantum mechanical effects as spin-dependent forces can be studied in scattering experiments by polarizing the particles of the incident beam or target, or both.

In the present section we will discuss only classical scattering and will consider the situation illustrated in Figure 3-1. The incident beam consists of monoenergetic particles such that I particles pass through unit area normal to

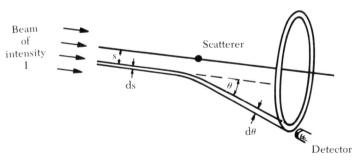

Figure 3-1 Schematic diagram of a scattering experiment.

⁹ E. Rutherford, *Phil. Mag.* **21,** 669 (1911).
* The de Broglie wavelength varies inversely with the momentum of the particle. This subject will be discussed in detail in section 4 of Chapter 4.

the beam per second. That is, the beam intensity is I particles per unit area per second. Consider a ray of the beam that would miss the scattering center O by the distance s if it were undeflected. We call s the *impact parameter*. Since there is cylindrical symmetry about the scattering axis, we will call ϕ the azimuthal angle about that axis, measured from some arbitrary reference. Then for a detector located at (θ, ϕ) which detects $dN(\theta, \phi)$ particles per second, we define the scattering cross section as

$$d\sigma = \frac{dN(\theta, \phi)}{I}. \qquad (3.3)$$

The scattering cross section must have the dimensions of area since the product $I \, d\sigma$ gives the number of particles per second arriving at the detector.

Because of the cylindrical symmetry, the total number of particles scattered into the conical wedge bounded by θ and $\theta + d\theta$ is equal to the number of particles incident on the washer of area $2\pi s \, ds$. That is,

$$dN(\theta) = 2\pi I s \, ds$$

or,

$$dN(\theta, \phi) = I s \, ds \, d\phi, \qquad (3.4)$$

since the integral of $d\phi$ is 2π. Combining (3.3) and (3.4) we have

$$d\sigma = s \, ds \, d\phi \qquad (3.5)$$

as the classical scattering cross section. In order to arrive at a cross section that is independent of the physical size of the detector, we define a *differential scattering cross section* obtained by dividing $d\sigma$ by the solid angle subtended by the detector. That is,

$$\frac{d\sigma}{d\Omega} = \frac{s \, ds \, d\phi}{\sin \theta \, d\theta \, d\phi} = \frac{s}{\sin \theta} \frac{ds}{d\theta}. \qquad (3.6)$$

For a specific interaction between particles one first obtains the relationship between s and θ; then $ds/d\theta$ can be found and the differential cross section can be calculated immediately. Experimentally, the number of particles counted per second per unit solid angle at the detector is

$$\frac{dN(\theta, \phi)}{d\Omega} = I \frac{d\sigma}{d\Omega},$$

or

$$\frac{d\sigma}{d\Omega} = \frac{1}{I} \frac{dN(\theta, \phi)}{d\Omega}. \qquad (3.7)$$

A comparison of theory and experiment can be made by noting the results obtained for Equations 3.6 and 3.7.

The total cross section, σ, may be obtained by integrating the differential cross section over all solid angles, that is,

$$\sigma = \int_\Omega \frac{d\sigma}{d\Omega} \, d\Omega \, . \tag{3.8}$$

PROBLEM 3-1

Consider the elastic scattering of a hard sphere of radius a from a stationary sphere of radius b. The incident sphere will rebound from the tangent plane at an angle equal to its angle of incidence as shown in Figure 3-2. Show that $d\sigma/d\Omega = \frac{1}{4}(a+b)^2$ and find σ.

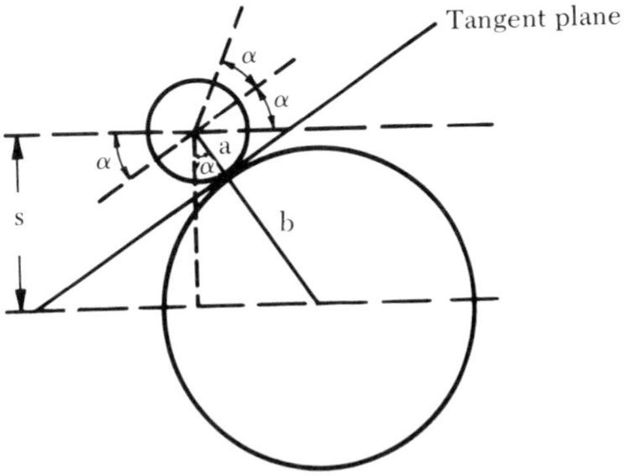

Figure 3-2 Classical scattering of hard spheres.

Now we will discuss the case of Coulomb or Rutherford scattering. Consider a projectile of mass m and charge ze incident upon a stationary scattering center of charge Ze. For z and Z both positive, a Coulomb repulsion occurs, and the scattered particle will have a hyperbolic trajectory as shown in Figure 3-3. The angular momentum about the target nucleus is constant for any central force. Hence,

$$mv_0 s = mr^2\dot{\theta} = \text{constant}.$$

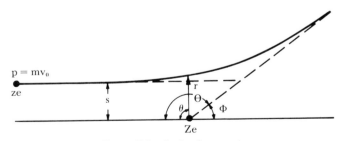

Figure 3-3 Coulomb scattering.

Then,

$$r(\dot\theta)^2 = \frac{v_0^2 s^2}{r^3}.$$ (3.9)

Using the radial part of the equation of motion,

$$\frac{zZe^2}{r^2} = m(\ddot r - r\dot\theta^2).$$ (3.10)

Letting $r = 1/u$,

$$\dot\theta = sv_0 u^2,$$

$$\dot r = \frac{dr}{du}\frac{du}{d\theta}\dot\theta = -sv_0\frac{du}{d\theta},$$ (3.11)

and

$$\ddot r = -sv_0\frac{d^2u}{d\theta^2}\dot\theta = -s^2v_0^2u^2\frac{d^2u}{d\theta^2}.$$

Substituting these expressions into Equation 3.10,

$$\frac{zZe^2}{ms^2v_0^2} = -u - \frac{d^2u}{d\theta^2}.$$

Define D, the collision diameter, as the closest distance of approach in the case of a head-on collision. That is, D is the distance from the target at which all of the kinetic energy of the projectile is converted to electrostatic potential energy. Then,

$$D = \frac{2zZe^2}{mv_0^2},$$

and the equation of the trajectory becomes

$$\frac{d^2u}{d\theta^2} + u = -\frac{D}{2s^2}.$$ (3.12)

A solution of Equation 3.12 is $u = A\cos\theta + B\sin\theta - D/2s^2$. Now as θ goes to zero, r becomes infinite and u goes to zero. Then $A = D/2s^2$. Also, as θ goes to zero, $\dot r$ becomes $-v_0$. From Equation 3.11 we have

$$\left(\frac{du}{d\theta}\right)_{\theta=0} = \frac{1}{s} = B.$$

Therefore our solution becomes

$$u = \frac{1}{s}\sin\theta + \frac{D}{2s^2}(\cos\theta - 1).$$ (3.13)

The scattering angle Φ corresponds to the situation when $\theta = \Theta$ and $u = 0$, since r becomes infinite here also. Then, from Equation 3.13,

$$s = \frac{D}{2}\left(\frac{1 - \cos \Theta}{\sin \Theta}\right) = \frac{D}{2} \tan \frac{\Theta}{2} = \frac{D}{2} \cot \frac{\Phi}{2}, \tag{3.14}$$

where Φ plays the role of θ in Equation 3.3 and those which follow.

This provides the relationship which is needed to calculate the differential scattering cross section in the center of mass system from Equation 3.6. Since

$$\frac{ds}{d\Phi} = \frac{D}{4} \csc^2 \frac{\Phi}{2},$$

then

$$\frac{d\sigma}{d\Omega} = \frac{D}{2} \cot \frac{\Phi}{2} (\sin \Phi)^{-1} \frac{D}{4} \csc^2 \frac{\Phi}{2} = \frac{D^2}{16}\left(\sin^4 \frac{\Phi}{2}\right)^{-1}. \tag{3.15}$$

This is the now famous Rutherford result. To find the fraction of particles scattered at a given angle Φ we use Equation 3.7 and write

$$\frac{dN(\Phi, \phi)}{I} = \frac{D^2}{16} \frac{1}{\sin^4 \frac{\Phi}{2}} d\Omega$$

per scattering center. Then for C scattering centers per unit area,

$$\frac{dN(\Phi)}{I} = \frac{D^2}{16} \cdot \frac{C}{\sin^4 \frac{\Phi}{2}} \cdot 2\pi \sin \Phi \, d\Phi = \frac{\pi D^2 C}{8} \cdot \frac{\sin \Phi \, d\Phi}{\sin^4 \frac{\Phi}{2}}. \tag{3.16}$$

For gold, the constant $C = \rho N_0 t / A = 5.92t \times 10^{22}$ atoms/cm^2, where ρ is the density in g/cm^3, t is the thickness of the film in cm, A is the atomic weight, and N_0 is Avogadro's number. For alpha particles having energies of the order of 5 MeV, D is about 2×10^{-11} cm. Thus the fraction of particles scattered into the cone bounded by Φ and $\Phi + d\Phi$, where

$$\Phi = 90°,$$

is

$$\frac{dN(90°)}{I} = \frac{\pi D^2}{2} \frac{\rho N_0 t}{A} d\Phi \sim 10^{-3} \, d\Phi,$$

for a foil one micron thick.

PROBLEM 3-2

What fraction of incident alpha particles having 2.5 MeV of kinetic energy will be scattered through an angle of 60° or more by a gold foil one micron thick?

Impact parameter

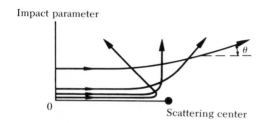

0   Scattering center

Figure 3-4 Relationship of impact parameter to scattered angle for Coulomb scattering.

PROBLEM 3-3

For alpha particles of energy 7.68 MeV scattered from a platinum foil two microns thick, find: (a) the collision diameter, (b) the differential scattering cross section, and (c) the fractional number of particles scattered through angles of 90° or more.

One of the important results of the scattering experiments of Rutherford and his collaborators was that an upper limit was obtained for the size of the nucleus. This figure is of the order of 10^{-12} centimeters. For impact parameters of the order of 5×10^{-13} cm, the scattering of alpha particles from light elements showed anomalies which were attributed to a force other than the pure Coulomb force.[10] This *nuclear force* was assumed to be an attractive, short-range force which dominates the Coulomb force at distances which are of the order of 10^{-13} cm. (This distance is now called the fermi. That is, $1 \text{ F} = 1 \times 10^{-13}$ cm.)

3. SCATTERING CROSS SECTIONS IN THE C AND L FRAMES

Since scattering experiments are performed in the laboratory frame, while the theory is conveniently worked out in the center of mass frame, it is necessary to be able to transform scattering results from one frame to the other. (See also section 9 of Chapter 1.) Consider a non-relativistic particle of mass m_1 and velocity v_1 moving toward a particle of mass m_2 and velocity $v_2 = 0$ in the laboratory frame (*L*-frame), as shown in Figure 3-5. The velocity of the zero momentum frame (*C*-frame) relative to the laboratory is $v_c = (m_1 v_1)/(m_1 + m_2) = \mu v_1/m_2$, where μ is defined as the reduced mass. The total kinetic energy in the *L*-frame is $T_L = \frac{1}{2} m_1 v_1^2$ and the kinetic energy in the *C*-frame is $T_C = \frac{1}{2} \mu v_1^2$. Since the kinetic energy of the center of mass of the

---

[10] E. Rutherford, *Phil. Mag.* **37**, 537 (1919).

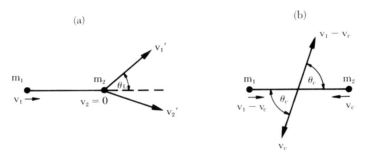

Figure 3-5 Scattering of mass m_1 from mass m_2 as viewed in the laboratory frame (a) and in the C-frame (b).

system with respect to the L-frame is $\frac{1}{2}(m_1 + m_2)v_c^2$, the reader should show that for elastic scattering this energy is equal to $T_L - T_C$. We then have the useful relationship,

$$\frac{T_C}{T_L} = \frac{\mu}{m_1},$$

(3.17)

for elastic scattering. Note that in the C-system each mass is scattered at the same angle, θ_c; thus, conservation of kinetic energy requires conservation of velocities and each mass has the same kinetic energy after the collision that it had before the collision. It is convenient to draw the following vector triangle:

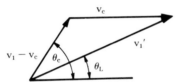

Then the following relations are obtained from the triangle:

$$(v_1 - v_c) \sin \theta_c = v_1' \sin \theta_L$$

$$v_c + (v_1 - v_c) \cos \theta_C = v_1' \cos \theta_L$$

and

$$\tan \theta_L = \frac{\sin \theta_c}{\dfrac{v_c}{v_1 - v_c} + \cos \theta_c} = \frac{\sin \theta_c}{\dfrac{m_1}{m_2} + \cos \theta_c} = \frac{\sin \theta_c}{r + \cos \theta_c},$$

(3.18)

where r is the ratio of the masses, m_1/m_2.

In order to relate the scattering cross sections in the two frames it is convenient to draw an auxiliary triangle with its sides labeled in accordance with the quantities in Equation 3.18.

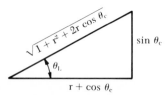

Using the fact that the total cross section is the same in each frame, we write,

$$\iint \left(\frac{d\sigma}{d\Omega}\right)_L \sin \theta_L \, d\theta_L \, d\phi_L = \iint \left(\frac{d\sigma}{d\Omega}\right)_C \sin \theta_C \, d\theta_C \, d\phi_C,$$

$$2\pi \int \left(\frac{d\sigma}{d\Omega}\right)_L \sin \theta_L \, d\theta_L = 2\pi \int \left(\frac{d\sigma}{d\Omega}\right)_C \sin \theta_C \, d\theta_C.$$

Then,

$$\left(\frac{d\sigma}{d\Omega}\right)_L = \left(\frac{d\sigma}{d\Omega}\right)_C \cdot \frac{\sin \theta_C}{\sin \theta_L} \cdot \frac{d\theta_C}{d\theta_L}. \tag{3.19}$$

Differentiating Equation 3.18,

$$\sec^2 \theta_L \cdot d\theta_L = \frac{1 + r \cos \theta_C}{(r + \cos \theta_C)^2} \, d\theta_C,$$

or

$$\frac{d\theta_C}{d\theta_L} = \frac{(r + \cos \theta_C)^2 \cdot \sec^2 \theta_L}{1 + r \cos \theta_C}. \tag{3.20}$$

From the triangle above,

$$\frac{\sec^2 \theta_L}{\sin \theta_L} = \frac{(1 + r^2 + 2r \cos \theta_C)^{\frac{3}{2}}}{(r + \cos \theta_C)^2 \sin \theta_C}. \tag{3.21}$$

Combining Equations 3.19, 3.20, and 3.21, we obtain our final result,

$$\left(\frac{d\sigma}{d\Omega}\right)_L = \left(\frac{d\sigma}{d\Omega}\right)_C \frac{(1 + r^2 + 2r \cos \theta_C)^{\frac{3}{2}}}{1 + r \cos \theta_C}. \tag{3.22}$$

An isotropic scattering cross section in the C-frame will thus be peaked in the forward direction in the laboratory by the factor $(1 + r)$.

PROBLEM 3-4

Sketch the differential scattering cross-section as a function of angle for the laboratory and zero momentum frames for the scattering of alpha particles by a carbon film.

4. BOHR'S THEORY OF ATOMIC SPECTRA

The nuclear atom proposed by Rutherford settled the problem associated with the scattering of alpha particles, but did not explain the stability of the atom. Since it is impossible to have a stable configuration of charges subject to electrostatic forces only, a dynamical system was proposed, analogous to a planetary system. Such a system could account for the fact that the nucleus is only of the order of 1×10^{-12} cm while the atom as a whole has an effective diameter of the order of 1×10^{-8} cm. However, a serious problem arose in connection with electromagnetic theory, namely, that a charge undergoing continuous centripetal acceleration should radiate continuously. If this were the case, the energy of the dynamical system would decrease continuously and the planetary charge would spiral into the nucleus after a nominal lifetime of about 10^{-8} second. That this does not occur is borne out by the infinite lifetimes of most elementary atoms and by the nature of their radiation spectra. Atoms do not radiate unless excited, and when radiation does occur its spectrum consists of discrete frequencies rather than the continuum of frequencies required by the classical theory of radiation.

In 1913, Bohr[11] proposed a theory which was successful in explaining the radiation spectra of one-electron atoms, although it is in direct disagreement with the classical theory of radiation. Bohr's postulates may be summarized as follows:

(1) The Coulomb force on a planetary electron provides the centripetal acceleration required for a dynamically stable circular orbit.
(2) The only permissible orbits are those in the discrete set for which the angular momentum of the electron equals an integer times $\hbar$, where $\hbar = h/2\pi$.
(3) An electron moving in one of these stable orbits does not radiate.
(4) Emission or absorption of radiation occurs only when an electron makes a transition from one orbit to another.

From the second postulate note that we now have angular momentum (as well as charge and energy) quantized in atomic systems. The third postulate rejects the troublesome claim that an accelerated charge must radiate in atomic systems, in spite of its validity in the macroscopic world. The fourth postulate provides the link with Planck's theory of radiation, since the frequency of the photon emitted or absorbed is given by the energy difference of the two states divided by h.

In order to appreciate the implications of the Bohr theory, let us consider a one-electron atom of nuclear mass M, nuclear charge Ze, electronic mass m_0 and electronic charge e. We will use the reduced mass μ given by

$$\mu = \frac{m_0 M}{M + m_0},$$
(3.23)

---

[11] N. Bohr, *Phil. Mag.* **26**, 1 (1913).

in order to neglect any motion of the nucleus. Then, from the first postulate,

$$\frac{\mu v^2}{r} = \frac{Ze^2}{r^2}, \tag{3.24}$$

and from the second,

$$\mu v r = n\hbar. \tag{3.25}$$

By combining Equations 3.24 and 3.25 we obtain the expression for the radius of the nth orbit,

$$r_n = \frac{n^2\hbar^2}{\mu Ze^2} \sim \frac{n^2 a_0}{Z}. \tag{3.26}$$

The quantity,

$$a_0 = \frac{\hbar^2}{m_0 e^2} \sim 0.53 \text{ Å},$$

is the radius of the first orbit of hydrogen calculated for a fixed nucleus, that is, for $\mu = m_0$. It is often simply called *the Bohr radius*. Note that Equation 3.26 specifies that the radii of the allowed orbits are proportional to the squares of the integers.

Using Equation 3.24 we may write the kinetic energy of the electron as

$$T = \tfrac{1}{2}\mu v^2 = \frac{Ze^2}{2r}.$$

Since the potential energy of the electron in the Coulomb field of the nucleus is

$$V = -\frac{Ze^2}{r},$$

then the total energy is

$$E = T + V = -\frac{Ze^2}{2r} = -\frac{\mu}{2}\frac{(Ze^2)^2}{n^2\hbar^2} \sim -\frac{w_0 Z^2}{n^2}, \tag{3.27}$$

where

$$w_0 = \frac{m_0 e^4}{2\hbar^2} \sim 13.6 \text{ eV}$$

is the magnitude of the ground state energy of hydrogen for a fixed nucleus. Equation 3.27 indicates that the quantization of energy has arisen as a result of assuming that the angular momentum is quantized. The frequency of a photon emitted by a transition from the nth level to the kth level, where $n > k$, is given by

$$h\nu = hc\left(\frac{1}{\lambda}\right) = hc\bar{\nu} = E_n - E_k = \frac{\mu}{2}\frac{(Ze^2)^2}{\hbar^2}\left(\frac{1}{k^2} - \frac{1}{n^2}\right).$$

Expressed in wave numbers, this becomes

$$\bar{\nu} = \frac{2\pi^2 \mu Z^2 e^4}{h^3 c}\left(\frac{1}{k^2} - \frac{1}{n^2}\right) = Ry\left(\frac{1}{k^2} - \frac{1}{n^2}\right), \tag{3.28}$$

where Ry is the Rydberg constant calculated from the reduced mass. For hydrogen $Ry = 109681$ cm^{-1}, whereas the experimental value from optical spectroscopy is $Ry = 109677.576 \pm 0.012$ cm^{-1}. The Rydberg constant calculated for a fixed nucleus (that is, for $\mu = m_0$) is

$$Ry_\infty = 109737.309 \text{ cm}^{-1}.$$

Furthermore, the experimentally established spectral series for hydrogen are all explained by Equation 3.28 by setting the value of k and letting n run through the integers such that $n > k$. For example, $k = 1$, $n = 2, 3, 4, \ldots$ corresponds to the Lyman series; $k = 2$, $n = 3, 4, 5, \ldots$ corresponds to the Balmer series; $k = 3$, 4, and 5 correspond to the Paschen, Brackett, and Pfund series, respectively. (See Figure 3-6). Equation 3.28 can also be used to calculate the ionization potential of hydrogen by letting $n \to \infty$. Then we obtain

$$V = \frac{chRy}{1.6 \times 10^{-12}} \sim 13.6 \text{ volts.}$$

The excellent agreement between the spectroscopic values for the Rydberg constant, the ionization potential, and the emission spectrum of hydrogen was

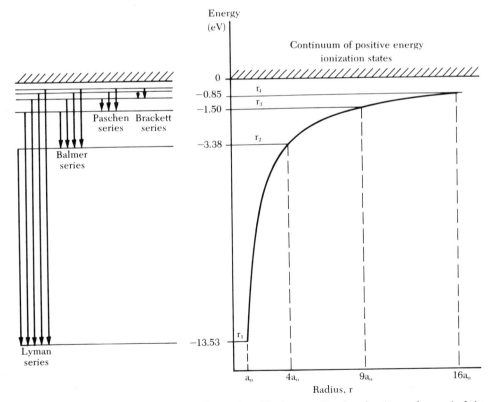

Figure 3-6 Bohr radii and allowed energies of hydrogen. The first few lines of several of the spectral series of hydrogen are shown at the left.

a great triumph for the Bohr theory. However, it was immediately apparent that the Bohr theory was good only for a one-electron atom, since it failed badly in the cases of helium and lithium, the next simplest neutral atoms. But for the spectra of singly ionized helium and doubly ionized lithium the Bohr theory is again quite good. We call such atoms "hydrogen-like" or "hydrogenic" since they are one-electron atoms but their nuclear masses and charges differ from that of hydrogen. The spectra of the alkali metals can also be calculated as if their single valence electron moves in an effective field of reduced Z due to the noble gas core of electrons. Although the gross features of these spectra are quite similar to that of hydrogen, there are some serious discrepancies due to the non-Coulomb nature of the electrostatic field and due to spin-orbit coupling.*

Interesting applications of the Bohr theory can be made to the short lived mesic atoms and to positronium. Mesic atoms are ordinary atoms which have captured an orbital negative mu meson in lieu of one of the orbital electrons. Although these atoms have short lifetimes,† the muon completes many millions of orbits before it decays or is captured by the nucleus. During its short lifetime, it can make transitions to or from excited states by absorbing or emitting quanta of radiation. The study of these transitions provides additional information about the nucleus.[12] Since the muon is about 207 times the mass of the electron, the radius of a Bohr orbit for a bound muon is about $\frac{1}{200}$ the radius of the corresponding orbit for an electron. Taking into account the difference in nuclear charge, a muon bound to a heavy element will have orbital radii $\sim 10^{-4}$ times the hydrogen radii, or $\sim 10^{-12}$ centimeter. According to Wheeler[13] a muon in its ground state orbit spends nearly half of its lifetime *within* the captor nucleus. Accordingly, we conclude that the nucleus is nearly transparent to the muon, which is to say that the muon interacts weakly with nucleons.

PROBLEM 3-5

(a) Use the Bohr theory to calculate the allowed energies and circular orbits for the two hydrogenic atoms positronium and muonium. Positronium consists of a positron and an electron, each of mass m_0; muonium consists of a proton of mass $1836\ m_0$ and a negative mu meson of mass $207\ m_0$. Each particle has a charge of magnitude e.

(b) Discuss the possible decay schemes of orthopositronium (parallel spins) and parapositronium (antiparallel spins) in the context of the discussion in section 10 of Chapter 1.

---

* This will be treated in section 2 of Chapter 8.
† The lifetime of the mesic atom is limited by the lifetime of the muon which decays into an electron and a neutrino. See problem 1-9.
[12] V. L. Fitch and J. Rainwater, *Phys. Rev.* **92,** 789 (1953).
[13] J. A. Wheeler, *Phys. Rev.* **92,** 812 (1953).

PROBLEM 3-6

(a) Using the Bohr model find the energy of the photon emitted when a mu meson makes a transition from the first excited state to the ground state of mu-mesic Pb^{208}.

(b) The $2P_{\frac{3}{2}}-1S_{\frac{1}{2}}$ transition energy for mu-mesic Pb^{208} is known to be 5.963 MeV. Suggest several reasons for the discrepancy between this value and your answer in (a).

PROBLEM 3-7

Determine the effect of nuclear mass on the first line of the Balmer series of hydrogen by calculating its wavelength shift for deuterium and tritium.
(Ans.: 1.79 Å; 2.38 Å.)

PROBLEM 3-8

Using the simple Bohr theory for circular orbits, calculate the first ionization potential of helium.
(Ans.: 54.4 volts.)

PROBLEM 3-9

If helium gas in a discharge tube is excited by a potential difference of 11 volts, which, if any, spectral lines corresponding to the Balmer series of hydrogen will be excited? (Use the simple Bohr theory and neglect spin.)

PROBLEM 3-10

An interesting application of the Bohr theory of the hydrogenic atom occurs in solid state physics. A phosphorus-doped silicon crystal may be regarded as a collection of hydrogenic atoms of unit net charge at each phosphorus site. The effect of the periodic potential, due to all of the silicon nuclei, on the motion of the orbital electron is taken into account by giving the electron an effective mass appreciably different from the electronic rest mass. In addition, the Coulomb force is diminished by the dielectric constant of silicon since the orbital radius is equivalent to many lattice spacings.

Calculate the energy and the radius of the first Bohr orbit if the dielectric constant of silicon is 11.7 and the effective electron mass is $0.2 \, m_0$.

(Ans.: -0.02 eV; 30 Å.)

5. THE FRANCK-HERTZ EXPERIMENT

The experiment of Franck and Hertz provided additional confirmation of the discrete energy states predicted by the Bohr theory.[14] In this experiment mercury vapor was bombarded with electrons of known kinetic energy. When this kinetic energy is less than the energy of the first excited state of atomic mercury, the only energy which the colliding electron loses is the small amount of kinetic energy (about one part in 10^5) that it can transfer to the massive mercury atom by an elastic collision. However, when the kinetic energy of the incident electron just exceeds the energy of the first excited state of mercury, then the electron gives up virtually all of its kinetic energy to the mercury atom in an inelastic collision. Thus, by comparing the kinetic energy of the electrons before and after the collision one can determine how much energy is transferred to the target atoms.

Franck and Hertz used a simple apparatus consisting of a mercury tube containing a cathode, a plate, and an accelerating grid which was located physically near the plate (see Figure 3-7). Both the plate and the grid were maintained at positive potentials with respect to the cathode, but the plate potential was slightly lower than the grid potential. This small retarding potential on the plate had the effect of preventing contributions to the plate current from electrons having negligible kinetic energy. Hence, if electrons were to lose most of their kinetic energy in inelastic collisions with mercury atoms, the retarding potential would prevent them from reaching the plate and a drop in plate current would result. Such a drop in plate current was found to occur at 4.9 volts, as shown in Figure 3-8. Furthermore, Franck and Hertz noted that at this voltage the 2536 Å spectral line of mercury appeared in the emission spectrum of the vapor. A simple calculation shows that the photon energy of the 2536 Å line corresponds to 4.86 electron-volts! At slightly higher voltages a large drop in plate current occurs and new lines appear in the

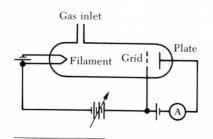

Gas inlet

Filament   Grid   Plate

Figure 3-7 Apparatus for Franck-Hertz experiment.

[14] J. Franck and G. Hertz, *Verhandl. deut. physik. Ges.* **16,** 457 and 512 (1914).

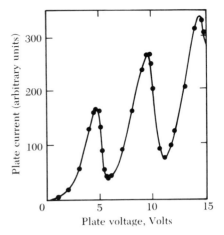

Figure 3-8 Current *vs.* voltage in the Franck-Hertz experiment. (After J. Franck and G. Hertz, *Verhandl. deut. physik. Ges.* **16**, 457 (1914).)

emission spectrum of the mercury vapor. This behavior is repeated at multiples of 4.9 volts as shown in the figure. We must conclude from this that the Bohr concept of discrete energy states is qualitatively correct.

6. X-RAY SPECTRA AND THE BOHR THEORY

When high-energy electrons are used to bombard a target made of a heavy metal, the electrons are decelerated and they will radiate. The resulting radiation produces a continuous spectrum called *Bremsstrahlung* which is characterized by a sharp cutoff at the short wavelength end. A typical spectrum is shown in Figure 3-9. This continuous spectrum can be accounted for classically by the theory of electromagnetic radiation[15] or quantum mechanically by invoking an inverse photoelectric effect. That is, the change in kinetic energy provides an upper limit on the photon frequency, whereas in the

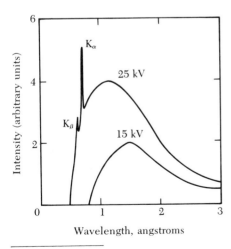

Figure 3-9 X-ray spectrum of molybdenum. The continuous spectrum is the bremsstrahlung radiation.

[15] F. K. Richtmyer, E. H. Kennard, and T. Lauritsen, *op. cit.*, p. 349.

photoelectric effect the photon frequency provides an upper limit on the change in kinetic energy. Since the initial kinetic energy of an electron is determined by the voltage on the x-ray tube, the short wavelength cutoff is given by

$$\lambda_c = \frac{hc}{eV} = \frac{12,400}{V},$$

(3.29)

where V is in volts and λ is in angstroms. Thus, a 12 kilovolt x-ray tube has a cut-off wavelength of about 1 angstrom. Not many photons having wavelength λ_c are emitted, because the majority of electrons perform mechanical work on the target when they stop and a large fraction of their kinetic energy is converted into heat. Equation 3.29 provides an accurate method[16] for determining the ratio e/h, since a plot of ν_c vs. V is a straight line whose slope is e/h.

Superimposed upon the continuous spectrum are a few discrete peaks whose wavelengths are characteristic of the target material. A few of these peaks are also shown in the figure. The characteristic x-ray spectra of many elements were studied by Moseley,[17] who showed that the frequency of a given line varies from element to element as the square of the atomic number. A plot of some of Moseley's data is shown in Figure 3-10. This is precisely the dependence predicted by the Bohr theory in Equation 3-28. Furthermore, the Bohr theory accounts for the discreteness and the order of magnitude of the photon energies of the characteristic spectrum by assuming that these photons are emitted as a result of transitions involving inner electron shells. If, for example, a high energy electron knocks out an electron from the first shell of a target atom, the vacancy may be filled by a transition from the second shell, or the third, or higher shells. Since the electrons of the first shell have been traditionally called K-electrons, the photon emitted by a transition from the second shell to the K shell is called K_α radiation, that due to a transition from the third shell to the K shell is called K_β radiation, and so on. The second

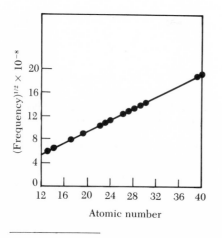

Figure 3-10 Moseley's results for the effect of atomic number on the frequency of a given x-ray line.

[16] J. DuMond and V. L. Bollman, *Phys. Rev.* **51,** 400 (1937).
[17] H. G. J. Moseley, *Phil. Mag.* **26,** 1024 (1913); **27,** 703 (1914).

shell is designated as the L shell, the third M, and so forth, so that L_α, L_β, . . . , M_α, M_β, . . . radiation is defined in a similar fashion.

Since Equation 3.28 shows that the photon energy is proportional to Z^2, a transition from the second to the first shell for copper ($Z = 29$) would result in the emission of a photon having an energy about 841 times the energy of the longest line of the Lyman series of hydrogen. This approximate calculation gives 1.44 Å for the K_α line of copper, whereas the correct value is 1.54 Å. The agreement is not nearly so good for transitions involving shells of higher order because of the shortcomings of the Bohr theory and the important screening effect of the nuclear charge by electrons in the innermost shells.

A process which generally competes with x-ray emission is the internal photoelectric effect or *Auger effect*. Here, the x-ray photon does not actually appear, but an equivalent amount of kinetic energy is given to an outer electron, which is in turn ejected from the atom.

PROBLEM 3-11

Calculate the wavelengths of the K_α and K_β lines of the characteristic spectrum of vanadium using the elementary Bohr theory. Compare these values with the accepted values.

PROBLEM 3-12

What is the kinetic energy of an Auger electron resulting from a radiationless transition from the L to the K shell of chromium? Use the elementary Bohr theory.

PROBLEM 3-13

(a) If the Bohr atom is interpreted as a classical oscillator whose frequency is $\nu_0 = v/2\pi r$, show that the classical oscillator frequency can be expressed as

$$\nu_0 = \frac{2w_0}{hn^3},$$

where w_0 is given in Equation 3-27.

(b) Show that in the limit as n gets very large, the Bohr frequency relation reduces to the expression in (a). This is an example of the *correspondence principle*, which states that a quantum theory result should agree with the equivalent classical solution in the limit of large quantum numbers.

7. NUCLEAR STRUCTURE AND SPECTROSCOPY

The nucleus of the Rutherford atom was at first thought to be composed of a number of protons equal to its atomic mass number A, and a number of electrons equal to $(A - Z)$, that is, the difference between the mass number and the atomic number. Although this composition accounted for the existence of the electrons that appeared in the case of β emitters, it was early rejected by Rutherford, who first proposed the existence of the neutron. The subsequent discovery of the neutron by Chadwick established it as a nuclear constituent. We now refer to both neutrons and protons as *nucleons* and call A either the mass number or nucleon number. The quantity $A - Z$ is often called the neutron number, N. Figure 3-11 shows a plot of N versus Z for some nuclei. There are now about 900 known nuclei, of which fewer than one–third are stable.

The nuclear radius was determined to be of the order of 10^{-12} centimeters by Rutherford. Subsequent experiments have led to the empirical relationship,

$$r = r_0 A^{\frac{1}{3}}, \tag{3.30}$$

which expresses the radius of any nucleus in terms of its nucleon number and

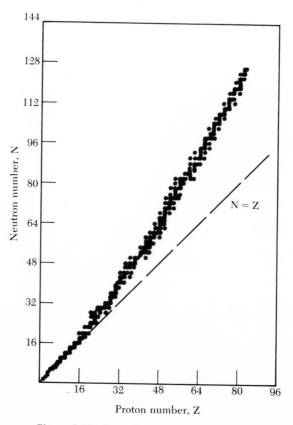

Figure 3-11   Plot of N vs. Z for stable nuclei.

the constant r_0. The accepted value of r_0 is 1.2×10^{-13} cm (or 1.2 F), which was obtained from measurements of the nuclear charge distribution by means of high-energy electron scattering.[18] Other methods of estimating nuclear radii are in good agreement with this result.[19,20] Equation 3.30 leads to two interesting conclusions. First, the *nuclear force* which holds the nucleus together must be charge independent, that is, the *n-n*, *p-p*, *n-p* interactions must be essentially the same since it is A, and not Z or N, which determines the nuclear radius. Second, the nuclear density is apparently the same for all nuclei, since it is given by the expression

$$\frac{A}{V} = \frac{A}{\frac{4\pi}{3} r_0^3 A} \sim \frac{1}{4 r_0^3}.$$

Thus, nuclear matter is roughly $(a_0/r_0)^3 = 10^{12}$ times as dense as ordinary matter.

The *binding energy* of a nucleus can be expressed in terms of the difference in rest mass energy between the constituent nucleons and the composite nucleus. Thus, the binding energy is the *mass defect* multiplied by c^2. For example, the binding energies of the deuteron and the alpha particle are as follows:

$$E_d = (M_p + M_n - M_d)c^2 = 2.2 \text{ MeV}$$
$$E_\alpha = (2M_p + 2M_n - M_\alpha)c^2 = 28.3 \text{ MeV}.$$

Figure 3-12 shows a plot of the binding energy per nucleon (that is, E/A) as a function of A. Note that the binding energy per nucleon is roughly 7 or 8 MeV

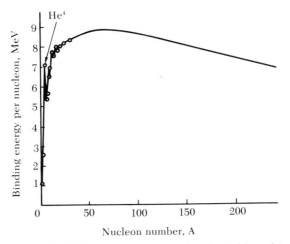

Figure 3-12 Binding energy per nucleon for stable nuclei.

[18] See, for example, Robert Hofstadter, *Annual Reviews of Nuclear Science* **7**, 231 (1957).
[19] L. R. B. Elton, *Introductory Nuclear Theory*. W. B. Saunders Co., Philadelphia, 1966.
[20] R. D. Evans, *The Atomic Nucleus*. McGraw-Hill Book Co., New York, 1955.

for most nuclei except the lighter nuclei. The sharp increase in binding energy for $A = 4$ accounts for the unusual stability of the alpha particle. The gradual fall-off of binding energy per nucleon for the heavy nuclei is due to the effect of the Coulomb repulsion, which is always present although it is dominated by the nuclear potential. This increase in Coulomb repulsion is partially compensated by an increased nuclear attraction resulting from an increase in the N/Z ratio as Z increases (see Figure 3-11). The nearly constant nuclear energy per particle contrasts sharply with the dependence of electronic binding energy per electron on Z (and hence A) in atomic potentials.

The study of positron emission from *mirror nuclei* provides an estimate of the Coulomb contribution to the binding energy as well as an approximate value of the nuclear radius. Mirror nuclei are pairs of nuclei having odd A such that Z and N differ by one. For example, $O^{15}(Z = 8, N = 7)$ and $N^{15}(Z = 7, N = 8)$ are mirror nuclei, each containing 15 nucleons. Thus, the binding energies of these nuclei should be the same except for the correction due to the difference in the Coulomb interactions. Using the classical expression for the Coulomb energy of a uniformly charged ball of charge Ze and radius r, the difference in Coulomb energy between nuclei of atomic numbers Z and $Z - 1$ is,

$$\Delta E = \frac{3e^2}{5r} [Z^2 - (Z - 1)^2] = \frac{3e^2}{5r} (2Z - 1). \tag{3.31}$$

PROBLEM 3-14

Find the mass defects of O^{15} and N^{15}. Compare this difference in rest mass energy with the Coulomb energy calculated from Equation 3.31.

By measuring the energy of β^+ decay from element Z to element $Z - 1$, r can be calculated. Studies of a number of such reactions confirm Equation 3.30 with a value of $r_0 \sim 1.5$ F. The fact that the binding energy of a nucleon is independent of its charge, except for the Coulomb correction, adds further confirmation to the validity of the assumption that the nuclear force itself is independent of the charge of the nucleon. The nucleon force is known to be an extremely short-range force, since the Coulomb law adequately describes the scattering of alpha particles for impact parameters greater than about 1×10^{-12} cm. A commonly used form for the nucleon potential is the co-called Yukawa potential,[21]

$$V \sim \frac{e^{-r/b}}{r}, \tag{3.32}$$

where b is the range of the nuclear force, $\sim 10^{-13}$ cm $= 1$ F. This expression is not valid for $r < 0.4$ F, where a repulsive term dominates. For $r = b = 1$ F,

---

[21] H. Yukawa, *Phys. Math. Soc. Japan* **17**, 48 (1935).

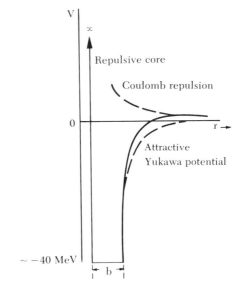

Figure 3-13 Schematic diagram of the nucleon–nucleon potential. Since the exact shape of the well is not known, it is approximated by a square well. For a neutron there is no Coulomb repulsion.

$V \sim 1$ to 10 MeV for a nucleon; for $r = 10b = 10$ F, $V \sim 10$ to 100 eV. A schematic diagram of the nucleon-nucleon potential is shown in Figure 3-13. Another property of the nuclear force, which is quite a departure from the behavior of Coulomb and gravitational forces, is that of *saturation*. That is, a nucleon seems to interact with only a limited number of other nucleons, analogous to nearest-neighbor interactions in solids or the saturation of chemical bonds in ligand theory. It is this characteristic that is evoked to explain the unusual stability of the alpha particle. There appears to be little or no attraction between an alpha and another nucleon and hence there is no stable nucleus of $A = 5$.

As a result of the discovery of natural and induced radioactivity, as well as the research leading to the identification of the "mysterious emanations" called α, β, and γ rays, it became evident that the nucleus itself must have an inner structure. From careful measurements of the energies of emitted particles and γ rays, schematic diagrams of the energy levels and decay schemes of many nuclei have been produced (see Figure 3-14). One striking difference between

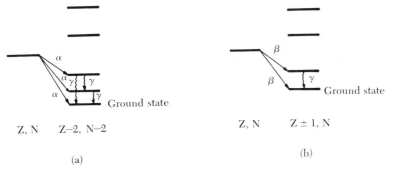

Figure 3-14 Schematic energy level structure for hypothetical nuclei. An α-emitter (a) or a β-emitter (b) may decay directly to the ground state or to an excited state of the daughter nucleus.

optical and nuclear spectroscopy is associated with the energies of the photons in each case. Recall that the energy of a photon in the visible region is of the order of a few eV, while an x-ray photon is generally in the keV regime. Since nuclear energies are of the order of MeV, γ photons are frequently in the MeV energy regime. However, there are no sharp demarcations between these energy regimes and accordingly, "hard" x-rays may equally well be called "soft" γ rays. The atom emitting a photon must suffer a recoil in order to conserve momentum as well as energy. For visible photons this recoil energy is often negligible, but it can become significant for energetic gammas.

PROBLEM 3-15

(a) Show that the energy shift of an emitted photon due to the recoil of the emitting atom may be expressed as

$$\Delta E \sim \frac{E^2}{2Mc^2}, \qquad (3.33)$$

where E is the transition energy and M is the mass of the atom.

(b) Calculate the recoil shift for the emission of the most energetic Lyman line from a free atom of hydrogen.

(c) Calculate the recoil shift for the emission of a 14.4 keV γ ray from a free atom of Fe^{57}.

The effect of such recoil on the light emitted from a source consisting of many atoms is to broaden the spectral line associated with the transition.* Of course, every spectral line has a natural width (there is no perfectly monochromatic source), but the recoil broadening can be many times greater than the natural line width. An important exception to this occurs in some crystals when the emitting atom is bound to the lattice with sufficient energy to prevent its recoil. In such cases the emitted line is very sharp since the mass in the denominator of Equation 3.33 becomes the mass of the whole crystal and $\Delta E \sim 0$. When recoilless emission occurs the linewidth is essentially its natural width, and the photon may be absorbed by an unexcited nucleus of the same species by the process of resonance absorption. This phenomenon is known as the *Mössbauer effect* after its discoverer.[22] It has become an important tool in many areas of physics, but particularly in solid state and nuclear physics.[23]

Brief mention should be made of the gamma emission process when recoil does occur. Here we may distinguish three separate cases. (1) If the free–atom

---

* See Chapter 2, section 9.

[22] R. Mössbauer, Z. *Physik* **151,** 124 (1958); *Naturwiss.* **45,** 538 (1958).

[23] See, for example, G. K. Wertheim, *Mössbauer Effect: Principles and Applications.* Academic Press, New York, 1964.

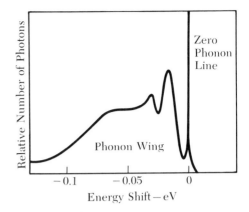

Figure 3-15 A hypothetical gamma ray spectrum for an atom in a solid at low temperature. The narrow zero-phonon line is the one normally used for Mössbauer spectrometry.

recoil energy, Equation 3.33, is greater than the binding energy of the atom in the solid (15 to 30 eV), the recoiling atom is dislodged from its lattice site. (2) If the free–atom recoil energy is less than the binding energy but large compared to the phonon* energy (~ 0.1 eV), the gamma-emitting atom will remain in its site and dissipate its recoil energy to the lattice as heat. (3) If the free–atom recoil energy is less than the phonon energy, then it is possible to observe either recoil-free gamma emission or gamma emission accompanied by the excitation of a discrete phonon of energy $\hbar\omega \sim 10^{-2}$ eV. This third case is the effect discovered by Mössbauer, and the gamma spectrum consists of a sharp peak corresponding to recoil-free gamma emission (the zero phonon peak) as well as a continuous spectrum on the low energy side (the phonon wing) corresponding to gamma emission accompanied by phonon excitation (see Figure 3-15). The reader should note that recoil-free gamma emission is analogous to the elastic scattering of photons as in the production of the unmodified line in the Compton effect (see section 7 of Chapter 2).

A frequently used source for Mössbauer studies is $_{27}\text{Co}^{57}$, which captures a K-electron and decays to an excited state of $_{26}\text{Fe}^{57}$. When Fe^{57} returns to its ground state it emits a recoilless photon having a linewidth of about 10^{-11} times its energy of 14.4 keV. Such a sharp line can easily excite an Fe^{57} atom in its ground state by the process of resonance absorption. Thus, a sample of Co^{57} provides a source of radiation of unbelievable spectral purity and unexcited Fe^{57} provides an equally sharp detector. To get some appreciation for the sharpness of this line, note that a relative velocity of only a few hundredths of a millimeter per second between source and detector will produce a Doppler shift great enough to prevent the resonance absorption.

PROBLEM 3-16

(a) Calculate the Doppler shift due to a relative velocity of 3 cm/sec.

(b) Find $\Delta E/E$, where $E = 14.4$ keV and ΔE is the Doppler shift obtained in (a).

---

* Phonons are the quantized lattice vibrations referred to in Problem 2-7.

(a)   What is the line broadening due to recoil of a Ni^{60} nucleus upon emission of a 1.33 MeV gamma ray?

(b)   What is the ratio $\Delta E/E$ for this case?

(Ans.: 15.8 eV; $\sim 10^{-5}$.)

Several nuclear models have been proposed to account for the observed properties of nuclei. Unfortunately, no single model satisfactorily explains all of the experimental facts, so that more than one model must be invoked unless one is concerned only with a specific property of the nucleus.

Perhaps the most popular model is the *shell model*. In this model each nucleon is regarded as moving in an orbital state under the influence of a nuclear field which represents the average effect of all the other nucleons. The allowed quantum states are found by using the *j-j* coupling scheme of combining angular momenta, which will be discussed in section 5 of Chapter 8. The noteworthy successes of the shell model are its ability to account for nuclear spins, the stability of the alpha particle and certain other nuclei (those associated with the so-called "magic numbers"), and for many features of nuclear spectra.

The shell model not only ignores such problems as the saturation of the nuclear force, the nearly constant nuclear density, and the binding energy per nucleon, but it gives the wrong results for the scattering of neutrons from nuclei. It turns out that neutrons interact so strongly with nuclei for certain discrete energies (sometimes spaced only a few hundred eV apart) that they are often trapped for a while before being ejected by the target nucleus. Such trapping events are called "resonances."

The model proposed to account for these resonances is the *liquid–drop model*. Here we regard the collective behavior of the nucleons as a many–body system analogous to the lattice vibrations of a solid. The energy states of the system are now the excitation energies of the collective system (analogous to the normal modes), which one would expect to be a set of discrete but closely spaced levels. Hence, a neutron having one of these energies will, in a collision with the nucleus, immediately share its energy with the whole system. A considerable time interval will elapse (compared to the transit time through a distance equal to the nuclear diameter) before a neutron accumulates enough energy to emerge as the "scattered" particle.

In addition to explaining the resonances, the liquid-drop model is useful in visualizing the saturation of nuclear forces and the phenomenon of fission. It has also been used to develop a phenomenological formula known as the semi-empirical mass formula.[24]

Another useful model of the nucleus is called the *optical model*. Here the nucleus may be regarded as a cloudy crystal ball, since it is essentially opaque to short wavelength neutrons (high energies) and it gets increasingly transparent as the neutron wavelength increases (lower energies). Its cloudiness is a measure

---

[24] See, for example, R. D. Evans, *op. cit.*, p. 366, or L. R. B. Elton, *op. cit.*, p. 118.

of the absorption probability and its index of refraction is related to the deflection of scattered neutrons. The phenomena of absorption and resonances can be accounted for by means of canceling and reinforcing traveling waves, respectively.

8. FUNDAMENTAL FORCES AND EXCHANGE PARTICLES

The historical problem of accounting for "action at a distance" was avoided in classical physics by postulating fields of force which pervaded all of space. Thus, the effect of body A on a remote body B can be explained by the fact that body B moves in the force field of A, and vice versa. The field has an energy density associated with it and it possesses momentum which can be transferred to and from massive particles. We have already seen in the case of the electromagnetic field that this energy density may be regarded as arising from a collection of quantized oscillators, each having a frequency corresponding to one of the allowed modes. The entity which possesses the quantum of energy (and the corresponding momentum) is the zero mass wavepacket which we know as the photon. Thus we may regard an electromagnetic field as a *photon gas*. To proceed a step further, the electromagnetic interaction between two charges is propagated by means of the exchange of *virtual* photons between the two charges. These photons are called virtual because their emission and reabsorption would constitute a violation of the conservation of energy were it not for the fact that they cannot be detected.

The uncertainty principle* permits a discrepancy in the energy of an amount ΔE provided that this non-conservation lasts for a time interval no greater than something of the order of $\hbar/\Delta E$. Thus, a charged particle can continually emit photons of energy ΔE and reabsorb them without violating the conservation of energy since the whole process is undetectable. By means of the same argument, a photon (not a virtual one) may be regarded as continually creating and annihilating virtual electron-positron pairs. These two processes are indicated schematically in Fig. 3-16.

When a virtual photon is emitted by one particle and absorbed by a different particle, momentum is transferred between the particles. The time rate of change of such a transfer of momentum is the force. There is a further complication, however, since the exchange of photons must account for both

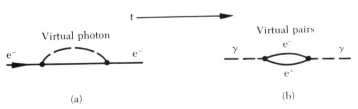

Figure 3-16 (a) Emission and reabsorption of a virtual photon by an electron. (b) Creation and annihilation of an electron–positron pair by a photon.

* See Section 9 of Chapter 2.

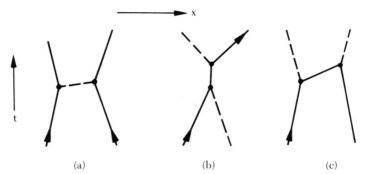

Figure 3-17 Feynman diagrams depicting (a) the interaction of two electrons, (b) the Compton effect, and (c) electron–positron annihilation. A dashed line represents a photon and a vertex indicates the absorption or emission of a photon.

attractive and repulsive forces. The interaction of two electrons by means of the exchange of virtual photons is symbolized in the Feynman diagram shown in Fig. 3-17(a). Neither the angles nor the lengths of the lines have any significance, but the intersections (or vertices) represent the physical processes. The Compton effect and (real) electron-positron annihilation are also shown in Fig. 3-17.

In a similar manner, the gravitational force may be regarded as being propagated by the exchange of virtual particles between gravitational masses. Since this force (the weakest of known forces) has an infinite range, we expect the exchange particle—which is called the *graviton*—to have zero rest mass. Although it is being hotly pursued at the present time,[25] the graviton is still undetected.

At the other extreme, the strongest force known is that between nucleons, the so-called nuclear force which was mentioned in the previous section. Unlike the weak electromagnetic interaction and the infinitesimally weak gravitational interaction, the nuclear force requires an exchange particle having a rest mass in order to account for its extremely short range. Yukawa[26] proposed a mass intermediate between that of the electron and the nucleon; hence, the name *meson* (intermediate) was given to this hypothetical exchange particle. Subsequently, a variety of mesons was discovered, but it is now generally believed that the *pi meson* or *pion* is the Yukawa particle. The charged pion has a rest mass that is 273 times the electron rest mass. It is illuminating to note how Yukawa was able to predict the mass of the nucleon exchange particle. We saw in the last section that the range of the nuclear force is $b \sim 1.4 \times 10^{-13}$ cm. If we take this to be the distance that the exchanged meson must move, then the transit time is at least $\Delta t \sim b/c$. Since the emission of a pion in the time Δt will produce an uncertainty in the energy of the nucleon of $\Delta E \sim M_\pi c^2$, then from the uncertainty principle we have

$$\Delta E \cdot \Delta t \sim M_\pi bc \sim \hbar,$$

[25] J. Weber, *Phys. Today* **21**, 34 (1968); *Phys. Rev. Letters* **21**, 395 (1968).
[26] H. Yukawa, *Phys. Math. Soc. Japan* **17**, 48 (1935).

or

$$M_\pi \sim \frac{\hbar}{bc} \sim \frac{\lambda_c}{b} m_0 \sim 270 \ m_0,$$

where λ_c is the Compton wavelength for the electron. Note the reciprocal relationship between the mass of the exchange particle and the range of the force. This explains why particles of zero rest mass are required for the inverse square forces which extend to infinity. Conversely, a force with a finite range requires an exchange particle with a non-zero rest mass.

There is at least one additional force which is involved in the so-called "weak interactions" of particle physics. This is the force that is responsible for nuclear β-decay, muon decay, and other lepton decay schemes. The exchange particle involved in this force has not yet been discovered.

9. ANGULAR MOMENTA AND MAGNETIC MOMENTS

In elementary electromagnetism one learns that a current loop has associated with it a magnetic moment whose magnitude is equal to the product of the current and the area enclosed by the loop, and whose direction is perpendicular to the plane of the loop. Then, for a circular Bohr orbit of radius r, we have

$$\vec{\mu}_\ell = \hat{r} \times \vec{i}A,$$

where $\vec{i}$ is in absolute units of current and the caret signifies a unit vector. But $A = \pi r^2$ and $\vec{i} = -e\vec{v}/2\pi cr$, where e is in esu. Then

$$\vec{\mu}_\ell = -\frac{e}{2c} \vec{r} \times \vec{v} = -\frac{e}{2mc} \vec{r} \times \vec{p} = -\frac{e\vec{\ell}}{2mc} . \tag{3.34}$$

According to Bohr's postulate, the angular momentum $\vec{\ell}$ is quantized and must be equal to an integer times $\hbar$, or $\vec{\ell} = m_\ell \hbar \hat{\ell}$. In particular, the angular momentum of the first Bohr orbit is given by $m_\ell = 1$, or $\vec{\ell} = \hbar \hat{\ell}$. Then

$$\vec{\mu}_\ell = -\frac{e\hbar}{2mc} \hat{\ell} = -\mu_B \hat{\ell}. \tag{3.35}$$

Here m is the mass of the electron. The unit vector $\hat{\ell}$ is included to show that the magnetic moment is directed antiparallel to the orbital angular momentum because of the negative charge of the electron. The quantity μ_B is called the *Bohr magneton*, and its value is approximately

$$\mu_B = \frac{e\hbar}{2mc} = 9.27 \times 10^{-21} \text{ erg/gauss}.$$

The ratio of the magnetic moment to the orbital angular momentum is called the classical gyromagnetic ratio (or the magnetomechanical ratio) and is expressed as

$$\gamma_\ell = \left| \frac{\vec{\mu_\ell}}{\vec{\ell}} \right| = \frac{e}{2mc} = \frac{\mu_B}{\hbar} . \tag{3.36}$$

In addition to its orbital momentum the electron also has an intrinsic spin angular momentum which cannot be explained classically. Its existence was postulated by Uhlenbeck and Goudsmit[27] in order to explain the fine structure of the hydrogen spectrum. It was later found that many of the details of the multiplet structure of many-electron atoms immediately became clear when spin was taken into account. Spin arises in a natural way in Dirac's relativistic quantum mechanics,[28] but it must be treated phenomenologically in the nonrelativistic Schrödinger formulation which will be used in this book. Therefore, we will postulate its existence and use it at will. Spin has a gyromagnetic ratio that is approximately twice the classical valué for orbital moments. That is,

$$\gamma_S = \left| \frac{\vec{\mu_S}}{\vec{S}} \right| = \frac{e}{mc} . \tag{3.37}$$

This value is called *anomalous* because it has no classical explanation.

Equations 3.36 and 3.37 are often combined by writing $\gamma = ge/2mc$. The quantity g is called the *spectroscopic splitting factor* because of its effect on the splitting of spectral lines under the influence of a magnetic field. In the case of spin only, the g-factor may be taken as approximately 2, although its experimental value is 2.0024. Since the g-factor is unity for orbital angular momenta, one might expect it to have non-integral values for states which are mixtures of orbital and spin angular momenta. Such cases will be discussed in section 4 of Chapter 8.

We have seen that spin is twice as effective as the orbital angular momentum in producing a magnetic moment. Another peculiarity of spin is that it can have half-integral units of angular momentum, in contrast to the integral multiples of $\hbar$ required for the component of orbital angular momentum along a given axis. Thus the spin of an electron is $\frac{1}{2}\hbar$. Protons and neutrons also have intrinsic spins of $\frac{1}{2}\hbar$, while photons have one unit of spin, namely, $\hbar$. The magnetic moment due to the spin of the electron is, however, also equal to one Bohr magneton because of the cancelling factors of 2. Thus

$$|\mu_S| = \gamma_S |S| = \frac{e}{mc} \cdot \frac{\hbar}{2} = \mu_B.$$

From the foregoing we note that the smallest unit of magnetic moment for the electron is the Bohr magneton, whether one considers orbital or spin angular momentum.

[27] G. E. Uhlenbeck and S. A. Goudsmit, *Naturwiss.* **13**, 593 (1925).
[28] P. A. M. Dirac, *Proc. Royal Society* (*London*) **117**, 610 (1928) and **118**, 351 (1928).

The intrinsic magnetic moment of the proton, which is also a spin-$\frac{1}{2}$ particle, and the deuteron, which is a spin-1 particle, were determined by Stern and his co-workers from beam deflection experiments on hydrogen and deuterium using the molecular species H_2, D_2 and HD.[29] Later, more accurate values were obtained for these nuclei by means of beam resonance and nuclear magnetic resonance techniques. These methods have now been applied to nearly all stable nuclei. The neutron magnetic moment was determined from neutron scattering experiments in magnetized iron.[30]

By analogy with the Bohr magneton we define the *nuclear magneton* as

$$\mu_N = \frac{e\hbar}{2M_pc} = 5.050951 \times 10^{-24} \text{ erg/gauss,}$$

where the mass of the proton has replaced the electronic mass. The current values of the proton, neutron and deuteron magnetic moments are:

$$\mu_p = 2.792782\,\mu_N$$

$$\mu_n = -1.9135\,\mu_N$$

$$\mu_d = 0.8576\,\mu_N.$$

Note that the deuteron magnetic moment is not the algebraic sum of the moment due to the neutron and proton. This discrepancy is believed to be evidence for a non-central contribution to the interaction between nucleons.

PROBLEM 3-18

(a) By means of a classical calculation, estimate the average flux density seen by an electron in the first Bohr orbit of hydrogen due to the proton revolving about it.

(b) Assuming that the spin moment of the electron experiences the magnetic field calculated in (a), find the energy required to flip the spin direction.

(c) What is the frequency of the photon having this energy?

PROBLEM 3-19

What is the magnetic field seen by an electron in the first Bohr orbit of hydrogen due to the intrinsic spin moment of the proton?

---

[29] For references and an excellent summary of the beam experiments, see N. F. Ramsey, *Molecular Beams*, Oxford University Press, N.Y. (1956), p. 102 cf.

[30] L. W. Alvarez and F. Bloch, *Phys. Rev.* **57**, 111 (1940).

The intrinsic spin angular momenta and associated magnetic moments of the elementary particles have provided the strongest arguments against the existence of electrons in the nucleus. The electron, having a magnetic moment roughly 1000 times greater than any known nuclear moment, could hardly remain undetected within a nucleus. Furthermore, it is known that nuclei with even A and odd Z and N have integral spin, whereas if such a nucleus contained A protons and N electrons its spin would have to be half-integral. On the other hand, the ˉnuclear shell model accounts for all known nuclear spins by regarding the nucleus to be comprised only of nucleons.

10. THE LARMOR THEOREM AND THE NORMAL ZEEMAN EFFECT

Let us now consider the effect of a weak magnetic field on an electron performing circular motion in a planar orbit such as that discussed in the previous section. In the classical treatment there is, of course, no electron spin, so our model simply consists of the interaction of a classical magnetic dipole $\vec{\mu}_\ell$ (arising from the orbital angular momentum $\vec{\ell}$) with the applied magnetic field, $\vec{\mathscr{B}}$, measured in gauss. Assume that the orbital angular momentum vector is oriented at the angle θ with respect to the z-axis along which the magnetic field is applied (see Figure 3-18). The torque on $\vec{\ell}$ is given by $\vec{\mu}_\ell \times \vec{\mathscr{B}}$, which is directed into the plane of the page in the ϕ-direction. Since, from elementary mechanics, the torque also equals the rate of change of the angular momentum, we have

$$\frac{d\vec{\ell}}{dt} = \vec{\mu}_\ell \times \vec{\mathscr{B}} = \gamma_\ell \vec{\ell} \times \vec{\mathscr{B}}.$$

But $|d\vec{\ell}| = \ell \sin \theta \, d\phi$, so we may write the scalar equation

$$\ell \sin \theta \cdot \frac{d\phi}{dt} = \gamma_\ell \ell \mathscr{B} \sin \theta.$$

Defining the precessional velocity by $\omega_L = d\phi/dt$, we obtain the result

$$\omega_L = \gamma_\ell \mathscr{B} = \frac{e\mathscr{B}}{2mc}. \tag{3.38}$$

The angular velocity ω_L is often called the *Larmor frequency*.

An important consequence of this, which is known as Larmor's theorem,[31] may be stated as follows: the effect of a weak magnetic field upon an atom can be removed (to first order in $\mathscr{B}$) by a mathematical transformation to a rotating

---
[31] J. H. Van Vleck, *The Theory of Electric and Magnetic Susceptibilities*, Oxford University Press, London, 1932, p. 22.

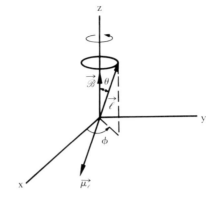

Figure 3-18 The precession of the angular momentum vector as a result of the torque produced by the action of a magnetic field on its associated magnetic moment.

coordinate system. Thus, if the energy levels of an atom are known (with no applied field) and then a magnetic field is turned on, a transformation to the coordinate system rotating at the Larmor frequency will restore the original field-free problem. It follows that the solution of a problem involving a static magnetic field can be written approximately as the superposition of the no-field solution and a rotation at the Larmor frequency. The exact expression which shows the quadratic term in the field is considered in Problem 3-20. Using the Planck relation, the energy associated with the Larmor frequency is

$$\Delta E = \pm \omega_L \hbar = \pm \frac{e\hbar \mathscr{B}}{2mc} = \pm \mu_B \mathscr{B}, \qquad (3.39)$$

where the signs refer to the sense of the rotation. This energy difference should be recognized as the potential energy of a magnetic dipole whose moment is one Bohr magneton, since the dipolar energy is given by

$$\Delta E = -\vec{\mu}_B \cdot \vec{\mathscr{B}}.$$

Hence, the plus sign (higher energy) in Equation 3.39 corresponds to anti-parallel alignment (as in Figure 3-18), while the minus sign (lower energy) indicates parallel alignment. If the energy of an electron having a moment μ_B is E_0 with no applied magnetic field, then it can take on one of the energies $E_0 \pm \mu_B \mathscr{B}$ in the magnetic field $\mathscr{B}$, provided that one can neglect quadratic terms in $\mathscr{B}$. It was found experimentally that in a collection of identical atomic systems of the type we have described above, a magnetic field actually produces a triplet of levels (called a Lorentz triplet) whose energies are $\{E_0, E_0 \pm \mu_B \mathscr{B}\}$. The existence of these levels can be confirmed by studying the optical transitions to and from these states. This phenomenon is known as the *normal Zeeman effect*.[32] The reason for the appearance of three levels will become apparent in the next section when we discuss spatial quantization. There we will see that the three energies correspond to the three allowed projections of the

[32] P. Zeeman, *Phil. Mag.* **5**, 43, 226 (1897).

angular momentum. However, the classical explanation invoked the Larmor frequency to account for the two shifted levels, and the unshifted level was attributed to the component of the motion parallel to the magnetic field, for which there would be no interaction.

The Zeeman effect is really more complicated than the above classical model would lead one to believe. We now know that electrons have spin and that this spin also has a magnetic moment associated with it. Thus, when a magnetic field is applied, both the spin and the orbital angular moments will find themselves acted upon by precessional torques. The resulting energy level splittings, however, cannot be interpreted from the classical theory because of the anomalous half-integral spin (in units of $\hbar$) of the electron and the spin g-value which is twice the classical value. As a consequence of this inexplicable behavior, the more general Zeeman effect, including spin, was historically misnamed the "anomalous Zeeman effect" in contrast with the "normal" effect, which is the less common variety. This topic will be discussed in considerable detail in section 4 of Chapter 8.

Before leaving this classical model the reader should note that the torque produced by the magnetic field acting on the magnetic dipole does *not* align the dipole with the field. Instead, the torque causes the dipole to precess in such a way as to maintain its initial polar angle θ (see Figure 3-18). In an ideal system (having no dissipative forces) this precession would continue as long as the field were to remain. Hence, it would be impossible to magnetize a piece of iron in such an ideal system, however great the applied field; the greater the field, the more rapid would be the precession. In the real world, however, some sort of "viscous" damping always exists. As energy is dissipated to the lattice, the polar angle of the precession decreases at the expense of the dipolar energy until complete alignment of the dipole occurs. The time interval required for complete alignment is known as the *longitudinal relaxation time*, or, in the case of a dipole moment due to spin, it is called the *spin-lattice relaxation time*. These relaxation times play an important role in magnetic resonance phenomena as well as in other areas of contemporary physics.[33]

PROBLEM 3-20

(a)   For an electron moving in a circular orbit about a nucleus of charge e in a magnetic field $\vec{\mathscr{B}}$, set up the equation of motion using the Lorentz force,

$$\vec{F} = e(\vec{\mathscr{E}} + \frac{1}{c}\vec{v} \times \vec{\mathscr{B}}).$$

(b)   Show that ω is given approximately by $e\mathscr{B}/2mc$ for small magnetic fields by estimating the relative magnitudes of the neglected terms.

[33] See, for example, C. P. Slichter, *Principles of Magnetic Resonance*, Harper and Row, New York, 1963; also, M. Sparks, *Ferromagnetic-Relaxation Theory*, McGraw-Hill, New York, 1964.

II. SPATIAL QUANTIZATION

The classical model of the previous section is often useful for describing certain aspects of resonance and relaxation experiments, but at the same time it can be misleading if taken too seriously. For example, the picture of a precessing angular momentum vector relaxing through a continuum of polar angles conflicts with a fundamental concept of quantum mechanics which is referred to as *spatial quantization*. Stated briefly, spatial quantization means that the projection of an angular momentum vector along any single axis in space can be only one of a discrete set of allowed values. At a given time there can be only one axis of quantization; if it is altered, say, by changing the direction of an applied uniaxial magnetic field, quantized states exist only along the new field direction. The close connection between space and angular momentum will be elaborated further in section 3 of Chapter 7.

The experiment proposed by Stern and performed by Gerlach and Stern[34] provided the first conclusive evidence of spatial quantization in 1922. Silver atoms were evaporated in an oven and collimated into a narrow beam which was passed through an inhomogeneous magnetic field as shown in Figures 3-19 and 3-20. Although a uniform field produces no net force on a magnetic dipole (only torque), a non-uniform field can deflect a dipole with a net translational force. This can readily be seen by considering a magnetic field $\mathscr{B}$ which acts in the positive z-direction, and which is also designed to have a gradient in the positive z-direction (Fig. 3-19). Assuming that all other derivatives of the magnetic field are zero, we find that the translational force on a dipole, oriented at an angle θ with the z-axis, is

$$\vec{F} = -\vec{\nabla}E = \vec{\nabla}(\vec{\mu}\cdot\vec{\mathscr{B}}) = \vec{\nabla}(\mu\mathscr{B}\cos\theta)$$

$$= \mu\cos\theta\,\frac{\partial\mathscr{B}}{\partial z}\,\hat{k}\,,$$

or
$$F_z = \mu\cos\theta\,\frac{\partial\mathscr{B}}{\partial z}\,.$$

Thus, a dipole aligned with the field is acted upon by an upward force of magnitude $\mu(\partial\mathscr{B}/\partial z)$, while a dipole aligned antiparallel to the field is acted upon by a downward force of the same magnitude. For arbitrary orientations the force could have any value between $-\mu(\partial\mathscr{B}/\partial z)$ and $+\mu(\partial\mathscr{B}/\partial z)$.

In the absence of spatial quantization, the beam of particles would contain a continuous distribution of angular orientations of the precessing dipoles, and the action of the non-uniform magnetic field would spread the narrow beam into a band at the detector screen. However, instead of a continuous band, Gerlach and Stern obtained two distinct lines whose breadth was due to the spread in particle velocities rather than to a continuum of dipole orientations.

This can only be explained if the dipoles are permitted just two orientational states, one state having a component in the $+z$-direction and the other

[34] O. Stern, Z. *Physik* **7,** 249 (1921); W. Gerlach and O. Stern, Z. *Physik* **8,** 110 and **9,** 349 (1922); *Ann. Physik* **74,** 673 (1924).

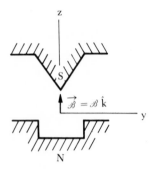

Figure 3-19 An inhomogeneous magnetic field having a large gradient in the field direction.

having a component in the $-z$-direction. The former are deflected toward the most intense region of the field while the latter are deflected toward the weakest region.

The quantization condition underlying the Stern-Gerlach experiment may be stated as follows: *The allowed orientations of the angular momentum are such that their projections on the axis of measurement differ successively by one unit of $\hbar$.* Thus, if the maximum projection is $1\hbar$, along the z-axis, there are three allowed orientations corresponding to z-components of 1, 0, and $-1\hbar$. In general, if the maximum component is $\ell\hbar$, there are $2\ell + 1$ allowed orientations of the angular momentum and its associated magnetic moment. Note that if the maximum component of the angular momentum is an integer there will always be an *odd* number of allowed orientations, and hence there should be an odd number of lines on the screen. However, for silver only two symmetric lines were observed in spite of the fact that the magnetic moment was found to be one Bohr magneton. In 1927, Phipps and Taylor repeated the experiment with hydrogen and again found just two lines.[35] The explanation, of course, is that the moments observed in these two cases were due to spin and *not* the orbital angular momentum. Because of its anomalous g-value, a spin projection of $\tfrac{1}{2}\hbar$ has a magnetic moment of one Bohr magneton. However, in order to satisfy the quantization condition, atoms with half-integral spin cannot have zero projections, so that the only allowed states correspond to $\pm\tfrac{1}{2}\hbar$. Hence, $2(\tfrac{1}{2}) + 1 = 2$, which accounts for the two lines in the original experiments. Thus, the experiment which first demonstrated spatial quantization also showed the existence of intrinsic spin angular momentum, although it was several years before the latter was recognized.

If the above experiment is repeated with particles having one unit of orbital or spin angular momentum, the beam is split into three distinct beams as required by spatial quantization. It is natural to inquire what would

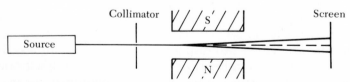

Figure 3-20 Particle trajectories in the experiment of Gerlach and Stern. The deflections of the beams are greatly exaggerated.

[35] T. E. Phipps and J. B. Taylor, *Phys. Rev.* **29**, 309 (1927).

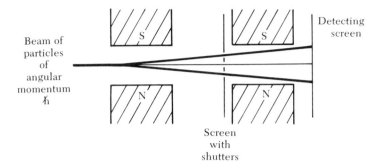

Figure 3-21   Apparatus for verifying the quantum states of particles having one unit of angular momentum.

happen if each of these three beams were, in turn, subjected to another apparatus identical with the first. In Figure 3-21 three such beams are shown entering a second magnet with the same field orientation as the first. If only one beam at a time is permitted to enter the second magnet, just one beam will appear at the detecting screen. Hence, no further splitting will occur, regardless of which of the three beams is allowed to enter the second apparatus. The second magnet produces an additional deflection of the upper and lower beams from which one can verify that the magnetic moments of the particles which comprise them correspond to $+\mu_B$ and $-\mu_B$, respectively. The center beam is undeflected and thus corresponds to zero projection of the magnetic moment along the vertical axis. From this result we draw the following conclusions. The first magnet establishes an axis of quantization and sorts the initial beam into three quantum states. The second magnet does not alter these quantum states but serves to confirm that the sorting of the beam into pure states has been preserved during the transit of the particles from one magnet to the other. On the other hand, if one were to alter Figure 3-21 by rotating the second magnet through an angle about the beam axis, the situation would be quite different. Each beam from the first magnet, which was a pure state in the first magnetic field, now appears as a mixture of the three quantum states in the new field orientation. Accordingly, if one beam at a time from the first magnet is allowed to enter the second, each will split into three components in the second apparatus.

PROBLEM 3-21

(a)   A beam of silver atoms is obtained by heating silver to a temperature of 2300°K. Assuming a Maxwell distribution of velocities, what is the most probable velocity of a silver atom?

(b)   Calculate the angular deflection of this beam in a Gerlach-Stern apparatus for a path length through the inhomogeneous magnetic field of 4 cm and an average field gradient of 5×10^4 gauss per cm.

SUMMARY

The Rutherford-Bohr atom is a dynamical atom in which nearly all of the atomic mass is concentrated in a nucleus having a diameter of about $10F$ ($\sim 10^{-4}$ Å). Charge neutrality is achieved by placing as many electrons in orbits about the nucleus as there are protons in the nucleus. The orbital radii and velocities of these electrons are such that the Coulomb attraction provides the appropriate centripetal acceleration to achieve dynamical stability. Bohr postulated that the orbital angular momentum associated with an electron is quantized in units of $\hbar = h/2\pi$, where h is Planck's quantum constant. This restriction on the possible values of the angular momentum results in an infinite but discrete set of allowed orbits and energies for each electron. Since the diameter of an allowed orbit is of the order of 1 to 10 Å, the "size" of an atom is roughly 10^5 times the size of the nucleus.

The Bohr theory predicts the energy levels of hydrogen and hydrogen-like atoms reasonably well, although serious discrepancies arise for non-hydrogenic atoms. An additional postulate of Bohr, namely, that the absorption or emission of a photon by an atom occurs when an electron makes a transition between two of the atomic energy levels, was an important link between the quantum theory of radiation and the quantum theory of matter. The magnetic moments associated with orbital angular momentum and with intrinsic particle spin are derived and the normal Zeeman effect is discussed. The phenomenon of spatial quantization, that is, that the angular momentum can have only discrete projections along a given axis, is illustrated by means of the Stern-Gerlach experiment.

SUGGESTED REFERENCES

American Institute of Physics, *Nuclear Structure, Selected Reprints*, 1965.

A. Beiser, *Perspectives of Modern Physics*. McGraw-Hill Book Co., Inc., New York, 1969.

C. H. Blanchard, C. R. Burnett, R. G. Stoner, and R. L. Weber, *Introduction to Modern Physics*, 2nd ed. Prentice-Hall, Inc. Englewood Cliffs, N.J., 1969.

M. Born, *Atomic Physics*, 6th ed. Hafner Publishing Co., New York, 1959.

R. M. Eisberg, *Fundamentals of Modern Physics*. John Wiley and Sons, Inc., New York, 1961.

L. R. B. Elton, *Introductory Nuclear Theory*. W. B. Saunders Co., Philadelphia, 1966.

Robley D. Evans, *The Atomic Nucleus*. McGraw-Hill Book Co., Inc., New York, 1966.

Richard P. Feynman, Robert B. Leighton, and Matthew Sands, *The Feynman Lectures on Physics*. Addison-Wesley Publishing Co., Inc., Reading, Mass., 1963.

R. B. Leighton, *Principles of Modern Physics*. McGraw-Hill Book Co., Inc., New York, 1959.

F. K. Richtmyer, E. H. Kennard, and J. N. Cooper, *Introduction to Modern Physics*, 6th ed. McGraw-Hill Book Co., Inc., New York, 1969.

Henry Semat, *Introduction to Atomic and Nuclear Physics*, 4th ed. Holt, Rinehart and Winston, New York, 1969.

Paul A. Tipler, *Foundations of Modern Physics*. Worth Publishers, Inc., New York, 1969.

F. W. Van Name, Jr., *Modern Physics*, 2nd ed. Prentice-Hall, Inc., Englewood Cliffs, N.J., 1962.

Hugh D. Young, *Fundamentals of Optics and Modern Physics*. McGraw-Hill Book Company, New York, 1968.

CHAPTER 4

THE DEVELOPMENT OF WAVE MECHANICS

I. INTRODUCTION

In this and the ensuing chapters we shall develop and use the formalism of non-relativistic quantum mechanics as it is understood today. We begin this chapter with a brief aside on the "old" quantum theory and its application to the hydrogen atom. Although this material may be omitted without hampering the reader's understanding of the current theory, it is included here for the following reasons. It represented a forward leap in that it attempted to incorporate quantum concepts into current theories. It correctly predicted a large body of experimental results from a few simple rules. It is of considerable historical importance because it occupied many of the greatest minds, and like the ether theory of pre-relativity days, it thus set the stage for the appearance of the modern theory.

The new quantum mechanics appeared in two forms which were later shown to be equivalent. One form is known as *wave mechanics* and is generally regarded as Schrödinger's formulation,[1] although it is based heavily on the work of de Broglie.[2] The other approach is called *matrix mechanics* and it is chiefly credited to Heisenberg, although Born and Jordan shared in its development.[3] Since wave mechanics is more easily grasped intuitively we shall begin our study with the development of the de Broglie theory of particle waves and the evolution of the Schrödinger equation. All of Chapter 5 is devoted to applications of the Schrödinger method to one-dimensional systems. The essentials of matrix mechanics will be introduced in Chapter 6. In the remaining chapters the wave and the matrix formulations will be used freely and interchangeably.

---

[1] E. Schrödinger, *Ann. Physik* (4) **79,** 361 (1925); **79,** 489 (1925); **80,** 437 (1926); **81,** 109 (1926).

[2] L. de Broglie, Nature **112,** 540 (1923); Thesis, Paris (1924); *Ann. Physique* (10) **2** (1925).

[3] W. Heisenberg, Z. *Physik* **33,** 879 (1925); M. Born and P. Jordan, Z. *Physik* **34,** 858 (1925); M. Born, W. Heisenberg and P. Jordan, Z. *Physik* **35,** 557 (1926).

2. THE OLD QUANTUM THEORY: WILSON-SOMMERFELD QUANTIZATION RULES

Wilson[4] and Sommerfeld[5] independently discovered a method of quantizing the action integrals of classical mechanics, and this method was subsequently applied to a number of physical systems. A necessary condition for the application of this method is that each generalized coordinate q_k and its conjugate momentum p_k must be periodic functions of time. Then the action integral taken over one cycle of the motion is quantized; that is,

$$\oint p_k \, dq_k = n_k h. \tag{4.1}$$

To illustrate the method, consider a one-dimensional simple harmonic oscillator whose equation of motion is

$$m\ddot{x} + kx = 0,$$

or

$$\ddot{x} + \omega^2 x = 0,$$

where

$$\omega^2 = \frac{k}{m}.$$

Then,

$$x = x_0 \sin \omega t$$

and

$$p_x = m\dot{x} = m\omega x_0 \cos \omega t.$$

Equation 4.1 becomes

$$nh = \oint p_x \, dx = \int_0^T m\omega^2 x_0^2 \cos^2 \omega t \, dt$$

$$= m\omega x_0^2 \int_0^{2\pi} \cos^2 \theta \, d\theta$$

$$= m\omega \pi x_0^2.$$

Therefore,

$$x_0^2 = \frac{nh}{\omega m \pi},$$

or the amplitudes are quantized. The energy states are

$$E = T + V = \tfrac{1}{2}m\dot{x}^2 + \tfrac{1}{2}kx^2$$

$$= \tfrac{1}{2}m\omega^2 x_0^2,$$

or,

$$E_n = n\omega\hbar = nh\nu. \tag{4.2}$$

[4] W. Wilson, *Phil. Mag.* **29**, 795 (1915).
[5] A. Sommerfeld, *Ann. Phys.* **51**, 1 (1916).

In both the classical theory and the old quantum theory, the ground state energy of an oscillator is incorrectly given as zero. However, the level spacings are correct in these older theories. From the oscillator energy levels obtained in Equation 4.2, there is no information about which transitions are most likely to occur, or in fact, whether any are forbidden. Information of this kind goes under the general heading of *selection rules* and is readily obtained in the new quantum mechanics. However, in the old theory, selection rules were inferred by comparing the system with the behavior of a classical system; that is, by employing what is called *Bohr's correspondence principle*. Thus, since a classical oscillator will emit only one frequency (and no harmonics), if a quantum mechanical oscillator is to correspond to the classical result in the limit of large n, then we must have the selection rule $\Delta n = \pm 1$. We had already assumed transitions between adjacent levels in our discussion of Planck's oscillators in section 3 of Chapter 2.

If we treat a two-dimensional harmonic oscillator as two independent one-dimensional oscillators in the x- and y-directions, the energy levels are

$$E_{n_x, n_y} = \hbar(n_x \omega_x + n_y \omega_y).$$

If the oscillator is isotropic $(k_x = k_y)$, then $\omega_x = \omega_y = \omega$, and

$$E_n = n\hbar\omega,$$

where $n = n_x + n_y$. But $n = 1$ now corresponds to the two states $(n_x = 0, n_y = 1)$ and $(n_x = 1, n_y = 0)$, which have the same energy. The two states are said to be *degenerate*. In general, the level of energy E_n is $(n + 1)$-fold degenerate. Similarly, the energy of the nth level of a three-dimensional iso-tropic oscillator is also given by Equation 4.2 with $n = n_x + n_y + n_z$. The degeneracy in this case is $\frac{1}{2}(n + 1)(n + 2)$.

Duane and Compton[6] applied the Wilson-Sommerfeld method to a corpuscular model for the diffraction of x-rays by a crystal. If the z-direction is normal to a set of identical atomic planes of separation d, then the quantum condition becomes

$$\oint p_z \, dz = \int_0^d p_z \, dz = p_z \, d = nh, \tag{4.3}$$

in the absence of forces. A photon of momentum h/λ incident at the angle θ, as shown in Figure 4-1, will be reflected at the same angle and will transfer an

Figure 4-1 Bragg reflection of a photon by a crystal.

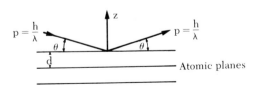

[6] W. Duane, *Proc. Nat. Acad. Sci.* **9**, 158 (1923); A. H. Compton, *ibid.* **9**, 359 (1923).

amount of momentum in the z-direction equal to $(2h/\lambda)\sin\theta$. But the momentum must satisfy the quantum conditions given by Equation 4.3, so we have

$$n\lambda = 2\,d\sin\theta, \qquad (4.4)$$

the well-known Bragg equation.

One can obtain the equivalent Bragg expression for electrons by letting $p_z = 2mv\sin\theta$ in Equation 4.3 to obtain

$$n\,\frac{h}{mv} = 2\,d\sin\theta. \qquad (4.5)$$

By comparing Equation 4.4 and Equation 4.5, it is apparent that the de Broglie wavelength, $\lambda = h/mv$, could also have been predicted by this theory.

PROBLEM 4-1

Use the Wilson-Sommerfeld method to obtain the energy levels of a rigid rotator of angular momentum, $p_\theta = I\omega$, where I is its moment of inertia about the rotation axis. (Ans.: $E_n = n^2\hbar^2/2I$.)

PROBLEM 4-2

Use the Wilson-Sommerfeld method to obtain the energy states of a perfectly elastic particle in a cubic box of edge a and with perfectly rigid walls. (Ans.: $E_n = n^2h^2/8ma^2$, where $n^2 = n_x^2 + n_y^2 + n_z^2$.)

PROBLEM 4-3

Find the energy states of a perfectly elastic ball bouncing in the gravitational field by applying the Wilson-Sommerfeld quantization condition. (Ans.: $E_n = (9g^2h^2mn^2/32)^{\frac{1}{3}}$.)

3. SOMMERFELD'S RELATIVISTIC THEORY OF THE HYDROGEN ATOM

Although the Bohr theory was quite successful in predicting the spectrum of hydrogen, there remained an unexplained fine structure or splitting of the lines. This splitting amounts to about one part in 10^4 and cannot be seen in spectrometers of low resolving power. Sommerfeld proposed that if elliptical

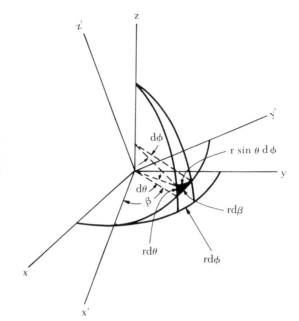

Figure 4-2 Relationship between the planar and spherical coordinates. The angle β is the azimuthal angle in the x'-y'-plane, ϕ is the azimuthal angle about the z-axis, and θ is the polar angle measured from the z-axis.

as well as circular orbits were allowed, the electron's velocity in an orbit of large eccentricity could become relativistic. He showed that the energy correction resulting from the relativistic treatment was of the proper order of magnitude to account for the fine structure splitting of hydrogen. Unfortunately, Sommerfeld's theory is not the correct explanation for the fine structure of atomic spectra. We now know that the fine structure is due to the fact that the electron has an intrinsic spin angular momentum which produces the so-called spin-orbit interaction. This will be treated in detail in the discussion of the quantum mechanical theory of radiation. In spite of the shortcomings of the Sommerfeld theory, it is of such great historical importance that it is worth sketching briefly here.

Consider an electron of reduced mass μ to be revolving about a fixed nucleus. Since the motion in a given orbit is confined to a plane, we will describe the position of the electron by the planar coordinates (r, β) measured in the plane of the orbit. In Figure 4-2, the x', y'-plane is the plane of the orbit and the unprimed system is used as the frame for a set of spherical polar coordinates. The relationship between the planar angle and the spherical coordinates is obtained by considering an infinitesimal angular displacement $d\beta$. Then the arc $r\, d\beta$ is the hypotenuse of an infinitesimal spherical triangle whose legs are $r\, d\theta$ and $r \sin \theta\, d\phi$. Then we have

$$(r\, d\beta)^2 = (r\, d\theta)^2 + (r \sin \theta\, d\phi)^2, \tag{4.6}$$

which we will need later in our discussion of the quantization rules.

The equation of the motion in the x', y'-plane is

$$\mu\ddot{r} = \mu r \dot{\beta}^2 - \frac{Ze^2}{r^2}.$$

Using the fact that $p_\beta = \mu r^2 \dot\beta = \ell = $ constant, we obtain

$$\mu\ddot{r} = \frac{\ell^2}{\mu r^3} - \frac{Ze^2}{r^2}.$$

Multiplying by $\dot{r}$ and integrating,

$$\mu\int \dot{r}\, d\dot{r} = \int \left(\frac{\ell^2}{\mu r^3} - \frac{Ze^2}{r^2}\right) dr,$$

or

$$\tfrac{1}{2}\mu\dot{r}^2 = -\frac{\ell^2}{2\mu r^2} + \frac{Ze^2}{r} + E. \qquad (4.7)$$

The constant of integration E is the total energy of the system. In order to get the path equation we eliminate the time by means of the substitution

$$\dot{r} = \frac{dr}{d\beta}\frac{d\beta}{dt} = \frac{\ell}{\mu r^2}\frac{dr}{d\beta}.$$

Substituting this into Equation 4.7 and multiplying by $2\mu/\ell^2$ results in

$$\left(\frac{1}{r^2}\frac{dr}{d\beta}\right)^2 = -\frac{1}{r^2} + \frac{2\mu Ze^2}{\ell^2 r} + \frac{2\mu E}{\ell^2}.$$

Introducing $u = 1/r$,

$$\frac{du}{d\beta} = -\frac{1}{r^2}\frac{dr}{d\beta} = \pm\sqrt{-u^2 + \frac{2\mu Ze^2}{\ell^2}u + \frac{2\mu E}{\ell^2}}$$

and

$$\pm\, d\beta = \frac{du}{\sqrt{\dfrac{2\mu E}{\ell^2} + \dfrac{2\mu Ze^2}{\ell^2}u - u^2}}.$$

Integrating,

$$u = \frac{1}{r} = \frac{\mu Ze^2}{\ell^2} + \sqrt{\frac{\mu^2 Z^2 e^4}{\ell^4} + \frac{2\mu E}{\ell^2}} \cdot \sin(\beta - \beta_0), \qquad (4.8)$$

for negative total energy. The equation of an ellipse of semiaxes a and b, eccentricity ϵ and orientation β_0 is

$$u = \frac{1}{r} = \frac{1 + \epsilon\sin(\beta - \beta_0)}{a(1 - \epsilon^2)} = \frac{a}{b^2} + \frac{\sqrt{a^2 - b^2}}{b^2}\sin(\beta - \beta_0),$$

where $b/a = \sqrt{1 - \epsilon^2}$. Comparing this with Equation 4.8 we immediately obtain the following results:

$$a = -\frac{Ze^2}{2E}$$

$$b = \frac{\ell}{\sqrt{-2\mu E}}$$

$$1 - \epsilon^2 = -\frac{2E\ell^2}{\mu Z^2 e^4}.$$

(4.9)

PROBLEM 4-4

Show that the time averages of T and V satisfy the same relation as that satisfied by circular Bohr orbits, namely, $\bar{E} = \frac{1}{2}\bar{V} = -\bar{T}$.

Applying the quantization condition given in Equation 4.1 to the planar angle and the three spherical coordinates, we have

$$\oint p_\beta \, d\beta = kh$$

(4.10a)

$$\oint p_\phi \, d\phi = mh$$

(4.10b)

$$\oint p_\theta \, d\theta = n_\theta h$$

(4.10c)

$$\oint p_r \, dr = n_r h$$

(4.10d)

Equation 4.10a can be integrated immediately since $p_\beta = \ell = $ constant, and the axis about which ℓ is measured is fixed in space. Then,

$$p_\beta = \ell = k\hbar.$$

(4.11a)

Similarly, p_ϕ is a constant and the z-axis about which it is measured is fixed, so

$$p_\phi = m\hbar.$$

(4.11b)

The quantum number m is called the *magnetic quantum number* because of the role it plays in distinguishing the energy levels of the atom in the presence of a magnetic field. Since the axis about which p_θ is measured is neither unique nor stationary, we must transform Equation 4.10c before it can be integrated.

If Equation 4.6 is multiplied by μ/dt, we obtain

$$\mu r^2 \dot{\beta} \, d\beta = \mu r^2 \sin^2 \theta \cdot \dot{\phi} \, d\phi + \mu r^2 \dot{\theta} \, d\theta,$$

or

$$p_\beta \, d\beta = p_\phi \, d\phi + p_\theta \, d\theta.$$

Substituting this into Equation 4.10c,

$$\oint (\, p_\beta \, d\beta - p_\phi \, d\phi) = n_\theta h = (k - m)h,$$

or,

$$k = n_\theta + m. \tag{4.11c}$$

The integer k is called the *azimuthal quantum number*. It can take on the values $1, 2, 3, \ldots$, with zero excluded.

Sommerfeld's integration of Equation 4.10d will not be repeated here but we will merely state his result, namely,

$$kh\left(\frac{1}{\sqrt{1 - \epsilon^2}} - 1\right) = n_r h,$$

or

$$\frac{a}{b} = \frac{n_r + k}{k} = \frac{n}{k}. \tag{4.11d}$$

The quantity n in the last expression is called the *total quantum number*, since it is defined as the sum of the radial and azimuthal quantum numbers, which is to say that

$$n = n_r + n_\theta + m.$$

Combining Equations 4.9, 4.11a, and 4.11d, we obtain the equalities

$$\frac{a}{b} = \frac{n\hbar}{\ell} = -\frac{Ze^2}{\ell}\sqrt{\frac{-\mu}{2E}},$$

from which the quantized orbits and energies are given by:

$$a = \frac{n^2 a_0}{Z},$$

$$b = \frac{ka}{n} = \frac{kna_0}{Z},$$

and

$$E_n = -\frac{w_0 Z^2}{n^2}, \tag{4.12}$$

where the constants a_0 and w_0 were defined in section 4 of Chapter 3. Note that the semimajor axis, a, is the counterpart of the radius of a circular orbit

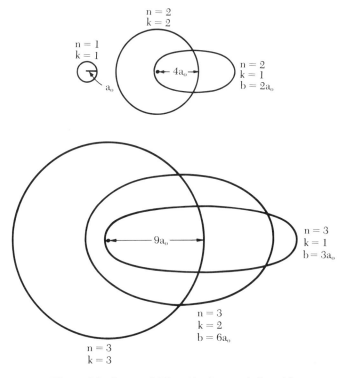

Figure 4-3 Sommerfeld's orbits for $n = 1$, 2, and 3.

in the Bohr theory. Furthermore, the energy of an orbit is independent of the value of the semiminor axis. This means that all of the orbits of different eccentricities associated with the same n value are degenerate. A few of these orbits are depicted in Figure 4-3.

It was at this point that Sommerfeld introduced a relativistic correction for the mass of the electron. For an orbit of large eccentricity the electron would pass close to the nucleus at a very high velocity such that the relativistic increase of mass would become noticeable. With this refinement to the theory the revised energies become

$$E_n = -\frac{w_0 Z^2}{n^2}\left[1 + \frac{\alpha^2 Z^2}{n}\left(\frac{1}{k} - \frac{3}{4n}\right)\right]. \tag{4.13}$$

The quantity

$$\alpha = \frac{e^2}{\hbar c} = 7.297 \times 10^{-3} \sim \frac{1}{137}$$

is called the *fine structure constant* since the term in Equation 4.13 in which α appears, correctly accounted for the fine structure splitting of the lines of the hydrogen spectrum.

Although the old quantum theory achieved many successes in atomic and molecular spectroscopy, it was an incomplete theory in the sense that none of its recipes for quantization were derived from first principles. Since it could not be applied to aperiodic systems, most collision and scattering problems were beyond its pale. Furthermore, it contained errors, contradictions, and ambiguities.[7] It did have some virtues, however. It predicted a large body of experimental results from a few simple rules, and it set the stage for the new quantum mechanics which soon replaced it. We will now proceed to discuss the wave mechanics of de Broglie and Schrödinger.

4. THE WAVE NATURE OF PARTICLES

The idea of associating both a wave and particle nature with the electron was first proposed by de Broglie in his doctoral thesis in 1925.[8] His work was motivated by the mystery of the Bohr orbits, which he attempted to explain by fitting a standing wave around the circumference of each orbit. Thus, de Broglie required that $n\lambda = 2\pi r$, where λ is the wavelength associated with the nth orbit and r is its radius. Combining this with Equation 3.25 we immediately obtain the result that

$$\lambda = \frac{h}{mv} = \frac{h}{p}.$$

Assuming the existence of a natural symmetry in the properties of matter and energy, he proposed that a material particle of total energy E and momentum p must be accompanied by a phase wave, analogous to that ascribed to the photon, whose wavelength is given by $\lambda = h/p$ and whose frequency is given by the Planck formula, $\nu = E/h$. The Planck and de Broglie relations may be expressed in the useful forms,

$$E = \hbar\omega$$

and   (4.14)

$$p = \hbar k,$$

where $\hbar = h/2\pi$, $k = 2\pi/\lambda$, and $\omega = 2\pi\nu$.

The physical nature of such a particle wave was not clearly described by de Broglie. Unlike a classical wave, the energy E of the particle wave is not thought of as spread out over the extent of the wave, but is regarded as localized with the particle. However, the accompanying wave is essential in order to account for the phenomena of interference and diffraction.

The concept of the de Broglie wavelength is one of the cornerstones of modern quantum theory, and the simple relationship

$$\lambda = \frac{h}{p}$$

holds for photons as well as for both relativistic and non-relativistic material particles, provided that the appropriate expression for p is used. On the other

---

[7] Albert Messiah, *Quantum Mechanics*, North-Holland Publishing Co., Amsterdam (1958), Chapter 1.

[8] L. de Broglie, *Ann. Phys. (Paris)* **3**, 22 (1925).

hand, the de Broglie frequency has not been a very useful concept and it comes into play only in the calculation of the phase velocity. Here a distinction appears between the relativistic and non-relativistic cases. If the rest mass energy is included in the total energy E (as in de Broglie's treatment), then the phase velocity of the wave becomes

$$u = \lambda \nu = \frac{E}{p} = \frac{1}{p} \sqrt{p^2 c^2 + (m_0 c^2)^2} = c \sqrt{1 + \left(\frac{m_0 c}{p}\right)^2} = \frac{c}{\beta},$$

or,

$$uv = c^2. \tag{4.15}$$

In obtaining Equation 4.15 we have expressed the relativistic momentum as $p = \gamma m_0 v$, where $v = \beta c$ is the particle velocity and $\gamma = (1 - \beta^2)^{-\frac{1}{2}}$. From our knowledge of waves* we identify the particle velocity v with the group velocity of the wave packet. Since special relativity requires that v be less than c, we note that Equation 4.15 calls for phase velocities greater than c. However, as no energy (that is, no signal or information) is transmitted at the phase velocity, the fact that $u > c$ constitutes no violation of the postulates of special relativity.

In non-relativistic quantum mechanics the rest mass term is neglected and the total energy E is merely the sum of the kinetic and potential energies. Accordingly, the phase velocity of the wave associated with a non-relativistic free particle ($V = 0$) is,

$$u = \frac{\omega}{k} = \frac{E}{p} = \frac{p}{2m} = \frac{\hbar k}{2m} = \frac{v}{2},$$

that is, one-half of the particle velocity. Thus the phase velocity is not of any physical significance. The group velocity of a particle wave, however, is given by

$$v = \frac{d\omega}{dk} = \frac{1}{\hbar}\frac{dE}{dk}.$$

5. THE DIFFRACTION OF PARTICLES

Whether or not particles of a given momentum will exhibit their wave characteristics will be determined by the relative magnitude of their de Broglie wavelength in comparison with the physical dimensions of the environment in which they are found. For wavelengths that are much smaller than the dimensions of apertures and obstacles, diffraction and other wave effects are not ordinarily observed. In such cases we can assume rectilinear propagation and problems can be treated by means of ray diagrams (for example, visible light in our everyday world). However, for wavelengths which approximate or exceed the dimensions of objects, diffraction effects become quite important and ray diagrams become meaningless (for example, audible sound in our everyday world). In order to get some insight into the kinds of behavior to

---

* See Chapter 2, section 8.

TABLE 4-1   de Broglie Wavelengths
Associated with Certain Particles.

144 volt electron	1 Å
1 volt electron	12 Å
1 volt proton	0.29 Å
1 volt alpha particle	0.15 Å

expect from particles, the de Broglie wavelengths for a few special cases are given in Table 4-1. Massive particles such as a bullet fired from a 0.22 rifle or a fast baseball have de Broglie wavelengths of the order of 10^{-23} or 10^{-24} angstrom.

Since the wavelengths associated with 10 to 100 volt electrons correspond to the atomic spacings of most crystalline solids, it was conjectured that electrons of these energies ought to be diffracted by crystals in the same manner as x-rays. This was confirmed experimentally in 1927 by the work of Davisson and Germer,[9] in which they scattered low energy electrons from a nickel crystal. Although electrons were scattered in all directions from the nickel crystal, a distinct peak in intensity occurred at an angle which corresponded to the first Bragg reinforcement for a wavelength of 1.65 Å. The size of this peak varied with the energy of the incident electrons and was found to reach a maximum of intensity for 54-volt electrons whose de Broglie wavelength is 1.67 Å! (See Figure 4-4.)

The Bragg equation, which gives the condition for intensity maxima in the reflected beams, may be expressed as

$$n\lambda = 2d \sin \phi = 2d \cos \theta, \qquad (4.16)$$

where n is the order of the diffraction, d is the distance between the reflecting planes, and the angles θ and ϕ are shown in Figure 4-5. For a mono-energetic beam of particles at normal incidence, the first order maximum will be found at an angle from the normal given by

$$2\theta = 2 \cos^{-1} \frac{\lambda}{2d}.$$

The higher ordered reflection maxima occur at progressively smaller angles from the incident beam.

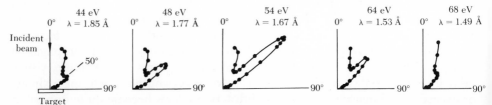

Figure 4-4   Angular plots of scattered intensity for low-energy electrons incident upon a nickel single crystal. The energies and de Broglie wavelengths of the electrons are given for each plot. (From C. Davisson and L. H. Germer, *Phys. Rev.* **30**, 705 (1927). Used with permission.)

[9] C. Davisson and L. H. Germer, *Phys. Rev.* **30**, 705 (1927).

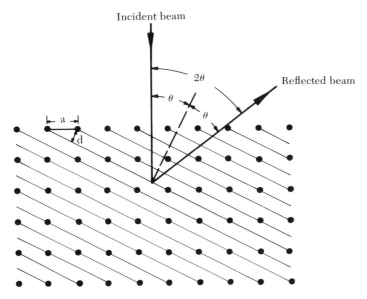

Figure 4-5 Bragg reflections of electron waves or x-rays for normal incidence on a crystal.

In addition to the Bragg reflection, the electron waves should undergo refraction at the crystal surface because of the abrupt change in potential energy at the surface. The index of refraction may be expressed as,

$$\mu = \frac{\lambda}{\lambda'},$$

where the prime refers to the wavelength inside the crystal. For electron energies that are small with respect to the rest-mass energy, the total energy and the momentum are related as follows:

$$E = \frac{p^2}{2m} + V.$$

Then,

$$p = \{2m(E - V)\}^{\frac{1}{2}},$$

and the index of refraction becomes,

$$\mu = \frac{\lambda}{\lambda'} = \frac{p'}{p} = \left(\frac{E - V'}{E}\right)^{\frac{1}{2}}, \tag{4.17}$$

where it has been assumed that $V = 0$ outside of the metal. If we take the potential well of nickel to be $V' \sim -5$ volts, then for incident electrons of 100 eV total energy we find that

$$\mu \sim \left\{\frac{105}{100}\right\}^{\frac{1}{2}} = \{1.05\}^{\frac{1}{2}} = 1.02.$$

Figure 4–6   Diffraction of 50 kV electrons from a disordered film of Cu_3Au alloy.   The alloy film was 400 Å   thick.   (Photograph kindly furnished by Dr. L. H. Germer.)

Figure 4–7   Diffraction of 300 volt electrons from a clean (110) surface of a tungsten single crystal.   (Photograph kindly furnished by Dr. L. H. Germer.   After L. H. Germer and J. W. May, *Surface Science* **4**, 452 (1966).   Used with permission of North-Holland Publishing Co., Amsterdam.)

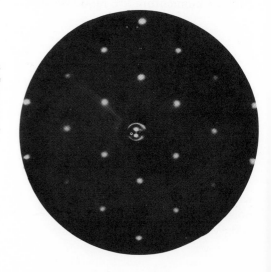

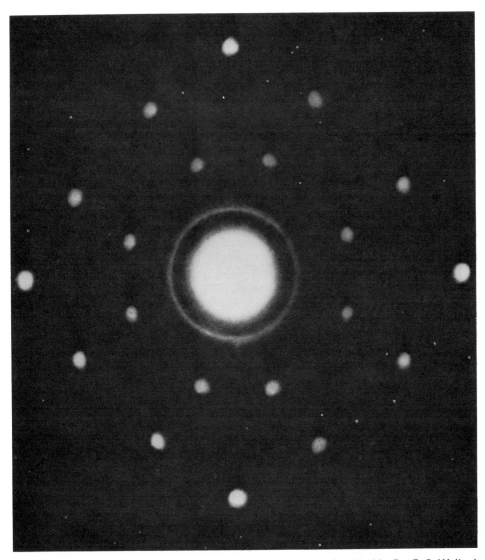

Figure 4-8 Neutron Laue photograph of NaCl. (Photograph kindly furnished by Dr. E. O. Wollan.)

PROBLEM 4-5

What is the wavelength of a photon whose energy equals the rest energy of an electron? What is the significance of this length? How does it compare with the classical radius of the electron?

PROBLEM 4-6

Using 2.15 Å for the lattice constant of nickel, calculate the angular positions of the first and second order maxima for 100 eV electrons incident normal to the surface.

Having established the fact that electrons have a wave nature, Davisson and Germer continued their work by looking at higher order Bragg reflections and by studying grazing angles of incidence. They were able to assign an index of reflection slightly greater than one to nickel for 100-volt electrons.

Pressing the analogy with x-rays a step further, Thomson proposed that electrons passing through a crystal should produce diffraction patterns similar to Laue patterns. He subsequently observed these from foils of gold, aluminum and other metals.[10] Later, diffraction patterns were observed for other elementary particles (see Figure 4-8) as well as for atoms and molecules.

6. THE WAVE FUNCTION FOR AN ELECTRON

As was seen in the previous sections, a beam of electrons behaves very much like a beam of photons for the proper choice of energies and physical dimensions. As a further illustration, consider the following particle analog of Young's double slit experiment in optics.* Let a mono-energetic beam of electrons be incident on two slits whose dimensions and spacing are chosen to be of the same order of magnitude as the de Broglie wavelength of the electrons so that diffraction effects can be detected. The detector may be regarded as an array of microscopic counters or as a fluorescent screen which can be photographed. If either slit is blocked off, the pattern observed on the screen would look like that shown in Figure 4-9(b). However, if both slits are open, the pattern resembles Figure 4-9(c), where the interference effects are strikingly evident at the points shown. If the electrons behaved like classical particles there would be no interference effects and the pattern on the screen would approximate that shown in Figure 4-9(d). By repeating the experiment with progressively lower beam intensities, the same pattern is observed after a sufficient time. Surprisingly enough, even if only one electron at a time were fired at the slits, the interference pattern of Figure 4-9(c) would be produced after a sufficiently large number of electrons were fired. (See Figure 4-10). This forces us to conclude that *each electron interacts with both slits at once* even though our classically trained intuition tells us that each electron can pass through only one of the slits. Thus, from the point of view of quantum mechanics, it is meaningless to ask which slit the electron goes through. Any attempt to determine experimentally *which* slit an electron goes through will destroy the interference pattern just as effectively as if the other slit had been blocked off!

It follows from the above discussion that it is impossible to predict at which point a given electron will strike the detecting screen. However, we can relate the relative height of the particle distribution curve at position x to the relative *probability* that an electron will strike the screen at position x. Drawing again upon the optical analogy, if monochromatic photons had been incident upon the slits of Figure 4-9(a), the curve shown in (c) would be a plot

---

[10] G. P. Thomson, *Proc. Roy. Soc. A.* **117**, 600 (1927); **119**, 651 (1928); **125**, 352 (1929); **133**, 1 (1931).

* For an elementary review of this subject see Hugh D. Young, *Fundamentals of Optics and Modern Physics*, McGraw-Hill Book Co., New York, 1968, Chapter 3.

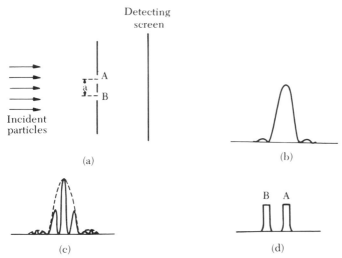

Detecting
screen

Incident
particles

(a)

(b)

(c)

(d)

Figure 4-9 (a) Double slit experiment with particles. (b) Distribution of particles recorded on the screen due to diffraction from either slit A or B. (c) Distribution of particles recorded on the screen due to diffraction with both slits A and B open. (d) Hypothetical distribution of particles recorded on the screen if wave effects are neglected.

(a) After 28 electrons

Figure 4-10 (a), (b), and (c). Computer-simulated growth of a two-slit interference pattern for electrons. (d) An actual photograph of a two-slit pattern produced by electrons. (Parts (a), (b), and (c) from E. R. Huggins, *Physics I*, W. A. Benjamin, Inc., New York, 1968. Part (d) is from C. Jönsson, *Zeitschrift für Physik*, **161**, 454 (1961). Used with permission.)

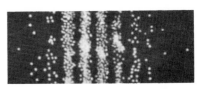

(b) After 1000 electrons

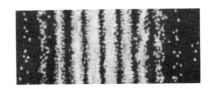

(c) After 10,000 electrons

(d) Two slit electron pattern

of relative light intensity versus position on the screen. The light intensity at each point is interpreted as the square of a wave amplitude vector which can be represented by the real or imaginary part of a function of the form,

$$\vec{\Psi} = \vec{A} e^{i(kx - \omega t)}.$$

Therefore, in quantum mechanics we define a wave amplitude function, or simply a *wave function*, for an electron, $\Psi(x, t)$, such that its modulus squared is proportional to the probability of finding the electron at position x at time t. This wave function is a complex function which we write as,

$$\Psi(x, t) = \Psi_0 e^{i(kx - \omega t)} = \Psi_0 e^{i(x p_x - Et)/\hbar},$$

where the de Broglie relation has been incorporated in the last expression in order to relate the frequency and wave number of the wave to the energy and momentum of the particle. Although this complex wave function is not directly observable (that is, measurable) its physical significance rests on the assumption that the quantity

$$|\Psi(x_1, t_1)|^2 \, dx = \Psi^*(x_1, t_1) \cdot \Psi(x_1, t_1) \, dx$$

is proportional to the probability of finding the electron in the element dx centered at x_1 at time t_1. Then the total probability for finding the electron *anywhere* in the space in question is proportional to the integral of $|\Psi(x, t)|^2$ over all of the space. Thus, the total probability, P, is

$$P \sim \int_{-\infty}^{\infty} |\Psi(x, t)|^2 \, dx,$$

for the one-dimensional space along the x-axis. This integral must be finite in order to represent a real particle. It is convenient to define the *probability density* for the particle as

$$\rho(x,t) = \frac{|\Psi(x,t)|^2}{\displaystyle\int_{-\infty}^{\infty} |\Psi(x,t)|^2 dx}, \tag{4.18}$$

where we then have

$$\int_{-\infty}^{\infty} \rho(x, t) \, dx = 1. \tag{4.19}$$

For physically acceptable wave functions it is always possible to introduce an appropriate factor in the wave function such that

$$\int_{-\infty}^{\infty} |\Psi(x, t)|^2 \, dx = \int_{-\infty}^{\infty} \Psi^*(x, t)\Psi(x, t) \, dx = 1. \tag{4.20}$$

When Equation 4.20 is true, the wave function Ψ is said to be *normalized*. It is evident from Equation 4.18 that when a wave function is normalized the probability density is simply the square of its absolute amplitude.

Furthermore, in order to account for interference effects we assume the validity of *the principle of superposition*. When superposition is valid in optics (that is, in the case of coherent light) we add the amplitudes at a point vectorially and square the resultant amplitude to obtain the intensity at the point. Thus,

$$I \sim (\vec{A}_1 + \vec{A}_2)^2 = A_1^2 + A_2^2 + 2\vec{A}_1 \cdot \vec{A}_2 \sim I_1 + I_2 + 2\vec{A}_1 \cdot \vec{A}_2,$$

where I represents the time average of the intensity over a full cycle. On the other hand, when light is incoherent (the relative phases of the different sources are washed out), the resultant intensity just goes as the sum

$$I \sim I_1 + I_2,$$

and the interference term $2\vec{A}_1 \cdot \vec{A}_2$ vanishes. Applying the principle of superposition to the experiment of Figure 4-9, we write:

$$\rho(x, t) \sim |\Psi(x, t)|^2 = |\Psi_A(x, t) + \Psi_B(x, t)|^2$$

$$\sim |\Psi_A|^2 + |\Psi_B|^2 + \Psi_A^* \Psi_B + \Psi_B^* \Psi_A$$

$$\sim \rho_A + \rho_B \pm |\Psi_A^* \Psi_B| \pm |\Psi_B^* \Psi_A|.$$

Here, the last two terms are the interference terms which depend upon the relative phases of the two waves. The plus and minus signs correspond to constructive and destructive interference. The phase factors in the wave functions Ψ_A and Ψ_B play a role analogous to that of vector addition in the above example from optics, thus indicating the importance of choosing complex wave functions.

The device of representing an electron by a complex wave function can be extended, of course, to other kinds of particles and even to atoms and molecules. In order to be physically admissible, however, a wave function which represents a particle must be finite, single-valued, and continuous. Furthermore, it must vanish suitably as $r \to \infty$ so that it can be normalized, as indicated by Equation 4.20. The latter requirement is necessary for the validity of the probabilistic interpretation of the modulus squared. The great significance of the principle of superposition is that a superposition of physically acceptable wave functions is itself acceptable for the representation of a real particle. We will now study the Fourier integral theorem. Its use will enable us to extend the mathematics of superposition from the simple case treated in section 8 of Chapter 2 to the formation of packets consisting of a continuum of frequencies.

PROBLEM 4-7

What are the phase and group velocities of an electron whose de Broglie wavelength is 0.01 Å? What is the kinetic energy of the electron?

(Ans.: $v_g = 0.925c$; $u_{ph} = 1.08c$; 0.836 MeV.)

PROBLEM 4-8

A particle is defined in the space, $-\infty < x < \infty$, by the wave function,

$$\Psi(x, t) = Ae^{-x^2}e^{i(kx-\omega t)}.$$

(a) Find the normalization constant A.
(b) What is the probability of finding the particle in the interval bounded by x and $x + dx$ at time t?
(c) What is the total probability of finding the particle somewhere between $-\infty$ and $+\infty$?

PROBLEM 4-9

A particle may be represented in the space, $-a \leq x \leq a$, by either of the wave functions
(a) $\Psi(x) = A \cos \pi x/2a$ or (b) $\Psi(x) = B \sin \pi x/a$.
Find the normalization constants A and B.

PROBLEM 4-10

A particle may be described in the space, $-a \leq x \leq a$, by a superposition of the two wave functions of the preceding problem, namely,

$$\Psi(x) = A \cos \frac{\pi x}{2a} + B \sin \frac{\pi x}{a}.$$

(a) Normalize this new function.
(b) Sketch the probability density as a function of x.

7. THE FOURIER INTEGRAL AND THE DELTA FUNCTION

Any periodic function, such as $f(t + T) = f(t)$, can be expanded in terms of sines and cosines provided that $f(t)$ is piecewise continuous and differentiable throughout the interval of the expansion.[11] If the interval of expansion

---

[11] R. Courant and D. Hilbert, *Methods of Mathematical Physics.* Interscience Publishers, Inc., New York, 1953, Chapter 2.

includes the end points of the period, then we add the assumption that the value of $f(t)$ at each end point is the arithmetic mean of its values at the right and left end points. Then, in the interval, $-(T/2) \leq t \leq T/2$,

$$f(t) = \frac{a_0}{2} + \sum_{n=1}^{\infty} a_n \cos \omega_n t + \sum_{n=1}^{\infty} b_n \sin \omega_n t, \qquad (4.21)$$

where

$$a_n = \frac{2}{T} \int_{-T/2}^{T/2} f(t') \cos \omega_n t' \, dt',$$

$$b_n = \frac{2}{T} \int_{-T/2}^{T/2} f(t') \sin \omega_n t' \, dt',$$

$$\omega_n = \frac{2\pi n}{T}.$$

This may be written in a more convenient form* by means of the Euler identity:

$$f(t) = \sum_{n=-\infty}^{\infty} c_n e^{i\omega_n t} \quad \text{and} \quad c_n = \frac{1}{T} \int_{-T/2}^{T/2} f(t') e^{-i\omega_n t'} \, dt', \qquad (4.22)$$

where $c_n = \frac{1}{2}(a_n - i b_n)$, $c_n^* = c_{-n}$ and $\omega_n = -\omega_{-n}$.

PROBLEM 4-11

Verify that Equations 4.21 and 4.22 are equivalent.

It is desirable to extend the interval of the expansion from T to ∞ so that non-periodic functions can also be represented by expansions in sines and cosines. To do this we must add the requirement that the integral $\int_{-\infty}^{\infty} |f(t)| \, dt$ exists. We rewrite Equation 4.22 in the form

$$f(t) = \sum_{n=-\infty}^{\infty} \frac{1}{T} \int_{-T/2}^{T/2} f(t') e^{i\omega_n (t-t')} \, dt', \qquad (4.23)$$

and note that we must eliminate the factor $1/T$ before going to the limit. Since n is restricted to integral values,

$$\Delta\omega = \omega_{n+1} - \omega_n = \frac{2\pi(n+1) - 2\pi n}{T} = \frac{2\pi}{T},$$

---

* Although the complex representation is introduced here as a mathematical device to represent real functions, its general utility will become evident when it is used to represent complex wave functions.

or

$$\frac{1}{T} = \frac{\Delta\omega}{2\pi}.$$ (4.24)

Substituting Equation 4.24 into 4.23,

$$f(t) = \frac{1}{2\pi} \sum_{n=-\infty}^{\infty} \Delta\omega \int_{-T/2}^{T/2} f(t')e^{i\omega_n(t-t')} dt'.$$

In the limit as $T \to \infty$, the frequencies are distributed continuously instead of discretely and

$$f(t) = \frac{1}{2\pi} \int_{-\infty}^{\infty} d\omega \int_{-\infty}^{\infty} f(t')e^{i\omega(t-t')} dt',$$

if the integrals are absolutely convergent. The Fourier integrals are written in a more nearly symmetric form by defining a function $g(\omega)$ as follows:

and

$$\left. \begin{array}{c} f(t) = \dfrac{1}{\sqrt{2\pi}} \displaystyle\int_{-\infty}^{\infty} d\omega\, g(\omega)e^{i\omega t} \\[4mm] g(\omega) = \dfrac{1}{\sqrt{2\pi}} \displaystyle\int_{-\infty}^{\infty} dt'\, f(t')e^{-i\omega t'} \end{array} \right\}.$$ (4.25)

The functions $f(t)$ and $g(\omega)$ are called Fourier transforms of one another. Any reasonably well-behaved function $f(t)$ can be represented by a super-position of harmonic functions with continuously varying frequency and weighting function $g(\omega)$. Conversely, $g(\omega)$ can be represented by a super-position of harmonic functions in time, each function multiplied by the weighting factor $f(t')$.

The above expansion is in the *time-frequency domain*. It is evident that an analogous expansion may be obtained in the position-wave-vector domain, or what is conveniently called the *coördinate-momentum domain*. Physically, this implies that our starting point was a *spatially* periodic function,

$$\psi(x + L) = \psi(x),$$

whose Fourier expansion is

$$\psi(x) = \sum_{n=-\infty}^{\infty} c_n e^{ik_n x},$$

where

$$c_n = \frac{1}{L} \int_{-L/2}^{L/2} \psi(x')e^{-ik_n x'} dx' \text{ and } k_n = \frac{2\pi n}{L}.$$

Following the same procedure as before, we eliminate the factor $1/L$ before taking the limit as $L \to \infty$ by the device that $\Delta k_n \sim 2\pi/L$ and $1/L \sim \Delta k_n/2\pi$.

The results are:

$$\left.\begin{array}{l} \psi(x) = \dfrac{1}{\sqrt{2\pi}} \displaystyle\int_{-\infty}^{\infty} dk\,\phi(k)e^{ikx} \\[4mm] \phi(k) = \dfrac{1}{\sqrt{2\pi}} \displaystyle\int_{-\infty}^{\infty} dx'\,\psi(x')e^{-ikx'} \end{array}\right\}. \tag{4.26}$$

Equations 4.26 are readily generalized to any number of dimensions. For example,

$$\left.\begin{array}{l} \psi(\vec{r}) = \left(\dfrac{1}{2\pi}\right)^{\frac{3}{2}} \displaystyle\iiint_{-\infty}^{\infty} d\vec{k}\,\phi(\vec{k})e^{i\vec{k}\cdot\vec{r}} \\[4mm] \phi(\vec{k}) = \left(\dfrac{1}{2\pi}\right)^{\frac{3}{2}} \displaystyle\iiint_{-\infty}^{\infty} d\vec{r}'\,\psi(\vec{r}')e^{-i\vec{k}\cdot\vec{r}'} \end{array}\right\}, \tag{4.27}$$

where $\vec{r}$ has components (x, y, z), $\vec{k}$ has components (k_x, k_y, k_z), $d\vec{r} = dx\,dy\,dz$ and $d\vec{k} = dk_x\,dk_y\,dk_z$. It is also possible to incorporate the time factor by writing

$$\Psi(\vec{r}, t) = \psi(\vec{r})e^{-i\omega t},$$

but the time will be omitted until it is specifically required.

The need for the Dirac delta function arises naturally from the Fourier integrals. Suppose we substitute $\phi(k)$ into $\psi(x)$ in Equations 4.26 as follows:

$$\psi(x) = \frac{1}{2\pi} \int_{-\infty}^{\infty} dk \int_{-\infty}^{\infty} dx'\,\psi(x')e^{ik(x-x')}.$$

Interchanging the order of integration,

$$\psi(x) = \frac{1}{2\pi} \int_{-\infty}^{\infty} dx'\,\psi(x') \int_{-\infty}^{\infty} dk\,e^{ik(x-x')}.$$

Now define the delta function,

$$\delta(x - x') = \frac{1}{2\pi} \int_{-\infty}^{\infty} dk\,e^{ik(x-x')}. \tag{4.28}$$

The delta function has the following properties:

$$\delta(x - x') = 0, \quad \text{if} \quad x' \neq x.$$
$$\int_a^b \delta(x - x')\,dx' = \begin{cases} 0, & \text{if} \quad x > b \text{ or if } x < a \\ 1, & \text{if} \quad a < x < b. \end{cases}$$

Since the delta function is zero everywhere but at the singular point, the only contribution to an integral occurs at that point. Then, using Equation 4.28

$$\psi(x) = \int_{-\infty}^{\infty} dx' \psi(x') \delta(x - x') \equiv \psi(x).$$

Thus the delta function may be thought of as a spike function having unit area but a non-zero amplitude at only one point, the point of singularity, where the amplitude becomes infinite. When integrated over all of space its effect is to yield the remaining factors of the integrand evaluated at the singular point.

It is often convenient to place the origin at the singular point, in which case the delta function may be written as

$$\delta(x) = \frac{1}{2\pi} \int_{-\infty}^{\infty} dk e^{ikx} .$$

An alternative definition of the delta function may be obtained by integrating as follows:

$$\delta(x) = \lim_{a \to \infty} \frac{1}{2\pi} \int_{-a}^{a} e^{ikx}\, dk = \frac{1}{2\pi} \lim_{a \to \infty} \left(\frac{e^{iax} - e^{-iax}}{ix} \right) = \lim_{a \to \infty} \frac{\sin ax}{\pi x}, \qquad (4.29)$$

where a is positive and real. Let us examine the behavior of this function for both small and large x. First, consider the limit as x goes to zero:

$$\lim_{x \to 0} \frac{\sin ax}{\pi x} = \frac{a}{\pi} \lim_{x \to 0} \frac{\sin ax}{ax} = \frac{a}{\pi}.$$

Thus, $\delta(0) = \lim_{a \to \infty} a/\pi \to \infty$, or the amplitude becomes infinite at the singularity. For large $|x|$, $\sin ax/\pi x$ oscillates with period $2\pi/a$, and its amplitude falls off as $1/|x|$. But in the limit as $a \to \infty$, the period gets infinitesimally narrow so that the function approaches zero everywhere except for the infinite spike of infinitesimal width at the singularity. It now remains to show that the integral of Equation 4.29 over all space is unity:

$$\int_{-\infty}^{\infty} \lim_{a \to \infty} \frac{\sin ax}{\pi x}\, dx = \lim_{a \to \infty} \frac{2}{\pi} \int_{0}^{\infty} \frac{\sin ax}{x}\, dx = \frac{2}{\pi} \cdot \frac{\pi}{2} = 1.$$

This establishes the validity of Equation 4.29 as a representation of the delta function.

PROBLEM 4-12

Show that the following two functions are valid representations of the delta function, where ϵ is positive and real:

$$\text{(a)} \quad \delta(x) = \frac{1}{\sqrt{\pi}} \lim_{\epsilon \to 0} \frac{1}{\sqrt{\epsilon}} e^{-x^2/\epsilon}$$

$$\text{(b)} \quad \delta(x) = \frac{1}{\pi} \lim_{\epsilon \to 0} \frac{\epsilon}{x^2 + \epsilon^2}.$$

PROBLEM 4-13

Verify the following properties of the delta function:
(a) $\delta(x) = \delta(-x)$.
(b) $x\delta(x) = 0$.
(c) $\delta'(-x) = -\delta'(x)$.
(d) $x\delta'(x) = -\delta(x)$.
(e) $c\delta(cx) = \delta(x), c > 0$.

8. PARTICLES AS WAVE PACKETS

The assumption that a particle can be represented in space and time by a complex wave packet is one of the fundamental premises of wave mechanics. We have seen how the concept of a complex wave function has resolved the classical wave-particle dualism through the de Broglie relations which connect the energy and momentum of the particle with the frequency and propagation constant of the wave. Also, in section 8 of Chapter 2, it was pointed out that in the classical limit the particle velocity v is associated with the group velocity of the packet.* Hence,

$$\frac{p}{m} = v = \frac{1}{\hbar}\frac{dE}{dk} = \frac{dE}{dp},$$

or

$$dE = \frac{p}{m}\,dp,$$

and

$$E = \frac{p^2}{2m} + V, \tag{4.30}$$

---

* This is another example of the *correspondence principle* which states that quantum mechanics must give the same result as classical mechanics in the appropriate classical limit. For particles, the classical limit occurs for very short de Broglie wavelengths.

where V may be regarded here as the integration constant. Experiments which depend upon particle-like properties can be regarded as observations of the behavior of the packet or wave-group as a whole, whereas experiments which depend upon interference and diffraction effects are observations of individual wavelength components of the packet. With the help of the Fourier integral theorem we are now able to construct and study some elementary packets composed of waves so closely spaced that they form a continuum of frequencies or k-values. This has the decided advantage of producing a truly localized packet having essentially zero amplitude except in one specific region of space. By way of contrast, the reader should recall that although a finite sum of waves whose frequencies differ by finite amounts can produce a modulation envelope that resembles a packet, this kind of packet repeats itself regularly throughout all of space. It is evident, then, that a packet which is used to represent a particle in free space is best constructed by means of the Fourier integral theorem.

As our first example of the use of this theorem, let us look at the effects of chopping a pure, infinite sine wave in both the time and the coordinate domains. Consider a sine wave of frequency ω_0 which is chopped to a lifetime of $2T$ seconds. That is,

$$f(t) = e^{i\omega_0 t}, \quad \text{for } -T \le t \le T,$$
$$f(t) = 0, \quad \text{for } |t| > T.$$

The chopping (or keying) process will introduce many new frequencies in varying amounts, given by the Fourier transform,

$$g(\omega) = \frac{1}{\sqrt{2\pi}} \int_{-\infty}^{\infty} dt\, f(t) e^{-i\omega t}$$
$$= \frac{1}{\sqrt{2\pi}} \int_{-T}^{T} e^{i(\omega_0 - \omega)t}\, dt$$
$$= \sqrt{\frac{2}{\pi}} \cdot T \cdot \frac{\sin(\omega_0 - \omega)T}{(\omega_0 - \omega)T}.$$

This function is plotted schematically in Figure 4-11. The attempt to chop in the time domain results in the mixing of an infinite number of frequencies, albeit in small amounts except for those frequencies close to ω_0. The principal contribution is still at the frequency ω_0. Note the similarity with the amplitude pattern for single slit diffraction or for a shutter, which is its analog in the space domain (the shutter chops an infinite wave front and produces a spread in propagation constants). Nature provides an example of this kind of chopping in the emission of photons during electronic and nuclear transitions in atoms.*

Notice that the central peak in Figure 4-11(a) has a breadth given by

$$\Delta\omega = \frac{2\pi}{T} \quad \text{or} \quad T \cdot \Delta\nu = 1.$$

----

* See section 9 of Chapter 2.

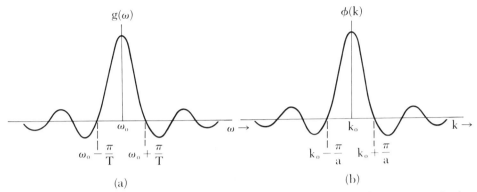

Figure 4-11   The spectral distribution resulting from chopping a pure, infinite sine wave in the time domain (a) and in the spatial domain (b).

Multiplying by h and replacing the time interval T by Δt, we have the statement that the minimum uncertainty product for this type of packet is

$$\Delta t \cdot \Delta E = h.$$

Therefore, we may state Heisenberg's uncertainty principle for simultaneous measurements of energy and time for such a packet as

$$\Delta t \cdot \Delta E \geq h. \tag{4.31}$$

The spatial counterpart of the above keying process can be illustrated by chopping an infinite, plane wave front with a shutter such that the length of the packet is $2a$. (Here $2a = 2cT$, where $2T$ is the time interval that the shutter is open.) Then,

$$\psi(x) = e^{ik_0 x}, \quad \text{for } -a \leq x \leq a,$$
$$\psi(x) = 0, \quad \text{for } |x| > a.$$

Then,

$$\phi(k) = \frac{1}{\sqrt{2\pi}} \int_{-\infty}^{\infty} \psi(x) e^{-ikx} \, dx$$

$$= \frac{1}{\sqrt{2\pi}} \int_{-a}^{a} e^{i(k_0 - k)x} \, dx$$

$$= \sqrt{\frac{2}{\pi}} \cdot a \cdot \frac{\sin (k_0 - k)a}{(k_0 - k)a}.$$

This function is also shown in Figure 4-11, but here it is the wave vector (or the momentum) that takes on a spread of values around k_0.* The breadth of

---

* Likewise, in single slit diffraction, we generally regard k_0 as a constant and allow its *direction* to take on a spread of values. This is equivalent to treating the slit as a source of Huygens wavelets.

the peak is $\Delta k = 2\pi/a$, from which we obtain the result

$$a \cdot \Delta p_x = h,$$

which enables us to write the uncertainty principle for simultaneous measurements of position and momentum as,

$$\Delta x \cdot \Delta p_x \geq h. \tag{4.32}$$

For the packet shape shown in Figure 4-11, the minimum value of the uncertainty product is h at $t = 0$. The packet shape that achieves the *absolute minimum* value of $\hbar/2$ is the Gaussian packet, which will be discussed later in this section. (See Problem 4-21.) However, as time progresses, any packet will broaden in coordinate space so that for a given Δp, the uncertainty product increases with time. (See Problem 4-16.) If we denote the time during which the packet retains its form as T_0, then

$$T_0 \sim \frac{m}{\hbar} (\Delta x)^2,$$

where Δx is its initial breadth in coordinate space. The more massive the particle, the more slowly its packet spreads with time. This is what we want in order to satisfy the correspondence principle. Thus, for an electron which is localized to 1 angstrom, $T_0 \sim 10^{-16}$ second, whereas for a .22 caliber bullet localized to 0.1 mm, $T_0 \sim 10^{24}$ seconds. It is evident that the concept of a classical trajectory, though meaningless for the electron, is quite appropriate for a macroscopic particle.

Now suppose that we wish to represent a free particle as a superposition of a continuum of plane wave states, namely,

$$\Psi(x, t) = \int \phi(k) \Psi_k(x, t) \, dk. \tag{4.33}$$

Each value of $\phi(k)$ serves as a weighting factor for the wave function it multiplies. That is, it tells how much of that particular wave function contributes to the packet at position x and time t. A plot of the values of $\phi(k)$ versus k is known as the spectral distribution of states, or simply as the *spectrum* of the states. The plane wave states of Equation 4.33 may be represented by

$$\Psi_k(x, t) = A e^{i(kx - \omega t)},$$

where A is the normalization factor. Since infinite plane waves are not well-behaved functions in the sense that they do not vanish at $x = \pm \infty$, special procedures must be used in order to normalize them. One such procedure is to use *delta-function normalization;* thus,

$$\int_{-\infty}^{\infty} \Psi_{k'}^{*}\Psi_k \, dx = |A|^2 \int_{-\infty}^{\infty} e^{ikx} e^{-ik'x} \, dx = |A|^2 \int_{-\infty}^{\infty} e^{ix(k-k')} \, dx.$$

Comparing this expression with Equation 4.28, written for the delta function in k,

$$\delta(k - k') = \frac{1}{2\pi} \int_{-\infty}^{\infty} e^{ix(k-k')} \, dx,$$

we see that by choosing $|A| = 1/\sqrt{2\pi}$ we can normalize a plane wave function to the delta function, so that the integral of its modulus squared over all space is unity.

It is now evident that the integration over a continuum of states is identical with the Fourier transform of Equation 4.26 with the time factor included. The function $\Psi(x, t)$ is said to be the wave function of the particle in the *coordinate representation* and $\Phi(k)$ is the wave function in the *momentum representation*. The two wave functions are Fourier transforms of one another. Consider a packet whose spectrum of k-values is a delta function in momentum space, that is,

$$\phi(k) = \delta(k - k_0).$$

Then,

$$\psi(x) = \frac{1}{\sqrt{2\pi}} \int_{-\infty}^{\infty} dk \, \delta(k - k_0) e^{ikx} = \frac{1}{\sqrt{2\pi}} e^{ik_0 x},$$

which is the normalized wave function for a packet containing a single momentum component $\hbar k_0$. Note that the infinite plane wave and the delta function are Fourier transforms of each other.

As a second example, consider a packet whose spectrum of k-values is the Gaussian function,

$$\phi(k) = Ae^{-k^2/2\sigma^2},$$

where σ is the standard deviation of the distribution and is a measure of the spread of the packet. (See Figure 4-12.) Then,

$$\psi(x) = \frac{A}{\sqrt{2\pi}} \int_{-\infty}^{\infty} dk e^{-k^2/2\sigma^2} e^{ikx}.$$

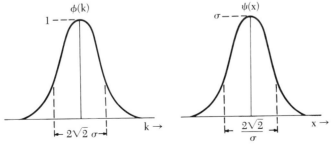

Figure 4-12 Gaussian packets. Each packet is the Fourier transform of the other. The breadth of the packet is arbitrarily taken to be its standard deviation, as shown.

In order to complete the square we multiply by $\exp\left[(x^2\sigma^2/2) - (x^2\sigma^2/2)\right] = 1$ and obtain:[12]

$$\psi(x) = \frac{A}{\sqrt{2\pi}} \int_{-\infty}^{\infty} dk \exp\left[-\frac{k^2}{2\sigma^2} + ikx + \frac{x^2\sigma^2}{2} - \frac{x^2\sigma^2}{2}\right]$$

$$= \frac{A}{\sqrt{2\pi}} e^{-x^2\sigma^2/2} \int_{-\infty}^{\infty} dk \exp\left[-\frac{1}{2\sigma^2}(k^2 - 2ikx\sigma^2 - x^2\sigma^4)\right]$$

$$= \frac{A}{\sqrt{2\pi}} e^{-x^2\sigma^2/2} \int_{-\infty}^{\infty} dk \exp\left[-\frac{1}{2\sigma^2}(k + ix\sigma)^2\right]$$

$$= \frac{\sqrt{2}A\sigma}{\sqrt{2\pi}} e^{-x^2\sigma^2/2} \int_{-\infty}^{\infty} e^{-u^2} du$$

$$= A\sigma e^{-x^2\sigma^2/2}.$$

Note that if $\Phi(k)$ is normalized by setting $A = \left(\dfrac{1}{\sqrt{2\pi}\,\sigma}\right)^{1/2}$, then $\psi(x)$ is also a normalized Gaussian function; that is, the Fourier transform of a Gaussian function is also a Gaussian. The spread of the packet in coordinate space is given by $1/\sigma$ and illustrates the reciprocal nature of Δk and Δx as required by the uncertainty principle.

PROBLEM 4-14

Show that $(d/dt)[\Phi(k)] = 0$.

PROBLEM 4-15

Show that if the coordinate wave function is normalized at $t = 0$, then both the coordinate and momentum wave functions remain normalized for all time.

PROBLEM 4-16

For a free particle show that a wave packet in coordinate space broadens with time (corresponding to increasing uncertainty in the position of the particle), while its representation in momentum space retains its shape and changes only in phase (since momentum is conserved).

---

[12] See J. L. Powell and B. Crasemann, *Quantum Mechanics*. Addison-Wesley Publ. Co., Inc., Reading, Mass. (1961), p. 475.

PROBLEM 4-17

What must be the frequency bandwidth of the detecting and amplifying stages of a radar system operating at pulse widths of 0.1 μsec? If the radar is used for ranging (distance measurements), what is the uncertainty in the range?

PROBLEM 4-18

(a)   Find the normalization constant N for the Gaussian wave packet,

$$\psi(x) = Ne^{-(x-x_0)^2/2K^2}.$$

(b)   Obtain the Fourier transform of $\psi(x)$ and verify that it is normalized.

PROBLEM 4-19

(a)   Use the uncertainty principle to obtain the uncertainty in the momentum of a particle of mass m constrained to the volume of a cubical box of side a.

(b)   What is the corresponding uncertainty in the kinetic energy of the particle?

(c)   How do you interpret the answer to (b)? (See Problem 4-2.)

9. THE SCHRÖDINGER WAVE EQUATION

In section 6 we defined a complex wave function, $\Psi(x, t)$, which is assumed to completely describe the dynamical state of a particle in the Schrödinger representation. Although the wave function $\Psi(x, t)$ cannot be measured directly, we interpret its "intensity," that is, its modulus squared, to correspond to the relative probability of detecting the particle at position x at time t. However, a knowledge of the state of the particle at one point in space and time is not enough, since we would also like to know how the particle is behaving at some later time. In order to describe the progress of the particle in space and time it is necessary to find a wave equation, preferably a differential equation, whose solutions will correspond to the motion of the particle. This equation cannot be derived, but must be postulated and then tested against experimental results. The equation should be linear in $\Psi(x, t)$ so that if $\Psi_1(x, t)$ and $\Psi_2(x, t)$ are solutions, any linear combination of Ψ_1 and Ψ_2 will also be a solution. This requirement insures the validity of the superposition principle. Furthermore, the differential equation must be of the first order with respect to time so that if $\Psi(x, t)$ is known at, say, t_0, then it will be uniquely specified at all later times. Our last requirement is that the wave equation must be consistent with de Broglie's hypothesis and the correspondence principle. For a free particle the total energy is

$$E = \frac{p_x^2}{2m},$$

(4.34)

which becomes in the de Broglie formulation, Equation 4.14,

$$\omega\hbar = \frac{\hbar^2 k^2}{2m}.$$

(4.35)

We have seen that such a free particle can be represented by an infinite plane wave of the form

$$\Psi(x, t) = \frac{1}{\sqrt{2\pi}} e^{i(k_x x - \omega t)}$$

(4.36)

Note that a single differentiation of Equation 4.36 with respect to time gives

$$\frac{\partial \Psi}{\partial t} = -i\omega\Psi,$$

(4.37)

whereas differentiating twice with respect to x results in

$$\frac{\partial^2 \Psi}{\partial x^2} = -k^2 \Psi.$$

(4.38)

From Equations 4.35, 4.37, and 4.38 we see that we can write,

$$i\hbar \frac{\partial \Psi}{\partial t} = -\frac{\hbar^2}{2m} \frac{\partial^2 \Psi}{\partial x^2},$$

(4.39)

which is Schrödinger's wave equation[13] for a free particle in one dimension. In three dimensions the wave function for a free particle is

$$\Psi(\vec{r}, t) = \frac{1}{(2\pi)^{\frac{3}{2}}} e^{i(\vec{k}\cdot\vec{r} - \omega t)}$$

(4.40)

and the corresponding Schrödinger equation is

$$i\hbar \frac{\partial}{\partial t} \Psi(\vec{r}, t) = -\frac{\hbar^2}{2m} \nabla^2 \Psi(\vec{r}, t),$$

(4.41)

---

[13] E. Schrödinger, *Ann. Phys.* **79,** 361 (1926).

where ∇^2 is the differential operator,

$$\frac{\partial^2}{\partial x^2} + \frac{\partial^2}{\partial y^2} + \frac{\partial^2}{\partial z^2} .$$

If the particle is not free but is acted upon by a force which can be expressed as the derivative of a scalar potential function, we merely add this potential to the right hand side of Equation 4.34. Thus,

$$E = \frac{p_x^2}{2m} + V,$$

and

$$\omega\hbar = \frac{\hbar^2 k^2}{2m} + V.$$

Equation 4.39 now becomes

$$i\hbar \frac{\partial}{\partial t} \Psi(x, t) = \left(-\frac{\hbar^2}{2m} \frac{\partial^2}{\partial x^2} + V \right) \Psi(x, t) = \mathscr{H}\Psi(x, t).$$

The operator, $\mathscr{H} = -(\hbar^2/2m)(\partial^2/\partial x^2) + V$, is called the *Hamiltonian operator* since, analogous to the classical Hamiltonian function, it is the sum of the kinetic and potential energy operators. In three dimensions the time-dependent Schrödinger equation is,

$$i\hbar \frac{\partial}{\partial t} \Psi(\vec{r}, t) = \mathscr{H}\Psi(\vec{r}, t), \tag{4.42}$$

where

$$\mathscr{H} = -\frac{\hbar^2}{2m} \nabla^2 + V(\vec{r}, t). \tag{4.43}$$

Equation 4.42 is a far more useful form than Equation 4.39 since all of the important physical problems do contain forces. In fact, one can almost go so far as to say that all of the interesting physics is contained in the potential term.

10. THE CONSERVATION OF PROBABILITY DENSITY

As in section 6 of this chapter, we define the probability of finding a particle in the volume element $d\tau = dx\, dy\, dz$ as,

$$\rho\, d\tau = \rho(\vec{r}, t)\, dx\, dy\, dz = \Psi^*(\vec{r}, t)\Psi(\vec{r}, t)\, dx\, dy\, dz, \tag{4.44}$$

where $\Psi(\vec{r}, t)$ is the *normalized* wave function associated with the particle. Now suppose that as time progresses, ρ decreases for one volume element in space. In order to be consistent with the fact that the total probability density over all of space must be unity, we must then have ρ increasing in some other volume element.* This shift or flow of probability density may be thought of as a "probability density current." We shall now derive the formal statement of the conservation law for probability density.

The time rate of change of probability density in any convenient volume is given by:

$$\frac{\partial}{\partial t} \int_\tau \rho \, d\tau = \int_\tau \left(\Psi \frac{\partial}{\partial t} \Psi^* + \Psi^* \frac{\partial}{\partial t} \Psi \right) d\tau.$$

Now, Ψ satisfies Equation 4.34,

$$\mathcal{H}\Psi = i\hbar \frac{\partial}{\partial t} \Psi,$$

while Ψ^* satisfies the conjugate wave equation,

$$\mathcal{H}\Psi^* = -i\hbar \frac{\partial}{\partial t} \Psi^*.$$

The reason for the fact that the conjugate wave functions require a different wave equation may be argued as follows. To claim that Ψ and Ψ^* satisfy the same equation is to say that $\Psi = \Psi^*$.

To go a step further, if Ψ represents momentum transfer to the right, then Ψ^* transfers momentum to the left; $\Psi^* = \Psi$ implies no momentum, that is, no dynamical state. Then,

$$\frac{\partial \Psi}{\partial t} = \frac{1}{i\hbar} \mathcal{H}\Psi = \frac{1}{i\hbar} \left(\frac{-\hbar^2}{2m} \nabla^2 + V \right) \Psi = \frac{i\hbar}{2m} \nabla^2 \Psi - \frac{i}{\hbar} V\Psi,$$

$$\frac{\partial \Psi^*}{\partial t} = \frac{-1}{i\hbar} \mathcal{H}\Psi^* = \frac{-1}{i\hbar} \left(\frac{-\hbar^2}{2m} \nabla^2 + V \right) \Psi^* = \frac{-i\hbar}{2m} \nabla^2 \Psi^* + \frac{i}{\hbar} V\Psi^*,$$

and

$$\frac{\partial \Psi^*}{\partial t} \Psi + \Psi^* \frac{\partial \Psi}{\partial t} = \frac{-i\hbar}{2m} [(\nabla^2\Psi^*)\Psi - \Psi^*(\nabla^2\Psi)].$$

Implicit in the last expression is the assumption that $V\Psi^*\Psi = \Psi^*V\Psi$. In the next section we will see that this is equivalent to assuming that the operator for the potential commutes with the wave function.

Now, by Green's theorem we have the vector identity

$$\nabla \cdot f(\nabla g) = (\nabla f) \cdot (\nabla g) + f(\nabla^2 g)$$

or

$$\nabla \cdot g(\nabla f) = (\nabla g) \cdot (\nabla f) + g(\nabla^2 f),$$

---

* This discussion assumes that matter is conserved. Therefore, these results are not valid for relativistic reactions where particles are created and annihilated.

where the latter merely interchanges the roles of the functions f and g. Subtracting,

$$\nabla \cdot [f(\nabla g) - g(\nabla f)] = f(\nabla^2 g) - g(\nabla^2 f).$$

Letting $f = \Psi$ and $g = \Psi^*$,

$$\nabla \cdot [\Psi(\nabla \Psi^*) - \Psi^*(\nabla \Psi)] = \Psi(\nabla^2 \Psi^*) - \Psi^*(\nabla^2 \Psi).$$

Then,

$$\frac{\partial}{\partial t} \int_\tau \rho \, d\tau = \int_\tau -\frac{i\hbar}{2m} \nabla \cdot [\Psi(\nabla \Psi^*) - \Psi^*(\nabla \Psi)] \, d\tau.$$

Now we define the *probability density current*,

$$\vec{S}(\vec{r}, t) = \frac{i\hbar}{2m} [\Psi(\nabla \Psi^*) - \Psi^*(\nabla \Psi)]. \tag{4.45}$$

Then

$$\frac{\partial}{\partial t} \int_\tau \rho \, d\tau = \int_\tau -\nabla \cdot \vec{S} \, d\tau,$$

or

$$\frac{\partial \rho}{\partial t} + \nabla \cdot \vec{S} = 0. \tag{4.46}$$

Equation 4.46 expresses the *conservation of probability density*. Its similarity to classical conservation laws in the absence of sources or sinks should be immediately recognized.

The probability density current may be written in a more convenient form by noting that $\Psi^*(\nabla \Psi)$ is the complex conjugate of $\Psi(\nabla \Psi^*)$. Thus the bracketed quantity in Equation 4.45 is equal to $2i$ times the imaginary part of $\Psi(\nabla \Psi^*)$, and Equation 4.45 becomes:

$$\vec{S}(\vec{r}, t) = \frac{-\hbar}{m} \text{Im} [\Psi(\nabla \Psi^*)]. \tag{4.47}$$

By way of illustration, consider a wave packet representing a free particle of momentum $\hbar k_0 = mv_0$ in the x-direction,

$$\Psi(x, t) = \frac{1}{\sqrt{2\pi}} e^{i(k_0 x - \omega t)}.$$

Then the probability density current in the x-direction associated with the motion of this particle is

$$S_x = \frac{-\hbar}{m} \text{Im} [\Psi(-ik_0 \Psi^*)] = \frac{\hbar k_0}{m} |\Psi \Psi^*| = \rho(x) v_0.$$

Hence, the probability density moves with the particle velocity.

PROBLEM 4-20

> If the Gaussian packet of Problem 4-18 represents a free particle, calculate the probability density current.

II. OBSERVABLES, OPERATORS, AND EXPECTATION VALUES

In the preceding sections we have seen how a particle can be represented by a complex wave function and we have obtained the wave equation which describes its progress in space and time. In addition, the probability density and its conservation law have been discussed. Yet the reader may justifiably wonder how all of this relates to the dynamical variables which can be measured in the laboratory. After all, this wave function, which contains all of the available information about the particle in the system under study, cannot even be directly observed!

In quantum mechanics the connecting link is that each dynamical observable is represented by a mathematical operator through the correspondence principle. When this operator operates on the wave function, a number is obtained which corresponds to a possible result of a physical measurement of that quantity. The operators we use in quantum mechanics to represent physical observables are *linear operators*. Thus, if P is a linear operator, then

$$P(a\psi_1 + b\psi_2) = aP\psi_1 + bP\psi_2,$$

where a and b are constants. The sum and the product of two linear operators also form linear operators. Although the sum of two operators is commutative, the product is, in general, not commutative.

We say that two quantities a and b commute under multiplication if $ab = ba$; that is, if the order of multiplication is immaterial. This property is taken for granted in ordinary algebra. However, you have already seen in vector algebra that the vector product $\vec{A} \times \vec{B} \neq \vec{B} \times \vec{A}$, although the scalar product $\vec{A} \cdot \vec{B} = \vec{B} \cdot \vec{A}$. Thus, the dot product is commutative, whereas the cross product is not.

In operator algebra we must not assume commutivity unless it is proven specifically for the operators in question. Consider the product of the operators x and d/dx operating on a function of x:

$$x\frac{d}{dx} f(x) = xf'(x).$$

But,

$$\frac{d}{dx} xf(x) = f(x) + xf'(x).$$

Subtracting,

$$\left[x\frac{d}{dx} - \frac{d}{dx}x\right]f(x) = -f(x).$$

Therefore, the equivalent value for the bracketed quantity is -1. This bracket is called the "commutator bracket" of x and d/dx and is usually written in the abbreviated form,

$$\left[x, \frac{d}{dx} \right] = -1.$$

Evidently,

$$\left[\frac{d}{dx}, x \right] = +1.$$

Therefore, *two operators commute if their commutator bracket is zero.*

Let us now consider a measurement of a position coordinate of a particle. The probability of finding the particle in the volume element $d\tau$ is, from Equation 4-18,

$$\rho(x, y, z, t) \, d\tau = \frac{\Psi^*(x, y, z, t)\Psi(x, y, z, t) \, d\tau}{\int_\tau \Psi^*(x, y, z, t)\Psi(x, y, z, t) \, d\tau}.$$

The fact that $\rho \, d\tau$ is called a *probability* implies that one would not necessarily expect to obtain the same result for two successive measurements on the system. We can define an *average value*, or *expectation value*, of the position variable by considering either a large number of measurements on the same system, or a single measurement on each of a large number of identical systems. Thus, the expectation value of, say, the variable x, is the weighted average,

$$\langle x \rangle = \int_\tau \rho x \, d\tau = \frac{\int_\tau \Psi^* x \Psi \, d\tau}{\int_\tau \Psi^* \Psi \, d\tau}. \tag{4.48}$$

The expectation value of a component of the momentum may be determined directly from the momentum wave functions in like manner. Thus:

$$\langle p_x \rangle = \frac{\int_{-\infty}^{\infty} \phi^* p_x \phi \, dk_x}{\int_{-\infty}^{\infty} \phi^* \phi \, dk_x}. \tag{4.49}$$

It is also possible to obtain the expectation value $\langle x \rangle$ by means of the momentum wave functions, provided that the appropriate operator is used. In order to find the correct operator expression, let us insert Equation 4-26, the Fourier transforms of Ψ^* and Ψ, into Equation 4-48, where we assume that Ψ is normalized. Then,

$$\langle x \rangle = \frac{1}{2\pi} \int\!\!\!\int\!\!\!\int_{-\infty}^{\infty} dx \, dk \, dk' \, \phi^*(k) e^{-ikx} \, x e^{ik'x} \, \phi(k')$$

$$= \frac{1}{2\pi} \int dx \int dk \, \phi^*(k) e^{-ikx} \int dk' \, x e^{ik'x} \, \phi(k').$$

Performing a partial integration over k', the last integral becomes

$$\left[-i\phi(k') e^{ik'x} \right]_{-\infty}^{\infty} + i \int e^{ik'x} \frac{\partial}{\partial k'} \phi(k') \, dk',$$

where the first term is zero since the wave function vanishes at both limits. Then,

$$\langle x \rangle = \frac{1}{2\pi} \int\!\!\!\int\!\!\!\int_{-\infty}^{\infty} dx \, e^{i(k'-k)x} \, dk \, dk' \, \phi^*(k) \, i\frac{\partial}{\partial k'} \, \phi(k')$$

$$= \int\!\!\!\int dk \, dk' \phi^*(k) \, i \frac{\partial}{\partial k'} \, \phi(k') \, \delta(k - k'),$$

where Equation 4-28 has been used. Performing the k' integration,

$$\langle x \rangle = \int dk \, \phi^*(k) \, i \frac{\partial}{\partial k} \, \phi(k) = \int dp_x \, \phi^*(p) \, i\hbar \frac{\partial}{\partial p_x} \, \phi(p).$$

Thus, the expectation value $\langle x \rangle$ can be obtained from the momentum wave functions by means of the operator, $i\hbar(\partial/\partial p_x)$. In like manner, it can be shown that the momentum operator in the coordinate representation is $-i\hbar(\partial/\partial x)$. That is,

$$\langle p_x \rangle = \int \Psi^* \left(-i\hbar \frac{\partial}{\partial x} \right) \Psi \, dx. \tag{4.50}$$

By comparing Equations 4-34 and 4-39 of section 9, it is evident that the requirement of the correspondence principle, namely,

$$\langle p_x^2 \rangle \leftrightarrow -\hbar^2 \frac{\partial^2}{\partial x^2},$$

is consistent with the last equation. Applying the correspondence principle to the left members of Equations 4-34 and 4-39, we see that the total energy operator may be represented by

$$E \leftrightarrow i\hbar \frac{\partial}{\partial t}.$$

A few of the commonly used operators are summarized in Table 4.2 for convenience.

Before closing this section, it is worth noting that the concept of the expectation value enables us to make a more precise statement of the uncertainty principle. The difference between a given measurement of x and the expectation value of x is called the *deviation* of x. We define the *uncertainty* of x as the root-mean-square deviation of x. Thus,

$$\Delta x = \sqrt{\langle (x - \langle x \rangle)^2 \rangle}$$
$$= \sqrt{\langle (x^2 - 2x\langle x \rangle + x^2) \rangle}$$
$$= \sqrt{\langle x^2 \rangle - 2\langle x \rangle \langle x \rangle + \langle x \rangle^2}$$
$$= \sqrt{\langle x^2 \rangle - \langle x \rangle^2}. \tag{4.51}$$

TABLE 4-2 Commonly Used Operators

Variable	Coordinate Representation	Momentum Representation
x	x	$i\hbar \dfrac{\partial}{\partial p_x}$
x^2	x^2	$-\hbar^2 \dfrac{\partial^2}{\partial p_x^2}$
p_x	$-i\hbar \dfrac{\partial}{\partial x}$	p_x
p_x^2	$-\hbar^2 \dfrac{\partial^2}{\partial x^2}$	p_x
E	$i\hbar \dfrac{\partial}{\partial t}$	$i\hbar \dfrac{\partial}{\partial t}$

By means of Equation 4.51, we are now able to define the uncertainty product given in Equation 4.32 as[14]

$$\Delta x \cdot \Delta p_x = \sqrt{\langle x^2 \rangle - \langle x \rangle^2} \cdot \sqrt{\langle p_x^2 \rangle - \langle p_x \rangle^2} \geq \frac{\hbar}{2}.$$

Equation 4.31, however, cannot be expressed in this manner since the time t is *not* a dynamical variable but is a parameter in the wave function. The meaning of Δt in Equation 4.31 is that of a time interval during which the system makes a transition from one state to another.[15]

PROBLEM 4-21

(a) For the Gaussian packet of Problem 4-18, calculate the expectation values $\langle x \rangle$, $\langle x^2 \rangle$, $\langle p_x \rangle$, and $\langle p_x^2 \rangle$ using the coordinate representation.

(b) Find the same expectation values using the momentum representation.

(c) Find the minimum uncertainty product $\Delta x \cdot \Delta p_x$. The Gaussian packet has the unique property of providing the theoretical minimum of uncertainty.

(d) Find the expectation value of the kinetic energy, $\langle T \rangle$.

[14] See A. Messiah, *Quantum Mechanics*. North-Holland Publishing Co., Amsterdam, 1958, Vol. 1, p. 133.
[15] *Ibid.*, p. 135.

Work out the following commutators:

$$[x, p_x], [x, p_y], [x, p^2], [p_x, p_y], [p_x, p^2].$$

Do the results just obtained depend upon whether the coordinate or the momentum wave functions are used in the calculation?

12. SEPARATION OF SPACE AND TIME IN THE SCHRÖDINGER EQUATION: ENERGY EIGENVALUES

There are many problems of physical interest in which the potential energy of the system is independent of the time. In such cases it is possible to simplify the Schrödinger equation (4.42) by regarding the wave function $\Psi(\vec{r}, t)$ as a product of two functions, one containing only spatial coordinates and the other containing only the time. Expressed mathematically, we assume that if $V = V(\vec{r})$, then

$$\Psi(\vec{r}, t) = \psi(\vec{r}) \cdot \chi(t).$$

Substituting this form of the wave function into Equation 4.42, we obtain

$$i\hbar \psi(\vec{r}) \frac{\partial}{\partial t} \chi(t) = \chi(t) \cdot \mathscr{H} \psi(\vec{r}),$$

since $\mathscr{H}$ is independent of the time. This may be written in the separated form,

$$\frac{i\hbar}{\chi(t)} \cdot \frac{\partial}{\partial t} \chi(t) = \frac{\mathscr{H} \psi(\vec{r})}{\psi(\vec{r})} \equiv E, \tag{4.52}$$

where the symbol E has been used for the separation constant, since we will soon see that it is readily identified as the total energy of the system. The equation involving the time can be immediately integrated, and its solution is

$$\chi(t) = Ce^{-(i/\hbar)Et}. \tag{4.53}$$

The spatial equation deriving from Equation 4.52 may be written in the form

$$\mathscr{H} \psi(\vec{r}) = E\psi(\vec{r}), \tag{4.54}$$

where $\mathscr{H} = -(\hbar^2/2m)\nabla^2 + V(\vec{r})$. Since each solution of Equation 4.54 will correspond to a definite value of the energy, let us write the nth solution as

$\psi_{E_n}(\vec{r})$. Thus, a particular solution of Equation 4.42 may be written as

$$\Psi_{E_n}(\vec{r}, t) = \psi_{E_n}(\vec{r}) \cdot e^{-(i/\hbar)E_n t}. \qquad (4.55)$$

Equation 4.54 is often called the *time-independent Schrödinger equation*. It belongs to the class of equations which is known collectively and traditionally as the *eigenvalue problem*. Before continuing with the application of the Schrödinger equation to physical problems, let us briefly discuss some of the properties of eigenequations.

Consider an operator Q which performs a mathematical operation on a function f. This is indicated by the product Qf. Suppose now that the operation Qf results in reproducing the *same function f* multiplied by a constant. That is,

$$Qf = \pm cf, \qquad (4.56)$$

where c is a positive constant.

When this occurs, f is called an *eigenfunction* of the operator Q, and $\pm c$ is called the *eigenvalue* associated with f. For example, for the function $f = \sin 2x$ and the operator $Q = (d^2/dx^2)$,

$$Qf = \frac{d^2}{dx^2}(\sin 2x) = -4 \sin 2x = -4f.$$

In this case the eigenvalue is -4 and the eigenfunction is $\sin 2x$. If there is a whole set of functions f_n, there is an eigenvalue of Q associated with each function. Thus $Qf_n = c_n f_n$. In the above example

$$f_n = \sin nx$$

and

$$c_n = -n^2.$$

Not all functions are suitable as eigenfunctions; for some functions there is *no* operator such that an eigenvalue exists. We say that an eigenfunction must be "well-behaved"; in particular, it must be single-valued and the integral of its modulus squared must be finite. In general, functions which are continuous, single-valued, and which are finite or vanish at infinity are suitable. It is evident that the suitability of a function depends somewhat on the nature of the operator, but generally the restrictions on eigenfunctions are the same as those that we require for wave functions which describe particles.

One of the main activities of quantum mechanics is that of solving an eigenvalue problem when only Q is known. That is, we want solutions of $Qf = \pm cf$, where the problem is to find both c and f. As an example, consider again the operator d^2/dx^2:

$$\frac{d^2 f}{dx^2} = \pm cf.$$

For positive c this has the solution $f = Ae^{\sqrt{c}x} + Be^{-\sqrt{c}x}$. This function diverges as $x \to \pm\infty$, so it is not a suitable eigenfunction. However, if $A = 0$ and $f_1 = Be^{-\sqrt{c}x}$, the function f_1 is suitable for $x > 0$ and has the eigenvalue c. For $x < 0$, the function $f_2 = Ae^{\sqrt{c}x}$ with $B = 0$ is suitable for the negative half-space. It also has the eigenvalue c.

To obtain well-behaved eigenfunctions for all-space, we must consider the eigenvalue $-c$. Then,

$$\frac{d^2f}{dx^2} = -cf$$

$$f = D \sin \sqrt{c}x + E \cos \sqrt{c}x.$$

This function is well-behaved over all of space, and an eigenfunction exists for Q for all eigenvalues $-c$. In this example there is a continuum of possible eigenvalues, whereas in many physical problems (such as standing waves on a string) the eigenvalues form a discrete set of numbers.

Returning now to the discussion of the energy eigenequation (the time-independent Schrödinger equation), which we will write as

$$\mathscr{H}\psi_{E_n}(\vec{r}) = E_n\psi_{E_n}(\vec{r}), \tag{4.57}$$

we wish to show that the energy eigenvalue E_n (which is a particular value of the separation constant for Equation 4.52) is a real number for the boundary conditions that we impose on the wave functions. From Equations 4.55 and 4.18, we have

$$\rho = \Psi^*(\vec{r}, t)\Psi(\vec{r}, t) = \psi_{E_n}^*(\vec{r})\psi_{E_n}(\vec{r})e^{-(i/\hbar)(E_n-E_n^*)t},$$

for $\psi_{E_n}(\vec{r})$ normalized. Now, making use of the conservation of probability, Equation 4.46,

$$-\frac{i}{\hbar}(E_n - E_n^*)\psi_{E_n}^*(\vec{r})\psi_{E_n}(\vec{r}) + \vec{\nabla} \cdot \vec{S}(\vec{r}, t) = 0.$$

Integrating over all of space,

$$(E_n - E_n^*)\int_\tau \psi_{E_n}^*(\vec{r})\psi_{E_n}(\vec{r})\,d\tau = \frac{\hbar}{i}\int_\tau \vec{\nabla} \cdot \vec{S}\,d\tau = \frac{\hbar}{i}\int_\sigma \vec{S} \cdot d\vec{\sigma},$$

where the latter is a surface integral. For a sufficiently large bounding surface, the probability flow through it must vanish, so we find that $E_n = E_n^*$, or E_n is real. This important result tells us that when space and time are separable, the time factor, Equation 4.53, is a harmonic function and not an exponentially rising or decaying function. $\Psi(\vec{r}, t)$ and $\psi_{E_n}(\vec{r})$ differ, then, only by a time-dependent phase factor. An immediate consequence of the reality of E_n is that the probability density is independent of time, since

$$\rho = \Psi^*(\vec{r}, t)\Psi(\vec{r}, t) = \psi_{E_n}^*(\vec{r})\psi_{E_n}(\vec{r}).$$

Following the pattern of Equation 4.48, for Ψ normalized, the expectation value of the energy is:

$$\langle E \rangle = \int_\tau \Psi^*(\vec{r}, t) i\hbar \frac{\partial}{\partial t} \Psi(\vec{r}, t) \, d\tau$$

$$= \int_\tau \psi^*_{E_n}(\vec{r}) e^{(i/\hbar) E_n t} i\hbar \left(-\frac{i}{\hbar} E_n \right) \psi_{E_n}(\vec{r}) e^{-(i/\hbar) E_n t} \, d\tau$$

$$= E_n \int_\tau \psi^*_{E_n}(\vec{r}) \psi_{E_n}(\vec{r}) \, d\tau$$

$$= E_n \int_\tau \rho \, d\tau$$

$$= E_n.$$

The meaning of the above results may be summarized as follows. When the Schrödinger equation is separable in space and time, each of its solutions corresponds to a definite energy which remains constant in time. The solutions that correspond to the spectrum of these energy eigenvalues are called *stationary states* or *eigenstates*. The probability density for an eigenstate is constant in time, and the expectation value for the energy of an eigenstate is precisely the energy of that state for all time. The reader can easily show that the expectation value of any operator which does not explicitly depend upon time is also independent of time for a stationary state.

We will devote all of Chapter 5 to the application of the Schrödinger eigenequation for the energy, Equation 4.57,

$$\mathcal{H} \psi_{E_n}(\vec{r}) = E_n \psi_{E_n}(\vec{r}),$$

to a number of one-dimensional systems. For unbound systems we will find that a continuum of eigenvalues and eigenfunctions exists, while for bound systems only discrete values of E_n exist. The state with the lowest value of E_n is taken as the ground state, and the states corresponding to higher values of E_n are called excited states. Since excited states generally have short lifetimes, one would suspect that the ground state is the only state which deserves to be called a "stationary state." In spite of this, however, we do approximate the states of a physical system (such as an atom) with the spectrum of absolute stationary states obtained by solving the energy eigenequation. To get a quantitative feeling for the validity of this approximation, recall that the nominal lifetime of an atomic state is 10^{-8} second (section 9 of Chapter 2) and that the period of an electron in a Bohr orbit is $\sim 10^{-15}$ second, since optical frequencies are $\sim 10^{15}$ sec^{-1}. Hence, the electron makes about 10^7 revolutions in the state before leaving it. Since this is comparable to 10 million years in a solar orbit, we must regard the electron's orbit in an excited state as relatively stable!

13. DIRAC BRACKET NOTATION

It is convenient to introduce Dirac's abbreviated notation for integrals of the type that are required for the normalization of wave functions and the calculation of expectation values. If ψ_α represents some definite state function, then the integral of its probability density over all of space is

$$\int_\tau \psi_\alpha^* \psi_\alpha \, d\tau.$$

We will now also represent this integral by the symbol

$$\langle \psi_\alpha \mid \psi_\alpha \rangle, \quad \text{or simply by } \langle \alpha \mid \alpha \rangle.$$

Note that there is no explicit designation of the complex conjugate or the variables of integration. It is understood that the complex conjugate of the function to the left of the vertical bar will be used when the integration is actually carried out. There is rarely any difficulty caused by omitting reference to the representation being used, but in case of ambiguity, the variables of integration may be easily included in the brackets as follows:

$$\langle \psi_\alpha(x, t) \mid \psi_\alpha(x, t) \rangle.$$

When finding the expectation value of an operator Q, the integral may be written as

$$\langle \psi_\alpha \mid Q\psi_\alpha \rangle \quad \text{or} \quad \langle \psi_\alpha \mid Q \mid \psi_\alpha \rangle, \tag{4.58}$$

where the form on the right implies that the operator is Hermitian. Since the Hermitian property will not be discussed until Chapter 6, we will restrict ourselves to the form on the left for the time being. By way of illustration, the expectation values given by Equations 4.48 and 4.49 now take the form:

$$\langle x \rangle = \frac{\langle \Psi \mid x\Psi \rangle}{\langle \Psi \mid \Psi \rangle} \quad \text{and} \quad \langle p_x \rangle = \frac{\langle \phi \mid p_x \phi \rangle}{\langle \phi \mid \phi \rangle}.$$

Although we will use Dirac's notation freely in the next chapter, its deeper significance will not become evident until we take up Chapter 6.

SUMMARY

The old quantum theory is briefly described and illustrated by the Sommerfeld treatment of the hydrogen atom. Its demise was brought about by Schrödinger's wave mechanics, which is developed in the remainder of the chapter. After a discussion of the de Broglie wavelength of a particle and particle diffraction, a complex wave function is defined. Although the wave function is not directly observable, its modulus squared is interpreted as the relative probability of finding the particle at a given point in space and time.

One of the fundamental assumptions of quantum mechanics is that a wave packet composed of a superposition of complex wave functions can be used to represent a localized particle. As required by the uncertainty principle, the more narrowly localized the packet in coordinate space, the greater the number of k-values required to form the packet. The Fourier integral theorem provides the mathematical tools necessary for constructing packets. A completely unlocalized particle may be represented by a single plane wave having one frequency and one k-value; that is to say, the Fourier transform of an infinite plane wave is a delta function in k-space.

The concept of the wave packet resolves the classical wave-particle dualism through the de Broglie relations, which connect the energy and momentum of the particle with the frequency and propagation constant of the wave. By means of this correspondence between packets and particles, a wave equation is constructed which enables us to describe the progress of a particle in space and time. Physical measurements are represented by linear mathematical operators which operate on the wave function. The average value of a large number of identical measurements is called the expectation value of the operator.

The principal activity of non-relativistic quantum mechanics is the solution of the Schrödinger equation for systems of fundamental particles or atoms. When the total energy of such a system is independent of the time, the solution of the Schrödinger equation reduces to the solution of the energy eigenequation. The total solution is then simply the energy eigenfunction ψ_{E_n} multiplied by the time factor, $\exp\left(-(i/\hbar)E_n t\right)$. The energy eigenstates are also referred to as stationary states, since the probability density for such a state is constant in time. Furthermore, the expectation value of any operator that is itself not a function of time is constant for a stationary state.

CHAPTER 4   APPENDIX A

APPENDIX A. NATURAL UNITS

The dimensionless constant α, which was named the fine structure constant, has a significance which goes beyond its role in Equation 4.13. Its magnitude may be regarded as a measure of the strength of the coupling of particles of charge $\pm e$ to electromagnetic fields. Indeed, one may call e^2 the "strength" of the interaction between particles of charge $\pm e$ and electromagnetic fields, but it is more "natural" to use the dimensionless quantity $\alpha = e^2/\hbar c$ for this purpose. Since α is independent of mass it can be used to describe the electromagnetic interactions of all particles of charge $\pm e$.

An appropriate set of natural units for describing atomic phenomena is as follows:

Variable	Natural Unit	Description
mass	m_0	electron rest mass
length	$\lambda_c = \dfrac{\hbar}{m_0 c}$	Compton wavelength divided by 2π
time	$\dfrac{\hbar}{m_0 c^2}$	reciprocal of the de Broglie frequency associated with electron rest mass energy
energy	$m_0 c^2$	electron rest mass energy
velocity	c	speed of light
angular momentum	$\hbar$	Planck's constant divided by 2π

In terms of the natural units and the coupling constant α, the relationships that were previously found for the hydrogen atom may be expressed as follows:

First Bohr orbit:
$$a_0 = \frac{\hbar^2}{m_0 e^2} = \frac{\lambda_c}{\alpha}$$

Ground state energy:
$$-w_0 = -\frac{m_0 e^4}{2\hbar^2} = -\tfrac{1}{2}\alpha^2 m_0 c^2$$

Rydberg constant:
$$Ry_\infty = \frac{\alpha}{4\pi a_0} = \frac{\alpha^2}{4\pi \lambda_c}$$

Orbital velocity:
$$v = \alpha c$$

Thus, when expressed in natural units, we may regard the hydrogen atom as being rather weakly bound, which is to say that the electromagnetic interaction is a weak interaction in the world of elementary particles. For instance, suppose that the coupling constant α were nearly unity. Then the Bohr radius would be nearly the Compton wavelength, and the ground state of hydrogen would be of the same order of magnitude as the rest mass energy of the electron. The density of matter then would be greater by the factor $(137)^3$.

It is evident from the form of the orbital velocity that α^2 is the appropriate correction for relativistic effects, which are always proportional to $(v/c)^2$.

Some form of simplified units is always used in advanced courses, although it may not be the system described here. A frequent practice in quantum electrodynamics is to define $\hbar = c = m = 1$, in order to simplify all equations. When the answer is obtained it is then multiplied by the appropriate combination of these universal constants which gives it the correct dimensions.

In a non-relativistic treatment of atomic systems, the "natural" units turn out to be the so-called atomic units wherein the unit of length is the Bohr radius a_0, the unit of mass is m_0, and energies are expressed in terms of w_0.

SUGGESTED REFERENCES

Sidney Borowitz, *Fundamentals of Quantum Mechanics*. W. A. Benjamin, Inc. New York, 1967.

R. H. Dicke and J. P. Wittke, *Introduction to Quantum Mechanics*. Addison-Wesley Publishing Co., Inc., Reading, Mass., 1960.

E. Merzbacher, *Quantum Mechanics*. John Wiley and Sons, Inc., New York, 1961.

Albert Messiah, *Quantum Mechanics*. North-Holland Publishing Co., Amsterdam, 1958.

D. Park, *Introduction to the Quantum Theory*. McGraw-Hill Book Co., Inc., New York, 1964.

L. Pauling and E. B. Wilson, *Introduction to Quantum Mechanics*. McGraw-Hill Book Co., Inc., New York, 1935.

J. L. Powell and B. Crasemann, *Quantum Mechanics*. Addison-Wesley Publishing Co., Inc., Reading, Mass., 1961.

F. K. Richtmyer, E. H. Kennard and J. N. Cooper, *Introduction to Modern Physics* (6th ed.). McGraw-Hill Book Co., New York, 1969.

V. Rojansky, *Introductory Quantum Mechanics*. Prentice-Hall, Inc., Englewood Cliffs, N.J., 1938.

D. S. Saxon, *Elementary Quantum Mechanics*. Holden-Day, Inc., San Francisco, 1968.

Leonard I. Schiff, *Quantum Mechanics*, 3rd ed. McGraw-Hill Book Co., New York, 1969.

Robert L. White, *Basic Quantum Mechanics*. McGraw-Hill Book Co., New York, 1966.

CHAPTER 5

SOLUTIONS OF SOME ONE DIMENSIONAL SYSTEMS

In this chapter the Schrödinger wave equation will be applied to a few one dimensional problems which will serve to illustrate the approach of quantum mechanics as well as some of the interesting predictions of the theory. Although the systems used here are idealized and are, consequently, somewhat artificial, this departure from reality serves to simplify the mathematics without destroying the essential physical features of the problems.

I. STEP POTENTIALS

We will begin with potentials that are constant in time and that are also constant throughout prescribed regions of space. The simplest of these is the step potential shown in Figure 5-1, in which the one dimensional space is divided into two regions such that $V = 0$ for $x < 0$, and $V = V_0$ for $x > 0$.

Let us first consider the case when the total energy E of a particle of mass m is greater than V_0. In the region where $x < 0$, which we will designate as region I, the potential is zero and the time-independent Schrödinger equation becomes

$$\mathscr{H}_{\mathrm{I}}\psi_{\mathrm{I}} = -\frac{\hbar^2}{2m}\frac{d^2\psi_{\mathrm{I}}}{dx^2} = E\psi_{\mathrm{I}}.$$

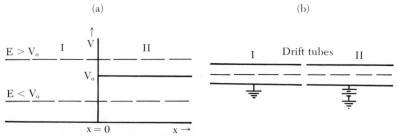

Figure 5-1. (a) A plot of energy vs position for a step potential. The dotted lines indicate possible particle energies. (b) A physical approximation of a step potential. The dotted line indicates a particle trajectory.

For a free particle in region I, the total energy of the particle is equal to its kinetic energy, that is,

$$\frac{p_1^2}{2m} = E,$$

or

$$p_1^2 = \hbar^2 k_1^2 = 2mE.$$

Thus, the Schrödinger equation for region I may be written as

$$\frac{d^2\psi_I}{dx^2} = -k_1^2\psi_I,$$

which has the general solution,

$$\psi_I = Ae^{ik_1x} + Be^{-ik_1x}. \tag{5.1}$$

Following the same procedure for region II, we obtain

$$p_2^2 = \hbar^2 k_2^2 = 2m(E - V_0)$$

$$\frac{d^2\psi_{II}}{dx^2} = -k_2^2\psi_{II},$$

and

$$\psi_{II} = Ce^{ik_2x} + De^{-ik_2x} \tag{5.2}$$

Thus far we have said nothing about the direction of the incident particle. It turns out that the continuous states for free particles are *doubly degenerate*, that is, there are two solutions for each value of the total energy. These solutions correspond physically to the fact that a particle of energy E may be traveling in either direction, $\pm x$. For convenience we will adopt the initial condition that the particle arrives from the left. Therefore, the coefficient A in Equation 5.1 represents the amplitude of the wave associated with the incident particle, B is the amplitude of a reflected wave at the potential discontinuity at $x = 0$, C is the amplitude of the transmitted wave in region II, and D is the amplitude of an incident wave from the right. Our initial condition requires that $D = 0$. Therefore,

$$\psi_{II} = Ce^{ik_2x}. \tag{5.3}$$

Equations 5.1 and 5.3 will together constitute a satisfactory solution of the problem only if they fulfill all of the requirements of good behavior which quantum mechanics imposes upon allowed wave functions. It was shown in an earlier section that plane wave solutions of this type can be normalized so as to satisfy the integral-square criterion. Although we will not normalize Equation 5.1 it is important to know that it *can* be normalized. Two other requirements are the continuity of the wave function itself and the existence of the first derivative everywhere. Both of these conditions will be met if we require ψ_I and ψ_{II} to be equal in magnitude and to have the same slope where they join at $x = 0$. That is, we impose the boundary conditions:

$$\left.\begin{array}{c} \psi_{\text{I}}(0) = \psi_{\text{II}}(0) \\ \psi_{\text{I}}'(x)]_{x=0} = \psi_{\text{II}}'(x)]_{x=0} \end{array}\right\} . \tag{5.4}$$

From the first condition we have

$$A + B = C,$$

and from the second we obtain

$$C = \frac{k_1}{k_2}(A - B).$$

After solving these equations we find:

$$\left.\begin{array}{c} \dfrac{B}{A} = \dfrac{k_1 - k_2}{k_1 + k_2} \\ \dfrac{C}{A} = \dfrac{2k_1}{k_1 + k_2} \end{array}\right\} . \tag{5.5}$$

Using the expression for the probability current in the previous chapter, we have

$$\left.\begin{array}{c} S_{\text{in}} = \dfrac{\hbar k_1}{m}|A|^2 \\[2ex] S_{\text{re}} = -\dfrac{\hbar k_1}{m}|B|^2 \\[2ex] S_{\text{tr}} = \dfrac{\hbar k_2}{m}|C|^2 \end{array}\right\} . \tag{5.6}$$

Then the transmission and reflection coefficients are:

$$\left.\begin{array}{c} T = \left|\dfrac{S_{\text{tr}}}{S_{\text{in}}}\right| = \dfrac{k_2}{k_1}\left|\dfrac{C}{A}\right|^2 = \dfrac{4k_1k_2}{(k_1 + k_2)^2} \\[2ex] R = \left|\dfrac{S_{\text{re}}}{S_{\text{in}}}\right| = \left|\dfrac{B}{A}\right|^2 = \left(\dfrac{k_1 - k_2}{k_1 + k_2}\right)^2 \end{array}\right\} . \tag{5.7}$$

Note that $T + R = 1$, as it must if probability is to be conserved. Figure 5-2

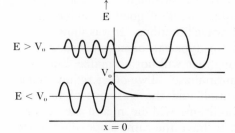

Figure 5-2. Possible eigenfunctions for the potential step. The amplitude of the wave is larger in the region where the velocity of the particle is lower, since the particle spends more time in that region. The wavelength is shorter in the region of higher kinetic energy.

illustrates schematically the behavior of solutions for the step potential. There will always be a reflected wave unless the discontinuity in the potential vanishes. Furthermore, Equation 5.7 indicates that the reflection coefficient is independent of whether the potential discontinuity is a step up or a step down. Here the behavior is quite different from that of a classical particle.

PROBLEM 5-1

Verify the probability currents given in Equation 5.6.

Now consider the case for a total particle energy less than V_0. Classically, the particle would rebound from the step, but we will find that in quantum mechanics there is a non-zero probability of finding the particle in region II. The analysis of region I is identical with the first case and results in the solution given by Equation 5.1. However, in region II we are now faced with the fact that the momentum is imaginary since the total energy is less than the potential energy. That is,

$$p_2^2 = \hbar^2 K_2^2 = -2m(E - V_0) = 2m(V_0 - E).$$

This is equivalent to defining the propagation constant as

$$K_2 = ik_2$$

in Equation 5.2. The general solution to Schrödinger's equation in region II then becomes

$$\psi_{II} = Ce^{-K_2 x} + De^{K_2 x}.$$

Instead of the harmonic solutions that we obtained previously, we now have two exponential terms, one increasing and the other decaying. Since the increasing exponential is not well-behaved for positive x we require D to be zero. Therefore, the solution in region II is

$$\psi_{II} = Ce^{-K_2 x}. \tag{5.8}$$

Application of the boundary conditions leads to

$$\left. \begin{array}{l} \dfrac{B}{A} = \dfrac{k_1 - iK_2}{k_1 + iK_2} \\[2ex] \dfrac{C}{A} = \dfrac{2k_1}{k_1 + iK_2}, \end{array} \right\} \tag{5.9}$$

and the result that

$$R = 1 \quad \text{and} \quad T = 0. \tag{5.10}$$

Here the reflection coefficient of unity, which agrees with the classical result, arises from the assumption that no absorption occurs in region II. Thus, the continuous damping results in continuous reflection until all of the incident energy is ultimately returned to region I.

Verify the results given in Equations 5.9 and 5.10.

It is of interest to point out that for a time-dependent solution the boundary conditions cannot be satisfied unless the same value for ω is used on both sides of the step. From our knowledge of the behavior of elastic and optical waves we know that when a wave passes from one medium to another, the frequency remains the same but the velocity and wavelength are both altered so that $\omega = kv = $ constant. This, of course, assures the conservation of energy since $E = \omega\hbar$. For a material particle this means that ω must be associated with the total energy of the particle including its rest mass, which is consistent with the original de Broglie hypothesis of Equation 4.14.*

2. THE FINITE POTENTIAL BARRIER

Consider the barrier shown in Figure 5-3, having height V_0 and width $2a$. Here there are three regions to be considered, but since the potential is zero for $x < -a$ and for $x > a$, these two regions have identical solutions.

For the case of $E > V_0$, following the same procedure as in the previous section, we obtain the solutions:

$$\psi_{\mathrm{I}} = Ae^{ik_1x} + Be^{-ik_1x}, \quad \text{for } x < -a$$

$$\psi_{\mathrm{II}} = Ce^{ik_2x} + De^{-ik_2x}, \quad \text{for } -a < x < a$$

$$\psi_{\mathrm{III}} = Fe^{ik_1x}, \quad \text{for } x > a,$$

where $\hbar k_1 = \sqrt{2mE}$ and $\hbar k_2 = \sqrt{2m(E - V_0)}$.

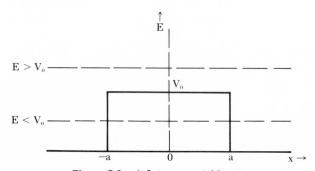

Figure 5-3. A finite potential barrier.

---

* This is confirmed experimentally by the coherent regeneration of short lifetime, neutral K mesons in absorbing material. See for example, Gunnar Källen, *Elementary Particle Physics*, Addison-Wesley Publishing Co., Reading, Mass., 1964, p. 438.

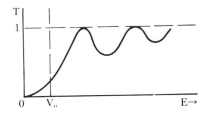

Figure 5-4. Transmission coefficient for a finite square barrier. For $E > V_0$ the curve is given by Eq. 5.11 and for $E < V_0$ it is given by Eq. 5.12.

Applying the boundary conditions (Equation 5.4) to ψ_I and ψ_{II} at $x = -a$ and to ψ_{II} and ψ_{III} at $x = a$, one can calculate all of the coefficients in terms of the coefficient A. Then, from the probability currents the transmission and reflection coefficients are determined to be

$$T = \left| \frac{S_{tr}}{S_{in}} \right| = \left| \frac{F}{A} \right|^2$$

$$R = \left| \frac{S_{re}}{S_{in}} \right| = \left| \frac{B}{A} \right|^2.$$

For the case of $E > V_0$ the transmission coefficient is an oscillating function as shown in Figure 5-4. An eigenfunction for this case is shown in Figure 5-5. Its functional form is given by

$$T = \frac{1}{\cos^2 (2k_2a) + \frac{1}{4}\left(\frac{k_2}{k_1} + \frac{k_1}{k_2}\right)^2 \sin^2 (2k_2a)}$$

or

$$T = \frac{1}{1 + \frac{V_0^2}{4E(E - V_0)} \sin^2 (2k_2a)}. \tag{5.11}$$

It is evident that the transmission coefficient is unity whenever an integral number of half wavelengths is contained inside the barrier, that is, when $2k_2a = n\pi$, where n is an integer. This oscillation is the analog of such phenomena as the optical transmission through a non-reflecting thin film, or the use of

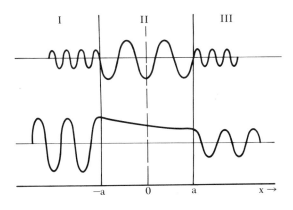

Figure 5-5. Real part of barrier eigenfunction. The function in region I is a superposition of the incident and reflected waves.

matching stubs in wave guides, or the Ramsauer-Townsend effect* in the scattering of low-energy electrons from noble gas atoms.

When $E < V_0$, corresponding to the classical case of no transmission, the momentum within the barrier is imaginary as in the potential step discussed previously. We will again find an exponentially damped wave within the barrier, but if the barrier is sufficiently narrow an appreciable transmitted wave can leak out into region III.

The functions ψ_I and ψ_{III} are the same as for $E > V_0$, but function ψ_{II} now becomes

$$\psi_{II} = Ce^{+K_2 x} + De^{-K_2 x},$$

where

$$\hbar K_2 = i\hbar k_2 = \sqrt{2m(V_0 - E)}.$$

The coefficients are obtained by satisfying the boundary conditions as before. The transmission coefficient for this case is:

$$T = \frac{1}{\cosh^2 (2K_2 a) + \frac{1}{4}\left(\frac{K_2}{k_1} - \frac{k_1}{K_2}\right)^2 \sinh^2 (2K_2 a)}$$

or,

$$T = \frac{1}{1 + \dfrac{V_0^2}{4E(V_0 - E)} \sinh^2 (2K_2 a)}. \tag{5.12}$$

It is of interest to evaluate the transmission coefficient for the case of $E \sim V_0$ when $\sinh (2K_2 a) \rightarrow (2K_2 a)$; then, Equation 5.12 becomes

$$T_{E \rightarrow V_0} = \frac{1}{1 + \dfrac{V_0^2}{4E(V_0 - E)} \cdot \dfrac{4a^2}{\hbar^2} \cdot 2m(V_0 - E)} = \frac{1}{1 + (k_1 a)^2}.$$

The same result is obtained by letting $E \rightarrow V_0$ in Equation 5.11.

Nature provides many examples of barrier penetration, although the barriers in the real world are not rectangular in shape and they are generally three dimensional. In spite of this, however, the idealized treatment given here forms the basis for understanding all of the more sophisticated tunneling processes and frequently provides a reasonably good order-of-magnitude answer. For example, alpha particle emission from heavy nuclei was first explained[1] by regarding it as a tunneling process through the nuclear barrier. Figure 5-6 shows a crude representation of a nuclear well which is about 40 MeV deep for a heavy nucleus. The Coulomb barrier is roughly 16 MeV high in this example and its breadth is of the order of 1×10^{-12} cm. A typical value for the energy of an emitted alpha particle is 6 MeV, which is considerably below the

---

* At bombarding energies of about 0.7 eV, rare gas atoms are almost transparent to electrons. This effect will be discussed in Chapter 10.

[1] G. Gamow, *Z. Physik* **51,** 204 (1928); R. W. Gurney and E. U. Condon, *Phys. Rev.* **33,** 127 (1929).

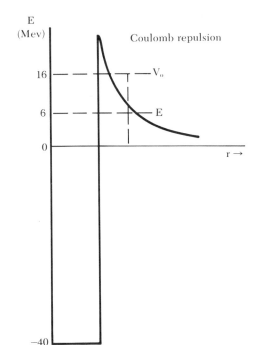

Figure 5–6. Approximation of the nuclear Coulomb barrier by a rectangular barrier.

summit of the barrier. Using Equation 5.12,

$$T = \cfrac{1}{1 + \cfrac{V_0^2}{4E(V_0 - E)} \sinh^2 (2K_2 a)}$$

$$= [1 + \tfrac{256}{240} \sinh^2 (2K_2 a)]^{-1}.$$

For $2K_2 a \gg 1$, we may write,

$$T \sim [\sinh (2K_2 a)]^{-2} \sim [\tfrac{1}{2}e^{2K_2 a}]^{-2} = 4e^{-4K_2 a}.$$

Now for the above example,

$$K_2 a = \frac{a}{\hbar} \sqrt{2m(V_0 - E)} \sim \frac{10^{-12}}{10^{-27}} \sqrt{2 \times 6.7 \times 10^{-24}(16 - 6) \times 1.6 \times 10^{-6}}$$

$$= 14.6$$

and

$$T \sim 4e^{-58.4} \sim 10^{-25}.$$

This means that an alpha particle incident on the barrier has about one chance in 10^{25} of tunneling through the barrier. Although this is a very small probability, we will see that a 6 MeV alpha particle trapped in the above nucleus bangs against the barrier about 10^{21} times per second (see Problem 5-15). Thus a tunneling event can be expected to occur with a probability of $10^{-25} \times 10^{21} = 10^{-4}$ per second, or the decay time is about 3 hours. To illustrate how

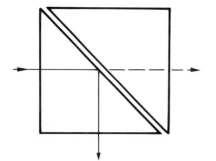

Figure 5-7. An optical example of barrier penetration. A light ray is totally reflected in the prism at the left. However, there is some penetration of the air barrier and when this barrier is made narrow enough a transmitted ray (dotted line) appears in the prism at the right.

critically the mean decay time depends upon the strength of the barrier, suppose that the barrier width were increased by only 20 percent, all other constants remaining the same. Then, $T \sim 10^{-30}$, and the mean decay time would be $\sim 10^9$ seconds or 100 years! Therefore, in spite of the crudeness of this model *it is able to account for the experimental fact that the mean lifetimes of alpha emitters vary from 1 microsecond to 10^{10} years!*

An elementary example of barrier penetration arises in optics in the case of total internal reflection from a glass–air interface (see Figure 5-7). If a second piece of glass is brought very near the reflecting surface, a weak transmitted wave will appear in it. The intensity of the transmitted wave is strongly dependent upon the thickness of the air barrier, and no transmission takes place for barriers more than one wavelength thick.

There are numerous examples of the tunneling of electrons and quasi-particles in solid state[2] and molecular physics. However, there is little to be gained by taking a detailed look at another example until we have developed the mathematical techniques for treating more realistic potentials than the rectangular potential. This will be done in Chapter 9.

PROBLEM 5-3

Derive Equation 5.11.

PROBLEM 5-4

Derive Equation 5.12.

PROBLEM 5-5

A beam of 5-volt electrons impinges on a square potential barrier of height 25 volts and width $a_0 = 0.52$ Å. What fraction of the incident beam gets through the barrier?

---

[2] C. B. Duke, ed., *Tunneling in Solids: Solid State Physics Supplement 10.* Academic Press, New York, 1969.

PROBLEM 5-6

For a 5-volt rectangular potential barrier what is the minimum barrier thickness for 100 percent transmission of 10-volt electrons?

3. THE SQUARE WELL IN ONE DIMENSION

A square well potential is shown in Figure 5-8. For particle energies greater than V_0 the eigenfunctions are similar to those of the barrier for $E > V_0$, and the transmission coefficients are again given by Equation 5.11. For energies less than V_0 we will find that bound states are possible. These bound states exist for only certain discrete values of E, in contrast to the continuum of states associated with all energies greater than V_0. In order to see how these discrete bound states arise we will briefly outline the mathematical steps.

The solutions for the three regions are:

$$\psi_{\text{I}} = Ae^{K_1 x} \qquad , \quad \text{for } x < -a$$
$$\psi_{\text{II}} = Ce^{ik_2 x} + De^{-ik_2 x}, \quad \text{for } -a < x < a$$
$$\psi_{\text{III}} = Fe^{-K_1 x} \qquad , \quad \text{for } x > a,$$

where

$$\hbar K_1 = i\hbar k_1 = \sqrt{2m(V_0 - E)}, \quad \text{and} \quad \hbar k_2 = \sqrt{2mE}.$$

Imposing the boundary conditions, Equation 5.4, at $x = -a$ one obtains:

$$\frac{C}{D} = \frac{k_2 - iK_1}{k_2 + iK_1} e^{2ik_2 a}. \tag{5.13}$$

At $x = a$, the boundary conditions give the following result:

$$\frac{C}{D} = \frac{k_2 + iK_1}{k_2 - iK_1} e^{-2ik_2 a}. \tag{5.14}$$

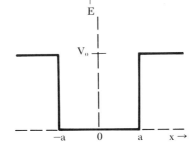

Figure 5-8. A square well potential.

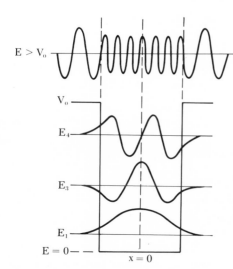

$E > V_0$

V_0

E_4

E_3

E_1

$E = 0$

$x = 0$

Figure 5-9. Eigenfunctions for the square well potential.

The product of Equations 5.13 and 5.14 is $C^2/D^2 = 1$, or

$$C = \pm D.$$

For $C = D$, the boundary conditions require $A = F$ and ψ_{II} becomes a cosine function; the eigensolution is then symmetric about the center of the well as illustrated by E_1 in Figure 5-9. For $C = -D$, $A = -F$, ψ_{II} becomes a sine function, and the eigensolution is antisymmetric about the axis of the well, such as E_4 in Figure 5-9. Note that the penetration of the walls increases with E.

In order to obtain specific information on the number of allowed solutions and the energy associated with each, it is convenient to transform the above equations in the following manner.

Let $\alpha = k_2 a$ and $\beta = K_1 a$. Then, setting $C/D = 1$ in either Equation 5.13 or 5.14, there results:

$$k_2(e^{i\alpha} - e^{-i\alpha}) = iK_1(e^{i\alpha} + e^{-i\alpha})$$

$$k_2 \sin \alpha = K_1 \cos \alpha$$

$$\tan \alpha = \frac{K_1}{k_2} = \frac{\beta}{\alpha}$$

$$\alpha \tan \alpha = \beta, \quad \text{for symmetric solutions.} \qquad (5.15)$$

In a similar fashion the condition $C/D = -1$ yields the equation

$$\alpha \cot \alpha = -\beta, \quad \text{for antisymmetric solutions.} \qquad (5.16)$$

Note that

$$\alpha^2 + \beta^2 = a^2(k_2^2 + K_1^2) = \frac{a^2}{\hbar^2}[2mE + 2m(V_0 - E)]$$

or,

$$\alpha^2 + \beta^2 = \frac{2mV_0a^2}{\hbar^2}. \tag{5.17}$$

Thus, the symmetric solutions can be obtained by solving Equations 5.15 and 5.17 graphically, and the antisymmetric solutions are obtained by solving Equations 5.16 and 5.17. These solutions are sketched in Figure 5-10. Note that the number of solutions depends upon the radius of the circle, Equation 5.17, which in turn depends upon the mass of the particle, the depth of the well and the width of the well. The product aV_0 is a measure of the "strength" of the well in terms of binding a particle. That is, the greater the depth and breadth of a well, the greater the number of bound states and the greater the probability of retaining a particle in the well. In particular, if

$$(n\pi)^2 < \frac{2mV_0a^2}{\hbar^2} < (n+1)^2\pi^2,$$

then there are $n + 1$ symmetric bound states. Similarly, if

$$(n - \tfrac{1}{2})^2\pi^2 < \frac{2mV_0a^2}{\hbar^2} < (n + \tfrac{1}{2})^2\pi^2,$$

then there are n antisymmetric bound states in the well. The total number of bound solutions exceeds by one the largest integer contained in $(a/\pi\hbar)\sqrt{2mV_0}$. Note that there is always at least one bound symmetric state, regardless of how shallow the well is.

In the limiting case of an infinitely deep well the energy eigenvalues are readily obtained without performing a graphical solution. As $V_0 \to \infty$, the radius of the circle, Equation 5.17, becomes infinite and will intersect Equations

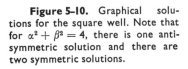

Figure 5–10. Graphical solutions for the square well. Note that for $\alpha^2 + \beta^2 = 4$, there is one antisymmetric solution and there are two symmetric solutions.

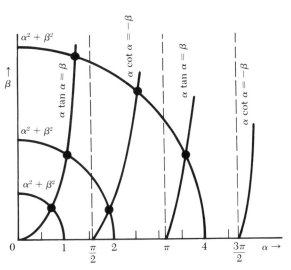

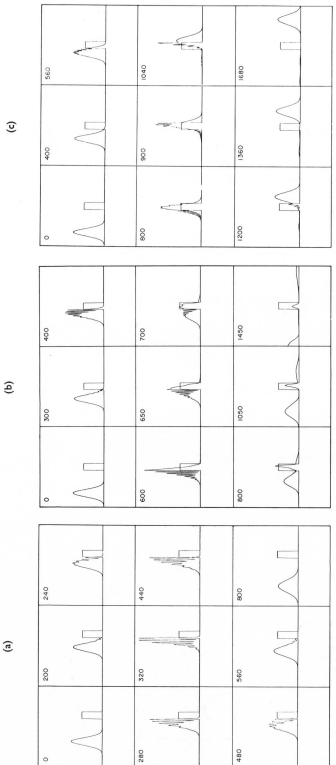

Figure 5-11. Gaussian wave packet scattering from a square barrier. The average energy is (a) one-half the barrier height, (b) equal to the barrier height, (c) twice the barrier height. Numbers denote the time in arbitrary units. (After A. Goldberg, H. M. Schey and J. L. Schwartz, Amer. J. Phys. 35, 177, 1967. Used with permission.)

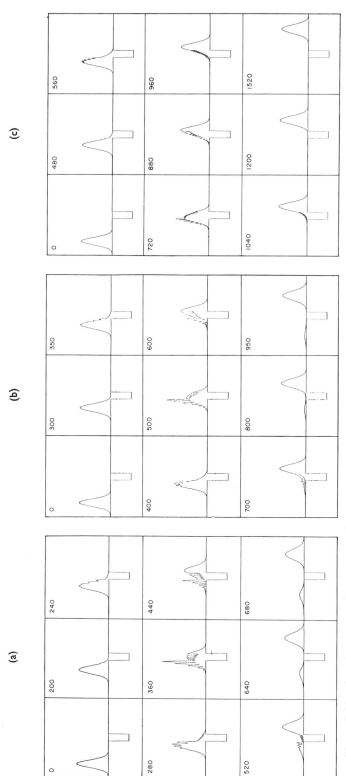

Figure 5-12. Gaussian wave packet scattering from a square well. The average energy is (a) one-half the well depth, (b) equal to the well depth, (c) twice the well depth. Numbers denote the time in arbitrary units. (After A. Goldberg, H. M. Schey and J. L. Schwartz, Amer. J. Phys. 35, 177, 1967. Used with permission.)

5.15 and 5.16 at their asymptotic limits. Then we have:

For symmetric solutions, $\alpha = \dfrac{\pi}{2}, \dfrac{3\pi}{2}, \dfrac{5\pi}{2}, \ldots$

For antisymmetric solutions, $\alpha = \pi, 2\pi, 3\pi, \ldots$

In general, $\alpha = n\pi/2$, where the lowest state is symmetric, the next antisymmetric, and so on, alternately. Since $\alpha = k_2 a$,

$$k_2 a = \frac{a}{\hbar} \sqrt{2mE_n} = \frac{n\pi}{2},$$

or

$$E_n = \frac{n^2\pi^2\hbar^2}{8ma^2}. \tag{5.18}$$

This will be recognized as the answer to the problem of a particle in a box with rigid walls as solved by the old quantum theory (Chapter 4, Problem 2). Note that $n = 0$ does not correspond to a physical state since it would require that $\psi = 0$ everywhere in the box.

The eigenfunctions for the infinite square well do not leak out of the walls of the well as they do in the finite potential well (Fig. 5-9). Since each solution must have zero amplitude at both walls, the eigenfunctions then become sine or cosine functions.

Before closing this section it is important to point out that a real particle is represented by a wave packet consisting of many k values instead of just a single value of k as we have assumed in the foregoing analysis. Furthermore, since the time dependence of each component differs from the others, a complete analysis of the interaction of a packet with a barrier or a well is quite complicated even in one dimension. Goldberg et al[3] have obtained computer solutions and computer-generated motion pictures of Gaussian packets scattering from one dimensional barriers and wells. Two of their scattering events are illustrated in Figures 5-11 and 5-12.

PROBLEM 5-7

Assuming an infinite square well of radius 2.8×10^{-13} cm, find the normalized wave functions and the energies of the four lowest states for a nucleon.

PROBLEM 5-8

Show that the energy of the lowest state in the infinite square well is consistent with the uncertainty in momentum as required by the uncertainty principle. Would the state $E = 0$ satisfy the uncertainty principle?

---

[3] A. Goldberg, H. M. Schey and J. L. Schwartz, *Amer. J. Phys.* **35,** 177 (1967).

PROBLEM 5-9

(a)   For the square well shown in Figure 5-8, find the function ψ_{II} defined in the region $-a < x < a$ for the case of $E > V_0$.

(b)   For what values of $E > V_0$ is the total probability within the well a maximum? A minimum?

(c)   Is the maximum value of $\int_{-a}^{a} \rho \, dx$ found in (b) greater than what it would be in the absence of the well? Physically, this means that particles are "trapped" for a short while in the well. Such events are called resonances.

PROBLEM 5-10

Calculate the minimum uncertainty product, $\Delta x \cdot \Delta p_x$, for the ground state and the first excited state of the infinite square well.

PROBLEM 5-11

Sketch the wave function $\psi_n(x)$ and the probability density $\rho_n(x)$ for each of the three lowest eigenfunctions for the infinite square well.

PROBLEM 5-12

Find the probability current density for the nth state of the infinite square well.

PROBLEM 5-13

Consider a state function which is a superposition of the two lowest eigenfunctions of the infinite square well, that is,

$$\psi = \frac{1}{\sqrt{a}} \cos \frac{\pi x}{2a} + \frac{2}{\sqrt{a}} \sin \frac{\pi x}{a}.$$

(a)   Normalize this wave function in the space $-a \le x \le a$.

(b)   Sketch the probability density function, $\rho(x)$.

(c)   Include the appropriate time factor in each term of the wave function and find the probability density function, $\rho(x, t)$.

(d)   Using the normalized wave function in (a), calculate the expectation value of the kinetic energy.

(Ans.: (d) $17\hbar^2\pi^2/40ma^2$)

PROBLEM 5-14

Use the uncertainty principle to give an argument against the existence of an electron in the nucleus of an atom. Take the nuclear diameter to be $\sim 10^{-12}$ cm.

PROBLEM 5-15

Approximately how many collisions per second would an alpha particle have with the walls of a nuclear well of radius 1×10^{-12} cm?

4. MULTIPLE SQUARE WELLS

If two identical square wells are isolated from each other and each has a particle in its ground state, the energy of the system is just the sum of the two ground state energies. Now, if the two wells are separated by only a very narrow barrier they will interact and the total wave function is a linear combination of the two single-well ground state functions as shown schematically in Figure 5-13. In (a) the total wave function is symmetric in the two wells, and in (b) it is antisymmetric. The energies of these two states are not the same, and the energy splitting increases as the width of the barrier decreases. This may be understood qualitatively by means of the following argument. In the limit as the barrier vanishes, the symmetric wave function corresponds to the ground state of a well having twice the width of the original wells; hence the energy *decreases*. As the wells coalesce, the antisymmetric function, on the other hand, corresponds to the first excited state of the well of double width. Thus, the energy decrease due to the increased width of the well is compensated by excitation of the particle to the first state above the ground state. The energy of the double-well system for these two cases is shown in Figure 5-14, where r is the width of the barrier.

The most important feature of this model is that the two wells, which have the same energy levels at large separations, interact more strongly as they are brought closer together, and the interaction splits each original level into two separate energy levels.

An important generalization of this model was made by Kronig and Penney[4] in treating the electron energies in solids. Here the number of

(a)   (b)

Figure 5-13. Total wave functions for two interacting square wells, (a) symmetric case, (b) antisymmetric case.

---

[4] R. de L. Kronig and W. G. Penney, *Proc. Roy. Soc.* (*London*) *A* **130**, 499 (1931).

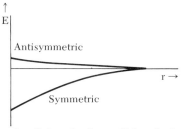

Figure 5-14. Energy versus barrier width r for two identical square wells.

interacting wells, N, is extremely large so that each of the single-well levels is split into N levels spaced so close together that they form nearly continuous energy bands. Figure 5-15 shows a schematic square-well representation of the potential seen by an electron due to a linear lattice of positive ions. Since the potential function has the period of the lattice, that is, $V(x + d) = V(x)$, we expect our solution to have the same periodicity, provided that we avoid the ends of the chain. This difficulty can be eliminated by imagining that the chain is joined into a ring containing a large number of atoms (of the order of Avogadro's number).

It was shown by Bloch[5] that the solution for such a periodic lattice may be written in the form of the product of a plane wave and a function having the periodicity of the lattice. Thus,

$$\psi(x) = e^{ikx}u(x),$$

where $u(x + d) = u(x)$ and $k = 2\pi nx/Nd$. Thus,

$$\psi(x + d) = e^{ikd}\psi(x).$$

The last result enables us to obtain the solution anywhere in the lattice once it is known for a single well. If we denote the width of the well by a, the width of the barrier by b and the height of the barrier by V_0, then we may define the propagation constants,

$$\hbar k_1 = \sqrt{2mE} \quad \text{and} \quad \hbar K_2 = i\hbar k_2 = \sqrt{2m(V_0 - E)},$$

as before. Although the algebra is laborious, the procedure is straightforward and one obtains:

For $E < V_0$,

$$\cos k_1 a \cdot \cosh K_2 b - \left(\frac{k_1^2 - K_2^2}{2k_1 K_2}\right) \sin k_1 a \cdot \sinh K_2 b = \cos \frac{2\pi n}{N},$$

Figure 5-15. A square-well representation of the potential seen by an electron due to a linear array of positive ions.

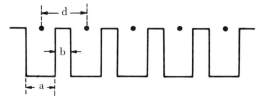

[5] F. Bloch, *Z. Physik* **52**, 555 (1928). This was first proven by Floquet in 1883.

and for $E > V_0$,

$$\cos k_1 a \cdot \cos k_2 b - \left(\frac{k_1^2 + k_2^2}{2k_1 k_2}\right) \sin k_1 a \cdot \sin k_2 b = \cos \frac{2\pi n}{N}.$$

Since the right hand side of each of these equations is restricted to the range of values between 1 and -1, solutions exist only for those energies for which the magnitude of the left hand side does not exceed unity. As a consequence of this restriction, the energies which correspond to "physically" realizable solutions form continuous bands which are separated by *forbidden* bands of energies for which no solutions exist.

Although the details of the energy band structure of a solid require a much more refined model than that used here, it is instructive to see that the general quantum mechanical behavior of the electrons can be predicted by just using the elementary concepts of wells and barriers developed in the previous sections.

5. THE HARMONIC OSCILLATOR: POLYNOMIAL SOLUTION[6]

The harmonic oscillator is one of the most important problems in modern physics. The overriding reason for its dominant rôle is that complex systems can be reduced, by means of Fourier analysis, to the solution of a collection of harmonic oscillators. Thus, the harmonic oscillator approximation has become one of the principal methods for studying the collective excitations of many-body systems such as collections of phonons and photons. We will use two different methods to solve the harmonic oscillator. In this section the polynomial method of Sommerfeld will be described, and later we will use the more elegant operator techniques which are based on the factorization method of Schrödinger and Dirac.

The classical Hamiltonian for a linear harmonic oscillator is given by the expression,

$$\mathscr{H} = \frac{p_x^2}{2m} + \tfrac{1}{2}\omega^2 m x^2,$$

and by means of the correspondence principle the Hamiltonian operator equivalent is

$$\mathscr{H} = -\frac{\hbar^2}{2m}\frac{d^2}{dx^2} + \tfrac{1}{2}\omega^2 m x^2.$$

Then the energy eigenstates of the oscillator will be given by the solutions of the time-independent Schrödinger equation,

$$\frac{d^2\psi}{dx^2} + (\beta - \alpha^2 x^2)\psi = 0, \tag{5.19}$$

---

[6] A. Sommerfeld, *Wave Mechanics*. Dutton, New York, 1929, p. 11.

where the parameters α and β are given by:

$$\alpha = \frac{\omega m}{\hbar} \quad \text{and} \quad \beta = \frac{2mE}{\hbar^2}.$$

It is convenient to transform Equation 5.19 to the dimensionless variable q, where

$$q = \sqrt{\alpha}\, x.$$

Substituting into Equation 5.19 the derivatives,

$$\frac{d\psi}{dx} = \frac{d\psi}{dq}\frac{dq}{dx} = \sqrt{\alpha}\,\frac{d\psi}{dq}$$

$$\frac{d^2\psi}{dx^2} = \frac{d}{dq}\left(\frac{d\psi}{dx}\right)\frac{dq}{dx} = \alpha\,\frac{d^2\psi}{dq^2},$$

there results

$$\frac{d^2\psi}{dq^2} + (\epsilon - q^2)\psi = 0, \tag{5.20}$$

where we have defined ϵ as the dimensionless eigenvalue

$$\epsilon = \frac{\beta}{\alpha} = \frac{2E}{\omega\hbar}.$$

In order to find acceptable solutions of Equation 5.20, we will first consider the form of the asymptotic solutions as q becomes infinite. Then it will be assumed that a satisfactory solution of Equation 5.20 can be expressed as the product of the asymptotic solution and a finite polynomial in q. That is,

$$\psi = \psi_a \cdot H(q),$$

where $H(q)$ is a power series in q which will later be terminated to the appropriate polynomial. In order to arrive at the asymptotic solution, we note that as q gets very large, Equation 5.20 may be approximated by

$$\frac{d^2\psi_a}{dq^2} - q^2\psi_a = 0, \tag{5.21}$$

which has the solution,

$$\psi_a = Ae^{q^2/2} + Be^{-q^2/2}.$$

We note that A must be zero in order to insure that ψ_a is a well-behaved function. If ψ_a is substituted into Equation 5.21 it is immediately obvious that the latter is not satisfied exactly. That is,

$$\frac{d^2\psi_a}{dq^2} - q^2\psi_a = \psi_a, \quad \text{where } \psi_a = Be^{-q^2/2}.$$

However, since ψ_a vanishes as $q \to \infty$, ψ_a satisfies both Equations 5.20 and 5.21 in the asymptotic limit. Our trial solution of Equation 5.20 may now be written as

$$\psi = e^{-q^2/2}H(q),$$

where the coefficient B has been absorbed in the polynomial. To determine $H(q)$ we now substitute ψ into Equation 5.20 and obtain

$$H''(q) - 2qH'(q) + (\epsilon - 1)H(q) = 0, \tag{5.22}$$

where the primes indicate differentiation with respect to q. Equation 5.22 is known as the Hermite equation. It can be solved by assuming a power series of the form

$$H(q) = \sum_{k=0}^{\infty} a_k q^k = a_0 + a_1 q + a_2 q^2 + a_3 q^3 + \cdots$$

$$H'(q) = \sum_{k=1}^{\infty} k a_k q^{k-1} = a_1 + 2a_2 q + 3a_3 q^2 + \cdots$$

and

$$H''(q) = \sum_{k=2}^{\infty} k(k-1) a_k q^{k-2} = 2a_2 + 2 \cdot 3a_3 q + 3 \cdot 4a_4 q^2 + \cdots$$

Substituting these functions into Equation 5.22 and setting the coefficients of each term of the resulting power series separately to zero, we obtain the following relations:

$$2a_2 + (\epsilon - 1)a_0 = 0$$

$$2 \cdot 3a_3 + (\epsilon - 1 - 2)a_1 = 0$$

$$3 \cdot 4a_4 + (\epsilon - 1 - 2 \cdot 2)a_2 = 0$$

$$\cdot \ \cdot \ \cdot \ \cdot \ \cdot \ \cdot \ \cdot \ \cdot \ \cdot \ \cdot \ \cdot \ \cdot \ \cdot$$

In general,

$$(k + 1)(k + 2)a_{k+2} + (\epsilon - 1 - 2k)a_k = 0,$$

or

$$a_{k+2} = - \frac{\epsilon - 1 - 2k}{(k + 1)(k + 2)} a_k \tag{5.23}$$

Equation 5.23, which is called the *recursion relation*, shows how all of the coefficients can be determined from a_0 and a_1. Thus, a_0 and a_1 are the two arbitrary constants required for the general solution of the second order differential equation. The solution of Equation 5.22 may be written as the sum of two series, one containing all odd powers of q and the other all even powers:

$$H(q) = a_0 \left(1 + \frac{a_2}{a_0} q^2 + \frac{a_4}{a_2} \cdot \frac{a_2}{a_0} q^4 + \frac{a_6}{a_4} \cdot \frac{a_4}{a_2} \cdot \frac{a_2}{a_0} q^6 + \cdots \right)$$

$$+ a_1 \left(q + \frac{a_3}{a_1} q_3 + \frac{a_5}{a_3} \cdot \frac{a_3}{a_1} q^5 + \frac{a_7}{a_5} \cdot \frac{a_5}{a_3} \cdot \frac{a_3}{a_1} q^7 + \cdots \right) \tag{5.24}$$

If the series does not terminate, the trial solution,

$$\psi = e^{-q^2/2} H(q),$$

will diverge as q becomes infinite and hence will be an unsatisfactory wave function. This can be seen by examining the behavior of $H(q)$ as k gets large. From Equation 5.23,

$$\frac{a_{k+2}}{a_k} \to \frac{2}{k}$$

Note that $e^{q^2} = 1 + q^2 + \frac{1}{2}q^4 + \cdots + 2^k q^{2k}/k! + \cdots$, and the ratio of two consecutive terms as k gets large is

$$\frac{b_{k+1}}{b_k} \to \frac{2}{k}.$$

Thus, $H(q)$ goes as e^{q^2}, except for a constant factor, and

$$\psi \sim e^{-q^2/2} \cdot e^{q^2} \sim e^{q^2/2}, \quad \text{which diverges.}[7]$$

From this we conclude that *the series solution of $H(q)$ must be terminated to a polynomial in order to obtain a physically acceptable wave function.* The recursion relation, Equation 5.23, tells us that when $k = n$, such that $\epsilon = 2n + 1$, one of the series will terminate with a_n, since a_{n+2} and all higher coefficients will be zero. The other series must be eliminated by setting a_0 equal to zero if n is odd, or a_1 equal to zero if n is even. As a consequence of the termination of the series we obtain the energy eigenvalues from the condition

$$\epsilon = \frac{2E}{\omega \hbar} = 2n + 1, \tag{5.25}$$

that is,

$$E = \omega \hbar (n + \tfrac{1}{2}).$$

We may now label the energy states and the wave functions by means of the index n which indicates the degree of the polynomial appearing in the solution. That is,

$$\psi_n(q) = e^{-q^2/2} H_n(q) \quad \text{and} \quad E_n = \omega \hbar (n + \tfrac{1}{2}).$$

Note that the energy levels are equally spaced at intervals of $\omega \hbar$ as in the old quantum theory. (See Figure 5-16.) However, the ground state is *not* zero, as in the older theory, but is

$$E_0 = \tfrac{1}{2}\omega \hbar.$$

This energy is the so-called zero-point energy, which accounts for the fact that

---

[7] P. M. Morse and H. Feshbach, *Methods of Theoretical Physics.* McGraw-Hill Book Co., Inc., New York, 1953, Part II, p. 1640.

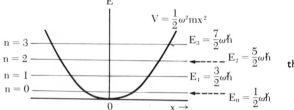

Figure 5-16. Energy levels of the harmonic oscillator.

elementary excitations can be present in a statistical system even when it is in thermodynamic equilibrium at absolute zero.

As in the case of the bound states of the square well, the harmonic oscillator provides another example of how the requirement of well-behaved wave functions results in energy quantization.

6. METHODS OF GENERATING THE HERMITE POLYNOMIALS

To illustrate how the Hermite polynomials can be obtained from the recursion relation, consider $H_3(q)$. From Equation 5.25 one obtains

$$\epsilon = 2(3) + 1 = 7.$$

Using $\epsilon = 7$ in Equation 5.23,

$$\frac{a_3}{a_1} = -\frac{7 - 1 - 2}{2 \cdot 3} = -\frac{2}{3}.$$

Then, from Equation 5.24,

$$H_3(q) = a_1(q - \tfrac{2}{3}q^3).$$

The value of a_1 is arbitrary and will ultimately be absorbed in the normalization factor of the wave function. However, a convenient convention is to choose a_1 (or a_0) such that the coefficient of the term of highest order is 2^n. In this example a_1 would then be -12.

A few of the Hermite polynomials normalized in this fashion are as follows:

$$H_0(q) = 1$$
$$H_1(q) = 2q$$
$$H_2(q) = 4q^2 - 2$$
$$H_3(q) = 8q^3 - 12q$$
$$H_4(q) = 16q^4 - 48q^2 + 12.$$

Note that $H_n(q)$ is odd or even as n is odd or even. In quantum mechanics the term *parity* is used to describe "even-ness" or "odd-ness." That is, $H_n(q)$ has even parity if n is even, and odd parity if n is odd. Since $e^{-q^2/2}$ is an even function, we note that $\psi_n(q)$ has the same parity as $H_n(q)$ and n.

In order to obtain two additional prescriptions for constructing the Hermite polynomials, let us first define the following function of two variables,

$$F(s, q) = \exp\left[q^2 - (s - q)^2\right] = \exp\left[s(2q - s)\right] = \sum_{n=0}^{\infty} A_n(q) \cdot \frac{s^n}{n!}. \quad (5.26)$$

The last expression is merely a Taylor expansion of $F(s, q)$ about $s = 0$. Thus, the $A_n(q)$ are given by

$$A_n(q) = \frac{\partial^n}{\partial s^n} F(s, q)\bigg]_{s=0} = e^{q^2} \frac{\partial^n}{\partial s^n} \left(e^{-(s-q)^2}\right)\bigg]_{s=0}.$$

But $\partial/\partial(s - q) = \partial/\partial s$, since q is held constant here. Then,

$$A_n(q) = e^{q^2} \frac{\partial^n}{\partial(s - q)^n} \left(e^{-(s-q)^2}\right)\bigg]_{s=0} = (-1)^n e^{q^2} \frac{\partial^n}{\partial q^n} \left(e^{-(s-q)^2}\right)\bigg]_{s=0}.$$

Or,

$$A_n(q) = (-1)^n e^{q^2} \frac{\partial^n}{\partial q^n} e^{-q^2}. \quad (5.27)$$

To show that the $A_n(q)$ are actually the Hermite polynomials, $H_n(q)$, we proceed as follows: Taking the partial derivatives of Equation 5.26,

$$\frac{\partial F}{\partial s} = 2(q - s)F \quad (5.28)$$

$$\frac{\partial F}{\partial q} = 2sF \quad (5.29)$$

$$\frac{\partial^2 F}{\partial q\, \partial s} = 2F + 2s\frac{\partial F}{\partial s}. \quad (5.30)$$

Adding Equations 5.28 and 5.29,

$$\frac{\partial F}{\partial s} + \frac{\partial F}{\partial q} = 2qF,$$

and differentiating this sum with respect to q,

$$\frac{\partial^2 F}{\partial q\, \partial s} + \frac{\partial^2 F}{\partial q^2} = 2F + 2q\frac{\partial F}{\partial q}. \quad (5.31)$$

Eliminating the mixed derivatives from Equations 5.30 and 5.31,

$$\frac{\partial^2 F}{\partial q^2} - 2q\frac{\partial F}{\partial q} + 2s\frac{\partial F}{\partial s} = 0. \quad (5.32)$$

Now, using the definition of F in Equation 5.26, we find $\partial F/\partial s$:

$$\frac{\partial F}{\partial s} = \sum_{n=0}^{\infty} A_n(q) \cdot \frac{ns^{n-1}}{n!} .$$ (5.33)

Substituting into Equation 5.32 we have

$$\sum_{n=0}^{\infty} \left(\frac{\partial^2 A_n}{\partial q^2} - 2q \frac{\partial A_n}{\partial q} + 2n\, A_n \right) \cdot \frac{s^n}{n!} = 0.$$

In order for this to hold for all s we must have

$$A_n'' - 2q\, A_n' + 2n\, A_n = 0,$$

where the primes signify differentiation with respect to the argument q. But the latter expression is identical with the Hermite equation, Equation 5.22, with ϵ replaced by its proper value of $2n + 1$. We therefore conclude that the A_n are the Hermite polynomials. Replacing A_n with H_n, Equations 5.26 and 5.27 then provide us with the following two new definitions of the Hermite polynomials:

$$F(s, q) = \exp\,[q^2 - (s - q)^2] = \exp\,[s(2q - s)] = \sum_{n=0}^{\infty} H_n(q) \cdot \frac{s^n}{n!}, \quad (5.34)$$

and

$$H_n(q) = (-1)^n e^{q^2} \frac{\partial^n}{\partial q^n}\, e^{-q^2}.$$ (5.35)

The first of these, Equation 5.34, says that if the generating function is expanded in a power series in s, the coefficient of s^n is the Hermite polynomial of order n divided by $n!$. The second, Equation 5.35, enables the calculation of the polynomial of any order by means of a sequence of differentiations.

PROBLEM 5-16

Calculate $H_3(q)$ and $H_4(q)$ by means of the three prescriptions given by Equations 5.24, 5.34, and 5.35.

Some useful identities may be easily derived which can provide a ready means of obtaining Hermite polynomials of other orders if $H_n(q)$ is known. By equating Equations 5.28 and 5.33 it follows that:

$$\sum_{n=0}^{\infty} H_n(q) \cdot \frac{s^{n-1}}{(n-1)!} + 2(s - q) \sum_{n=0}^{\infty} H_n(q) \cdot \frac{s^n}{n!} = 0,$$

or,

$$\sum_{n=0}^{\infty} H_n(q) \cdot \frac{s^{n-1}}{(n-1)!} + 2 \sum_{n=0}^{\infty} H_n(q) \cdot \frac{s^{n+1}}{n!} - 2q \sum_{n=0}^{\infty} H_n(q) \cdot \frac{s^n}{n!} = 0.$$

Collecting terms in like powers of s,

$$\sum_{n=0}^{\infty}\left[H_{n+1}(q)\cdot\frac{1}{n!}+\frac{2H_{n-1}(q)}{(n-1)!}-\frac{2qH_n(q)}{n!}\right]\cdot s^n=0.$$

In order for this to hold for arbitrary s, we must have

$$H_{n+1}(q)+2nH_{n-1}(q)-2qH_n(q)=0. \tag{5.36}$$

In like manner, by equating the expressions for $\partial F/\partial q$ from Equations 5.26 and 5.29, it follows that:

$$\sum_{n=0}^{\infty}H_n'(q)\cdot\frac{s^n}{n!}=2s\sum_{n=0}^{\infty}H_n(q)\cdot\frac{s^n}{n!},$$

from which we obtain

$$H_n'(q)=2nH_{n-1}(q). \tag{5.37}$$

The latter may be extended to higher derivatives as follows:

$$H_n''(q)=2\cdot 2n(n-1)H_{n-2}(q)=4n(n-1)H_{n-2}(q)$$

$$H_n'''(q)=8n(n-1)(n-2)H_{n-3}(q),\text{ and so on.}$$

PROBLEM 5-17

(a) Derive the following additional identities:

$$H_{n+1}(q)=2qH_n(q)-H_n'(q)$$

$$H_n(-q)=(-1)^nH_n(q).$$

(b) Verify the four identities by substituting the polynomials and their derivatives into the above expressions (for $n=4$).

7. NORMALIZATION OF THE HARMONIC OSCILLATOR WAVE FUNCTIONS

Letting N_n be the normalization coefficient to be determined, we write the wave function for the nth state of a linear harmonic oscillator as

$$\psi_n(q)=N_ne^{-q^2/2}H_n(q).$$

Then we require that

$$\int_{-\infty}^{\infty}|\psi_n(q)|^2\,dq=1.$$

At this point we will use the Dirac notation for such integrals over all of space (see section 13 of Chapter 4):

$$\langle \psi_n(q) \mid \psi_n(q) \rangle \equiv \int_{-\infty}^{\infty} \psi_n^*(q) \psi_n(q) \, dq,$$

where it is understood that the complex conjugate of the quantity to the left of the vertical bar is used in performing the integration. Then we may write the normalization integral as

$$\langle \psi_n \mid \psi_n \rangle = |N_n N_{n'}| \int_{-\infty}^{\infty} e^{-q^2} H_n(q) H_{n'}(q) \, dq = |N_n N_{n'}| \langle H_n(q) \mid H_{n'}(q) \rangle,$$

where it should be noted that the factor e^{-q^2} is understood in the last integral.

The condition $n' = n$ may be imposed at a later time. The integral may be evaluated in two ways. First, Equation 5.35 may be integrated by parts n times, assuming $n' > n$, to obtain

$$\langle H_n(q) \mid H_{n'}(q) \rangle = (-1)^{n'-n} 2^n n'! \int_{-\infty}^{\infty} H_0(q) \frac{d^{n'-n}}{dq^{n'-n}} e^{-q^2} \, dq = 0,$$

since e^{-q^2} and all of its derivatives vanish[8] for infinite q. For $n' = n$, the integral becomes

$$\langle H_n(q) \mid H_n(q) \rangle = 2^n n! \int_{-\infty}^{\infty} H_0(q) e^{-q^2} \, dq = 2^n n! \sqrt{\pi}.$$

Therefore, we note that the Hermite polynomials are *orthogonal* with respect to the weighting factor e^{-q^2}, since the integral vanishes for $n' \neq n$. For $n' = n$ the normalization factor is determined by the result,

$$|N_n|^2 \, 2^n n! \sqrt{\pi} = 1.$$

The normalized wave functions for the linear harmonic oscillator are then

$$\psi_n(q) = \left(\frac{1}{2^n n! \sqrt{\pi}} \right)^{\frac{1}{2}} e^{-q^2/2} H_n(q). \tag{5.38}$$

Normalization may also be accomplished by means of the definition in Equation 5.34. Define two functions F and G such that:

$$F(q, s) = \sum_n \frac{H_n(q)}{n!} s^n = e^{q^2 - (s-q)^2}$$

$$G(q, s') = \sum_{n'} \frac{H_{n'}(q)}{n'!} s'^{n'} = e^{q^2 - (s'-q)^2}.$$

---

[8] R. Courant and D. Hilbert, *Methods of Mathematical Physics.* Interscience Publishers, Inc., New York, 1953, Vol. I, p. 92.

Then,

$$\int_{-\infty}^{\infty} FGe^{-q^2}\, dq = \sum_{n,n'} s^n s'^{n'} \int_{-\infty}^{\infty} \frac{H_n(q)H_{n'}(q)}{n!\,n'!} e^{-q^2}\, dq = e^{2ss'} \int_{-\infty}^{\infty} e^{-(q-s-s')^2}\, dq.$$

Defining $u = q - s - s'$, the integral on the right becomes

$$\int_{-\infty}^{\infty} e^{-u^2}\, du = \sqrt{\pi}.$$

Thus,

$$\sum_{n,n'} \langle H_n(q) \mid H_{n'}(q) \rangle \frac{s^n}{n!} \cdot \frac{s'^{n'}}{n'!} = \sqrt{\pi}\, e^{2ss'}$$

$$= \sqrt{\pi}\left[1 + 2ss' + \frac{(2ss')^2}{2!} + \cdots + \frac{(2ss')^n}{n!} + \cdots \right].$$

$$= \sqrt{\pi} \sum_n \frac{(2ss')^n}{n!}.$$

It is immediately obvious that there can be no solution unless $n' = n$, since the two series expansions can be equal only when this condition is met. For $n' = n$ we have

$$\sum_n \left[\left(\frac{1}{n!}\right)^2 \langle H_n(q) \mid H_n(q) \rangle - \frac{2^n \sqrt{\pi}}{n!} \right] (ss')^n = 0.$$

Therefore,

$$\langle H_n(q) \mid H_n(q) \rangle = 2^n \sqrt{\pi}\, n!, \qquad (5.39)$$

and for $n' \neq n$, the integrand is zero as before.

It is sometimes convenient to use the normalization over x rather than the dimensionless parameter q. In this case the normalized wave function becomes

$$\psi_n(x) = \left(\frac{\sqrt{\alpha}}{2^n n! \sqrt{\pi}}\right)^{\frac{1}{2}} e^{-\alpha x^2/2} H_n(\sqrt{\alpha}\, x). \qquad (5.40)$$

Before closing this section it will be instructive to evaluate an integral of the type that is used to calculate expectation values and transition probabilities for the harmonic oscillator. Such integrals are called *matrix elements* in the matrix formulation of quantum mechanics. Consider the following matrix element of the operator x:

$$\langle \psi_n(x) \mid x\psi_{n'}(x) \rangle = \int_{-\infty}^{\infty} \psi_n^*(x) x\psi_{n'}(x)\, dx = \frac{1}{\sqrt{\alpha}} \int_{-\infty}^{\infty} \psi_n^*(q) q\psi_{n'}(q)\, dq$$

$$= \frac{1}{\sqrt{\alpha}} \langle \psi_n(q) \mid q\psi_{n'}(q) \rangle,$$

where the wave functions in Equation 5.40 are used for the integral on the left and those in Equation 5.38 are used for the integral on the right. Using the latter and the definitions in Equation 5.26, we write:

$$\langle \psi_n(q) \mid q\psi_{n'}(q) \rangle = \sum_{n,n'} \frac{s^n}{n!} \cdot \frac{s'^{n'}}{n'!} \langle H_n(q) \mid qH_{n'}(q) \rangle$$

$$= e^{2ss'} \int_{-\infty}^{\infty} e^{-(q-s-s')^2} (q - s - s') dq + (s + s') e^{2ss'} \int_{-\infty}^{\infty} e^{-(q-s-s')^2} dq$$

$$= \sqrt{\pi}(s + s')e^{2ss'}.$$

Then,

$$\sum_{n,n'} \left[\langle H_n(q) \mid qH_{n'}(q) \rangle \frac{s^n}{n!} \cdot \frac{s'^{n'}}{n'!} \right] = \sqrt{\pi} \left[\left(s + 2s^2s' + \frac{2^2 s^3 s'^2}{2!} + \cdots \right. \right.$$

$$\left. + \frac{2^n s^{n+1} s'^n}{n!} + \cdots \right) + \left(s' + 2ss'^2 + \frac{2^2 s^2 s'^3}{2!} + \cdots + \frac{2^n s^n s'^{n+1}}{n!} + \cdots \right) \Bigg].$$

By comparing series we see that the first series of the right member can be equated to the left member only when $n' = n - 1$; the second series on the right can be equated to the left only when $n' = n + 1$; for $n' = n$ the integral must vanish. The results are summarized below:

$$\left. \begin{array}{ll} \langle H_n \mid qH_{n+1} \rangle = 2^n(n + 1)!\sqrt{\pi} & \langle \psi_n \mid x\psi_{n+1} \rangle = \sqrt{\dfrac{n + 1}{2\alpha}} \\[2mm] \langle H_n \mid qH_n \rangle = 0 & \langle \psi_n \mid x\psi_n \rangle = 0 \\[2mm] \langle H_n \mid qH_{n-1} \rangle = 2^{n-1}n!\sqrt{\pi} & \langle \psi_n \mid x\psi_{n-1} \rangle = \sqrt{\dfrac{n}{2\alpha}} \end{array} \right\} \qquad (5.41)$$

The following example will illustrate further the method of evaluating expectation values by means of the orthogonality properties of the Hermite polynomials.

EXAMPLE

Calculate the expectation values $\langle x \rangle$, $\langle p_x \rangle$, $\langle T \rangle$, $\langle V \rangle$ and $\langle E \rangle$ for the linear harmonic oscillator. Utilizing the identities 5.36 and 5.37, we have

$$H_{n+1}(q) - 2qH_n(q) + 2nH_{n-1}(q) = 0. \qquad (5.42)$$

Multiplying by $H_n(q)e^{-q^2}$ and integrating over all of space,

$$\langle H_n \mid H_{n+1} \rangle - 2\langle H_n \mid qH_n \rangle + 2n\langle H_n \mid H_{n-1} \rangle = 0.$$

From the orthogonality property of the Hermite polynomials, it follows that

$\langle H_n | H_{n+1} \rangle = \langle H_n | H_{n-1} \rangle = 0.$ Therefore,

$$\langle H_n | q H_n \rangle = 0 \quad \text{and} \quad \langle x \rangle = \frac{1}{\sqrt{\alpha}} \langle H_n | q H_n \rangle = 0.$$

Now multiply Equation 5.42 by $H_n q e^{-q^2}$ and integrate:

$$\langle H_n | q H_{n+1} \rangle - 2 \langle H_n | q^2 H_n \rangle + 2n \langle H_n | q H_{n-1} \rangle = 0.$$

Substituting the values for $\langle H_n | q H_{n+1} \rangle$ and $\langle H_n | q H_{n-1} \rangle$ from Equation 5.41, we obtain

$$\langle H_n | q^2 H_n \rangle = 2^{n-1}(n+1)! \sqrt{\pi} + n \cdot 2^{n-1} n! \sqrt{\pi} = (2n+1) \cdot 2^{n-1} n! \sqrt{\pi}.$$

Then

$$\langle x^2 \rangle = \frac{1}{\alpha} \langle \psi_n | q^2 \psi_n \rangle = \frac{1}{2^n n! \sqrt{\pi \alpha}} \langle H_n | q^2 H_n \rangle = \frac{2n+1}{2\alpha}.$$

To find the expectation value of the momentum, we use the expression

$$\langle p_x \rangle = -i\hbar \sqrt{\alpha} \left\langle \psi_n \left| \frac{\partial \psi_n}{\partial q} \right. \right\rangle = +i\hbar \sqrt{\alpha} \langle \psi_n | q \psi_n \rangle - i\hbar \sqrt{2n} \sqrt{\alpha} \langle \psi_n | \psi_{n-1} \rangle,$$

where the identity given by Equation 5.37 was used in the last expression. Since the first integral has been already shown to be zero, we have

$$\langle p_x \rangle \sim \langle H_n | H_{n-1} \rangle = 0,$$

from the orthogonality relation. Now, to find $\langle p_x^2 \rangle$, recall that

$$p_x^2 = -\hbar^2 \frac{\partial^2}{\partial x^2} = -\alpha \hbar^2 \frac{\partial^2}{\partial q^2}.$$

Utilizing Equation 5.20 we note that

$$\langle p_x^2 \rangle = \alpha \hbar^2 \langle \epsilon - q^2 \rangle.$$

Utilizing our results above,

$$\langle p_x^2 \rangle = \alpha \hbar^2 \left(2n + 1 - \frac{2n+1}{2} \right) = \alpha \hbar^2 \left(\frac{2n+1}{2} \right).$$

Then,

$$\langle T \rangle = \frac{1}{2m} \langle p_x^2 \rangle = \frac{\alpha \hbar^2}{2m} (n + \tfrac{1}{2}) = \frac{\omega \hbar}{2} (n + \tfrac{1}{2}),$$

$$\langle V \rangle = \frac{\omega^2 m}{2} \langle x^2 \rangle = \frac{\omega^2 m}{2\alpha} (n + \tfrac{1}{2}) = \frac{\omega \hbar}{2} (n + \tfrac{1}{2}),$$

and

$$\langle E \rangle = \langle T \rangle + \langle V \rangle = \omega \hbar (n + \tfrac{1}{2}).$$

The correspondence between the quantum mechanical expectation values and their classical counterparts is again illustrated by the above example. A few of the normalized harmonic oscillator wave functions are:

$$\psi_0(q) = \left(\frac{1}{\sqrt{\pi}}\right)^{\frac{1}{2}} e^{-q^2/2}$$

$$\psi_1(q) = \left(\frac{2}{\sqrt{\pi}}\right)^{\frac{1}{2}} q e^{-q^2/2}$$

$$\psi_2(q) = \left(\frac{1}{2\sqrt{\pi}}\right)^{\frac{1}{2}} (2q^2 - 1) e^{-q^2/2}$$

$$\psi_3(q) = \left(\frac{1}{3\sqrt{\pi}}\right)^{\frac{1}{2}} (2q^3 - 3q) e^{-q^2/2}$$

$$\psi_4(q) = \frac{1}{2} \left(\frac{1}{6\sqrt{\pi}}\right)^{\frac{1}{2}} (4q^4 - 12q^2 + 3) e^{-q^2/2}$$

$$\psi_5(q) = \frac{1}{2} \left(\frac{1}{15\sqrt{\pi}}\right)^{\frac{1}{2}} (4q^5 - 20q^3 + 15q) e^{-q^2/2}$$

Some of these functions are sketched in Figure 5-17.

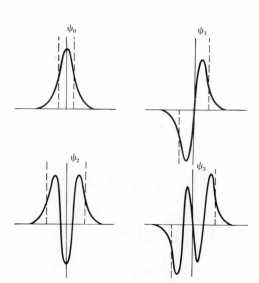

Figure 5-17. The first four harmonic oscillator wave functions. The dashed lines indicate the classical limits of motion.

8. APPLICATIONS TO MOLECULAR VIBRATIONS

An obvious application of the harmonic oscillator solutions is to the vibrational spectra of linear diatomic molecules. A schematic diagram of the potential energy versus internuclear distance for such a molecule is shown in Figure 5-18. Note that this potential may be approximated by the parabolic harmonic oscillator potential, at least for small oscillations about the equilibrium position, r_0. Therefore, to the extent that this approximation is valid, the vibrational energy states are given by

$$E_n = \omega \hbar (n + \tfrac{1}{2}),$$

where $\omega^2 = k/\mu$, the force constant divided by the reduced mass of the two atoms. A better fit to the real potential is obtained by using the so-called *Morse potential*,

$$V(r) = V_0[1 - e^{-a(r-r_0)}]^2 - V_0.$$

Using this potential, the energy levels become[9]

$$E_n = \omega \hbar (n + \tfrac{1}{2}) - \left[\frac{\omega \hbar}{2\sqrt{V_0}} (n + \tfrac{1}{2})\right]^2.$$

The effect of the second term is to depress the levels as n increases. Hence, the interval between successive levels diminishes instead of remaining constant as in the harmonic oscillator. This is shown schematically in Figure 5-19.

Another model that sometimes proves useful in molecular problems is the double oscillator with a barrier between the two oscillator wells (see Figure 5-20). A particle having energy $E < V_0$ can tunnel from one oscillator to the

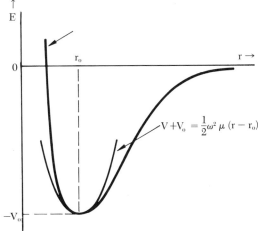

Figure 5-18. Potential energy versus interatomic distance for a diatomic molecule.

$-V + V_0 = \frac{1}{2}\omega^2 \mu (r - r_0)^2$

[9] See L. Pauling and E. B. Wilson, *Introduction to Quantum Mechanics*, McGraw-Hill Book Co., 1935, p. 271.

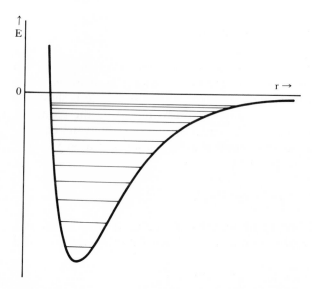

Figure 5-19. The vibrational levels for the Morse potential.

other at a rate determined by the height and breadth of the barrier and the total energy of the particle. This model can be used to describe the oscillation of the nitrogen atom between two stable configurations of the ammonia molecule. The geometrical arrangement of this molecule is that of a tetrahedron whose basal plane contains the three hydrogen atoms at the vertices of an equilateral triangle. The nitrogen atom is at the fourth vertex, either above or below the basal plane (see Figure 5-21). The oscillation of the nitrogen atom between these two equilibrium positions is a tunneling process whose qualitative features are similar to those of the double square well of section 4. A rigorous solution of the double oscillator shows that the energy splitting of the two lowest levels, for example, is[10]

$$\Delta E = 4\epsilon_0 \sqrt{\frac{V_0}{\pi\epsilon_0}} \, e^{-V_0/\epsilon_0} ,$$

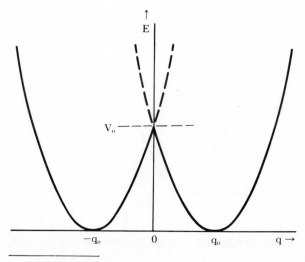

Figure 5-20. The double oscillator. Each energy level of a single oscillator is split into two levels. For $E < V_0$ tunneling occurs.

---

[10] Merzbacher, *Quantum Mechanics*, John Wiley and Sons, New York, 1961, p. 64 cf.

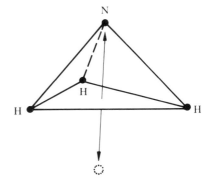

Figure 5-21. The ammonia molecule. The nitrogen atom oscillates between two stable configurations by means of tunneling. In each configuration it experiences a potential that is approximately harmonic.

where $V_0 \gg \epsilon_0 = \frac{1}{2}\omega\hbar$, the ground state energy of either single oscillator. This corresponds to a tunneling frequency of $\Delta E/h$. The ratio of the tunneling time τ to the period of either oscillator, T, is

$$\frac{\tau}{T} \sim \frac{1}{2}\sqrt{\frac{\pi\epsilon_0}{V_0}} \cdot e^{V_0/\epsilon_0}.$$

It is evident that the tunneling process is greatly affected by the barrier height in the exponential factor.

PROBLEM 5-18

(a) Calculate the vibrational energy states for the CO molecule in the harmonic oscillator approximation. Take the force constant to be $k = 1.87 \times 10^5$ dynes/cm. Will any of these states be excited at room temperature?

(b) Assuming a rigid bond length of 1.13 Å, calculate the moment of inertia of the molecule about an axis perpendicular to its symmetry axis. Using the result for the quantization of the rigid rotator, find the energies of the first few states for this rotational mode. Will these states be excited at room temperature?

(c) Estimate the energies of the first few rotational states about the symmetry axis of the molecule. Will any of these states be excited at room temperature?

(Ans.: (a) $E_1 = 0.13$ eV; (b) $E_1 \sim 4 \times 10^{-4}$ eV; (c) $E_1 \sim 4$ eV.)

9. THE HARMONIC OSCILLATOR: OPERATOR METHOD

In this section we will solve the oscillator equation (Eq. 5.20), by the method of Dirac.[11] The method is a powerful one, frequently used for algebraic

[11] P. A. M. Dirac, *Quantum Mechanics*. Clarendon Press, Oxford, 1958, 4th edition, section 34.

equations involving non-commuting operators, and is commonly called the factorization method. Since the harmonic oscillator has already been studied in great detail in the previous sections, it provides an excellent subject for illustrating this important operator method.

For convenience let us define the dimensionless position and momentum operators,

$$q = \sqrt{\alpha}x,$$

$$p = \frac{1}{\hbar\sqrt{\alpha}} p_x = -\frac{i}{\sqrt{\alpha}}\frac{\partial}{\partial x} = -\frac{i}{\sqrt{\alpha}}\frac{\partial}{\partial q}\frac{\partial q}{\partial x} = -i\frac{\partial}{\partial q},$$

$$p^2 = -\frac{\partial^2}{\partial q^2}.$$

In terms of these operators, Equation 5.20 becomes

$$(p^2 + q^2)\psi = \epsilon\psi. \tag{5.43}$$

Since p and q are operators associated with x and p_x (see Problem 4-22), we do not expect them to commute. In particular,

$$[p, q] = \frac{1}{\hbar}[p_x, x] = -i. \tag{5.44}$$

If the commutator in 5.44 were zero, Equation 5.43 could be factored into $(q + ip)(q - ip)\psi = \epsilon\psi$. However, because of the non-commutivity of p and q we can write:

$$(q + ip)(q - ip) = q^2 + p^2 + 1$$
$$(q - ip)(q + ip) = q^2 + p^2 - 1.$$

Adding,

$$p^2 + q^2 = \tfrac{1}{2}[(q + ip)(q - ip) + (q - ip)(q + ip)].$$

If we define two new operators,

$$a = \frac{1}{\sqrt{2}}(q + ip)$$

$$a^\dagger = \frac{1}{\sqrt{2}}(q - ip),$$

then

$$p^2 + q^2 = aa^\dagger + a^\dagger a,$$

and the Schrödinger equation, Equation 5.20, may be written as

$$(aa^\dagger + a^\dagger a)\psi = \epsilon\psi. \tag{5.45}$$

The operators a and $a^\dagger$ satisfy the following commutation relation:

$$[a, a^\dagger] = aa^\dagger - a^\dagger a = 1.$$

Using the latter result, the Schrödinger equation (5.45) can be written in the following two additional but equivalent forms:

$$aa^\dagger \psi = \left(\frac{\epsilon}{2} + \frac{1}{2}\right)\psi \qquad (5.46)$$

$$a^\dagger a\psi = \left(\frac{\epsilon}{2} - \frac{1}{2}\right)\psi. \qquad (5.47)$$

Before actually solving the Schrödinger equation, we will show that we can construct an infinite set of eigenfunctions if we can somehow find one solution of any of Equations 5.45, 5.46, or 5.47. Operating on Equation 5.46 with $a^\dagger$ from the left, we have

$$a^\dagger aa^\dagger \psi = \left(\frac{\epsilon}{2} + \frac{1}{2}\right)a^\dagger \psi. \qquad (5.48)$$

Using the commutation relation, $a^\dagger a = aa^\dagger - 1$,

$$(aa^\dagger - 1)a^\dagger \psi = \left(\frac{\epsilon}{2} + \frac{1}{2}\right)a^\dagger \psi$$

$$aa^\dagger (a^\dagger \psi) = \left(\frac{\epsilon}{2} + \frac{3}{2}\right)a^\dagger \psi. \qquad (5.49)$$

That is, if ψ is an eigenfunction of $aa^\dagger$ corresponding to the energy eigenvalue $\epsilon/2 + \frac{1}{2}$, then $a^\dagger \psi$ is also an eigenfunction of $aa^\dagger$, but it has the new eigenvalue $\epsilon/2 + \frac{3}{2}$. Since $a^\dagger \psi$ is also an eigenfunction of $a^\dagger a$, from Equation 5.45 we see that

$$(aa^\dagger + a^\dagger a)(a^\dagger \psi) = (\epsilon + 2)(a^\dagger \psi),$$

where we have used Equations 5.48 and 5.49.

Similarly, operating on Equation 5.49 with $a^\dagger$, we obtain

$$a^\dagger (aa^\dagger)(a^\dagger \psi) = (a^\dagger a)(a^\dagger)^2\psi = (aa^\dagger - 1)(a^\dagger)^2\psi = \left(\frac{\epsilon}{2} + \frac{3}{2}\right)(a^\dagger)^2\psi.$$

Then,

$$aa^\dagger (a^\dagger)^2\psi = \left(\frac{\epsilon}{2} + \frac{5}{2}\right)(a^\dagger)^2\psi,$$

and $(aa^\dagger + a^\dagger a)(a^\dagger)^2\psi = (\epsilon + 4)(a^\dagger)^2\psi$.

This process may be continued indefinitely, since the harmonic oscillator has an infinite set of states. Since each successive operation with $a^\dagger$ raises the oscillator to the next higher energy state, $a^\dagger$ is called the *raising operator*. In like manner we will show that a is a *lowering operator* for the same set of functions such that the following relations will hold:

$$a^\dagger \psi_n = C_+ \psi_{n+1} \quad \text{and} \quad a\psi_n = C_- \psi_{n-1}, \tag{5.50}$$

where C_+ and C_- are constants of proportionality which will be determined later.

Let a operate from the left on Equation 5.47:

$$aa^\dagger a\psi = (1 + a^\dagger a)a\psi = \left(\frac{\epsilon}{2} - \frac{1}{2}\right)a\psi$$

or

$$a^\dagger a(a\psi) = \left(\frac{\epsilon}{2} - \frac{3}{2}\right)a\psi.$$

That is, if ψ is an eigenfunction of $a^\dagger a$ with eigenvalue $(\epsilon/2 - \frac{1}{2})$, then $a\psi$ is also an eigenfunction of $a^\dagger a$ but with the new eigenvalue $(\epsilon/2 - \frac{3}{2})$. The lowering process can also be repeated by successive application of the lowering operator a as shown in Table 5-1. However, in contrast with the raising operation, the

TABLE 5-I Eigenfunctions Obtained by Means of the Raising and Lowering Operators

The lowering operator is subject to the constraint, $\epsilon \geq 2n' + 1$

Oscillator Wave function	Eigenvalue of $aa^\dagger$	Eigenvalue of $a^\dagger a$	Eigenvalue of $p^2 + q^2$
ψ	$\frac{\epsilon}{2} + \frac{1}{2}$	$\frac{\epsilon}{2} - \frac{1}{2}$	ϵ
$a^\dagger \psi$	$\frac{\epsilon}{2} + \frac{3}{2}$	$\frac{\epsilon}{2} + \frac{1}{2}$	$\epsilon + 2$
$(a^\dagger)^2 \psi$	$\frac{\epsilon}{2} + \frac{5}{2}$	$\frac{\epsilon}{2} + \frac{3}{2}$	$\epsilon + 4$
...	...	...	...
$(a^\dagger)^n \psi$	$\frac{\epsilon}{2} + \frac{2n+1}{2}$	$\frac{\epsilon}{2} + \frac{2n-1}{2}$	$\epsilon + 2n$
...	...	...	...
$a\psi$	$\frac{\epsilon}{2} - \frac{1}{2}$	$\frac{\epsilon}{2} - \frac{3}{2}$	$\epsilon - 2$
$a^2\psi$	$\frac{\epsilon}{2} - \frac{3}{2}$	$\frac{\epsilon}{2} - \frac{5}{2}$	$\epsilon - 4$
...	...	...	...
$a^{n'}\psi$	$\frac{\epsilon}{2} - \frac{2n'-1}{2}$	$\frac{\epsilon}{2} - \frac{2n'+1}{2}$	$\epsilon - 2n'$

lowering operator can be applied only a finite number of times because the eigenvalue, $\epsilon/2 - (2n' + 1)/2$, cannot become negative. The physical reason for this is that the energy states of the harmonic oscillator must always correspond to positive energies. This, of course, was a well-known classical result and could be accepted in quantum mechanics on the grounds of the correspondence principle. However, we can readily show its validity in quantum mechanics by examining the expectation value of the energy. Thus,

$$\langle \epsilon \rangle = \int_{-\infty}^{\infty} \psi^* \epsilon \psi \, dq = \int_{-\infty}^{\infty} \psi^* (p^2 + q^2) \psi \, dq = -\int_{-\infty}^{\infty} \psi^* \frac{d^2 \psi}{dq^2} \, dq + \int_{-\infty}^{\infty} \psi^* q^2 \psi \, dq.$$

Integrating the first integral by parts,

$$\langle \epsilon \rangle = -\left[\psi^* \frac{d\psi}{dq} \right]_{-\infty}^{\infty} + \int_{-\infty}^{\infty} \frac{d\psi^*}{dq} \frac{d\psi}{dq} \, dq + \int_{-\infty}^{\infty} |\psi|^2 \, q^2 \, dq.$$

The first term vanishes since ψ must be well behaved. Then,

$$\langle \epsilon \rangle = \int_{-\infty}^{\infty} \left(\left| \frac{d\psi}{dq} \right|^2 + q^2 \, |\psi|^2 \right) dq \geq 0,$$

since the integrand is positive definite.

If we designate the ground state wave function of the oscillator as ψ_0, then we must have

$$a\psi_0 = 0, \tag{5.51}$$

since there can be no lower state. Then, from Equation 5.47, we have

$$a^\dagger a \psi_0 = 0 = \left(\frac{\epsilon}{2} - \frac{1}{2} \right) \psi_0,$$

$$\epsilon = 1 = \frac{2E_0}{\omega \hbar}$$

or

$$E_0 = \tfrac{1}{2}\omega \hbar,$$

the ground state energy of the oscillator.

We will now see how easily the ground state wave function is obtained by this method. Inserting the explicit form of the lowering operator,

$$a = \frac{1}{\sqrt{2}} \left(q + \frac{\partial}{\partial q} \right),$$

into Equation 5.51, we obtain the first order differential equation,

$$\frac{d\psi_0}{dq} + q\psi_0 = 0.$$

This may be integrated immediately to obtain the ground state wave function,

$$\psi_0 = C_0 e^{-q^2/2},$$

where the constant of integration, C_0, is chosen so as to normalize ψ_0. Then each higher state can be obtained by successive operations with the raising operator, $a^\dagger$. Thus,

$$\psi_1 = \frac{C_1}{C_0} a^\dagger \psi_0$$

$$\psi_2 = \frac{C_2}{C_1} a^\dagger \psi_1 = \frac{C_2}{C_0} (a^\dagger)^2 \psi_0$$

. . . .

$$\psi_n = \frac{C_n}{C_0} (a^\dagger)^n \psi_0. \tag{5.52}$$

It will be shown below that the normalization constant C_n, when divided by $(2^n)^{\frac{1}{2}}$, becomes the same factor as that indicated by N_n in the discussion preceding Equation 5.38. It should be pointed out that Equation 5.52 provides still another method of obtaining the Hermite polynomials, since

$$\psi_n = N_n e^{-q^2/2} H_n(q) = C_n (a^\dagger)^n e^{-q^2/2} = \frac{C_n}{\sqrt{2^n}} \left(q - \frac{\partial}{\partial q} \right)^n e^{-q^2/2}.$$

That is,

$$H_n(q) = e^{q^2/2} \left(q - \frac{\partial}{\partial q} \right)^n e^{-q^2/2}. \tag{5.53}$$

Let us now evaluate the normalization coefficients above. We require first that

$$\langle \psi_0 \mid \psi_0 \rangle = 1 = |C_0|^2 \int_{-\infty}^{\infty} e^{-q^2} dq = |C_0|^2 \sqrt{\pi}.$$

Then, $C_0 = (1/\sqrt{\pi})^{\frac{1}{2}}$.

To determine the ratio C_n/C_0 we use Equation 5.52 and form the integral

$$\langle \psi_n \mid \psi_n \rangle = \left| \frac{C_n}{C_0} \right|^2 \langle (a^\dagger)^n \psi_0 \mid (a^\dagger)^n \psi_0 \rangle.$$

Recall from the definitions of a and $a^\dagger$ that $a^* = a^\dagger$; that is, the operators are complex conjugates. They are, in fact, Hermitian conjugates or adjoints of one another. The full meaning of this property will not be evident until one uses matrix representations, but its usefulness here lies in the fact that it permits us "to move an operator past the vertical bar" as follows:[12]

$$\langle a^\dagger \psi \mid a^\dagger \psi \rangle = \langle \psi \mid a a^\dagger \psi \rangle = \langle a a^\dagger \psi \mid \psi \rangle.$$

---

[12] The equivalence of these expressions may be established by performing partial integrations.

Using the property of the adjoint,

$$\langle \psi_n \mid \psi_n \rangle = \left| \frac{C_n}{C_0} \right|^2 \langle \psi_0 \mid (aa^\dagger)^n \psi_0 \rangle.$$

This expression can be greatly simplified by using the following identity:

$$a(a^\dagger)^n = (a^\dagger)^n a + n(a^\dagger)^{n-1} \qquad (5.54)$$

Its validity can be readily deduced from the commutation relation, that is,

$$
\begin{aligned}
a(a^\dagger)^n &= aa^\dagger(a^\dagger)^{n-1} = (1 + a^\dagger a)(a^\dagger)^{n-1} = (a^\dagger)^{n-1} + a^\dagger a(a^\dagger)^{n-1} \\
&= (a^\dagger)^{n-1} + a^\dagger[(a^\dagger)^{n-2} + a^\dagger a(a^\dagger)^{n-2}] \\
&= 2(a^\dagger)^{n-1} + (a^\dagger)^2[(a^\dagger)^{n-3} + a^\dagger a(a^\dagger)^{n-3}] \\
&= 3(a^\dagger)^{n-1} + (a^\dagger)^3 a(a^\dagger)^{n-3} \\
&\quad \cdots \cdots \cdots \cdots \cdots \\
&= n(a^\dagger)^{n-1} + (a^\dagger)^n a \ .
\end{aligned}
$$

The result of applying Equation 5.54 to ψ_0 is:

$$a(a^\dagger)^n \psi_0 = n(a^\dagger)^{n-1} \psi_0,$$

since $a\psi_0 = 0$. Then,

$$
\begin{aligned}
a^2(a^\dagger)^n \psi_0 &= na(a^\dagger)^{n-1} \psi_0 = n(n-1)(a^\dagger)^{n-2} \psi_0 \\
a^3(a^\dagger)^n \psi_0 &= n(n-1)a(a^\dagger)^{n-2} \psi_0 = n(n-1)(n-2)(a^\dagger)^{n-3} \psi_0.
\end{aligned}
$$

$$\cdots \cdots \cdots \cdots$$

$$a^n(a^\dagger)^n \psi_0 = n! \psi_0. \qquad (5.55)$$

Using Equation 5.55,

$$\langle \psi_n \mid \psi_n \rangle = \left| \frac{C_n}{C_0} \right|^2 n! \langle \psi_0 \mid \psi_0 \rangle = \left| \frac{C_n}{C_0} \right|^2 n! \, |C_0|^2 \sqrt{\pi}.$$

Therefore,

$$C_n = \left(\frac{1}{n! \sqrt{\pi}} \right)^{\frac{1}{2}}$$

and

$$N_n = \left(\frac{1}{2^n n! \sqrt{\pi}} \right)^{\frac{1}{2}}.$$

The general form for the wave function for the nth state is then

$$\psi_n(q) = \left(\frac{1}{2^n n! \sqrt{\pi}} \right)^{\frac{1}{2}} e^{-q^2/2} H_n(q)$$

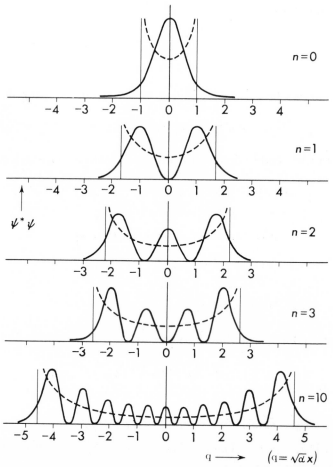

Figure 5-22. Probability densities for a few harmonic oscillator states. The dotted curves show the classical probabilities corresponding to the same energies. (From Chalmers W. Sherwin, *Introduction to Quantum Mechanics*, Holt, Rinehart and Winston, Inc., 1959. Used with permission.)

or

$$\psi_n(x) = \left(\frac{\sqrt{\alpha}}{2^n n! \sqrt{\pi}}\right)^{\frac{1}{2}} e^{-q^2/2} H_n(q), \qquad (5.56)$$

where

$$H_n(q) = (-1)^n e^{q^2} \frac{\partial^n}{\partial q^n} e^{-q^2}$$

and

$$\alpha = \frac{\omega m}{\hbar} = \frac{\sqrt{km}}{\hbar}.$$

PROBLEM 5-19

(a)   Consider a particle of mass m performing simple harmonic motion along the x-axis. Show that the classical probability of finding the particle in the element dx is

$$\rho(x)\, dx = \frac{dx}{\pi\sqrt{A^2 - x^2}},$$

where A is the amplitude of the motion.

(b) Show that the classical amplitudes corresponding to the quantized oscillator energies are given by

$$A_n^2 = \frac{(2n + 1)}{\alpha},$$

where $\alpha = \omega m/\hbar$.

(c) Verify the fact that the classical limit on the motion of the particle in the state $\psi_n(q)$ is $q = \sqrt{2n + 1}$.

PROBLEM 5-20

Using the normalized, linear harmonic oscillator wave functions perform the necessary integrations to verify the following:

(a) that ψ_0 and ψ_1 are normalized.

(b) that ψ_0 and ψ_1 are orthogonal.

(c) that $\langle \psi_0 | q^2 | \psi_0 \rangle = \frac{1}{2}$.

(d) that $\langle \psi_1 | q^2 | \psi_1 \rangle = \langle \psi_1 | p^2 | \psi_1 \rangle = \frac{3}{2}$.

(e) that $\langle \psi_0 | q | \psi_1 \rangle = \langle \psi_1 | q | \psi_0 \rangle = \dfrac{\sqrt{2}}{2}$.

PROBLEM 5-21

(a) Carry out the necessary differentiations and integrations to find the following expectation values for the first excited state of the harmonic oscillator: $\langle x \rangle$, $\langle x^2 \rangle$, $\langle p_x \rangle$, and $\langle p_x^2 \rangle$.

(b) Using the results of (a), find $\langle T \rangle$, $\langle V \rangle$, and $\langle E \rangle$ for the same state.

(c) Calculate the minimum uncertainty product $\Delta x \cdot \Delta p_x$ for this state.

PROBLEM 5-22

A harmonic oscillator has a time-dependent wave function which is a superposition of its ground state and first excited state eigenfunctions, that is

$$\Psi(x, t) = \frac{1}{\sqrt{2}}[\Psi_0(x, t) + \Psi_1(x, t)].$$

Find the expectation value of the energy.

PROBLEM 5-23

Where is the maximum of the probability of $\psi_0(x)$, the ground state harmonic oscillator wave function? How does this compare with the classical value?

PROBLEM 5-24

Find the minimum uncertainty product, $\Delta x \cdot \Delta p_x$, for the nth state of the harmonic oscillator.

10. EXPECTATION VALUES IN THE OPERATOR FORMALISM

It is useful to evaluate the proportionality constants in Equation 5.50 before obtaining the matrix elements and expectation values of section 7 by the operator method. Using Equation 5.52,

$$\psi_{n+1} = \frac{C_{n+1}}{C_n} a^\dagger \psi_n,$$

or

$$a^\dagger \psi_n = \frac{C_n}{C_{n+1}} \psi_{n+1} = \left(\frac{(n+1)!}{n!}\right)^{\frac{1}{2}} \psi_{n+1} = \sqrt{n+1}\, \psi_{n+1}. \qquad (5.57)$$

Also, operating on Equation 5.52 with the lowering operator,

$$a\psi_n = \frac{C_n}{C_0} a(a^\dagger)^n \psi_0.$$

Using the identity (Eq. 5.54),

$$a\psi_n = \frac{C_n}{C_0}\left[(a^\dagger)^n a + n(a^\dagger)^{n-1}\right]\psi_0$$

$$= n\frac{C_n}{C_0}(a^\dagger)^{n-1}\psi_0, \quad \text{(since } a\psi_0 = 0\text{)},$$

$$= n\frac{C_n}{C_0}\cdot\frac{C_0}{C_{n-1}}\psi_{n-1}, \quad \text{(from Eq. 5.52)},$$

$$= n\left(\frac{(n-1)!}{n!}\right)^{\frac{1}{2}}\psi_{n-1}.$$

Thus,

$$a\psi_n = \sqrt{n}\psi_{n-1}. \qquad (5.58)$$

It is convenient to transform p and q by means of the definitions of a and $a^\dagger$. Thus,

$$\left.\begin{aligned} q &= \frac{1}{\sqrt{2}}\,(a + a^\dagger) \\[2mm] p &= \frac{i}{\sqrt{2}}\,(a^\dagger - a). \end{aligned}\right\} \tag{5.59}$$

Now,

$$\langle a \rangle = \langle \psi_n \mid a\psi_n \rangle = \sqrt{n}\,\langle \psi_n \mid \psi_{n-1} \rangle = 0$$

$$\langle a^\dagger \rangle = \langle \psi_n \mid a^\dagger \psi_n \rangle = \sqrt{n+1}\,\langle \psi_n \mid \psi_{n+1} \rangle = 0$$

$$\langle aa^\dagger \rangle = \langle \psi_n \mid aa^\dagger \psi_n \rangle = \left(\frac{\epsilon_n}{2} + \frac{1}{2} \right) \langle \psi_n \mid \psi_n \rangle = n + 1$$

$$\langle a^\dagger a \rangle = \langle \psi_n \mid a^\dagger a\psi_n \rangle = \left(\frac{\epsilon_n}{2} - \frac{1}{2} \right) \langle \psi_n \mid \psi_n \rangle = n.$$

From this it readily follows that:

$$\langle q \rangle = \frac{1}{\sqrt{2}}\,(\langle a \rangle + \langle a^\dagger \rangle) = 0$$

$$\langle p \rangle = \frac{i}{\sqrt{2}}\,(\langle a^\dagger \rangle - \langle a \rangle) = 0$$

$$\langle q^2 \rangle = \tfrac{1}{2}(\langle aa \rangle + \langle aa^\dagger \rangle + \langle a^\dagger a \rangle + \langle a^\dagger a^\dagger \rangle)$$

$$= \tfrac{1}{2}(2n + 1)$$

$$= n + \tfrac{1}{2}$$

$$\langle p^2 \rangle = \frac{-1}{2}\,(\langle a^\dagger a^\dagger \rangle - \langle a^\dagger a \rangle - \langle aa^\dagger \rangle + \langle aa \rangle)$$

$$= n + \tfrac{1}{2}.$$

Therefore,

$$\langle x \rangle = \frac{1}{\sqrt{\alpha}}\,\langle q \rangle = 0$$

$$\langle p_x \rangle = \hbar\sqrt{\alpha}\,\langle p \rangle = 0$$

$$\langle x^2 \rangle = \frac{1}{\alpha}\,\langle q^2 \rangle = \frac{1}{\alpha}\,(n + \tfrac{1}{2})$$

$$\langle p_x^2 \rangle = \alpha\hbar^2\langle p^2 \rangle = \alpha\hbar^2(n + \tfrac{1}{2})$$

$$\langle T \rangle = \frac{1}{2m}\,\langle p_x^2 \rangle = \frac{1}{2}\,\omega\hbar(n + \tfrac{1}{2})$$

$$\langle V \rangle = \frac{\omega^2 m}{2}\,\langle x^2 \rangle = \tfrac{1}{2}\omega\hbar(n + \tfrac{1}{2})$$

$$\langle E \rangle = \frac{\omega\hbar}{2}\,\langle \epsilon \rangle = \frac{\omega\hbar}{2}\,(\langle p^2 \rangle + \langle q^2 \rangle) = \omega\hbar(n + \tfrac{1}{2}).$$

The reader should compare these results with those obtained in section 7 using the algebra of the Hermite polynomials.

PROBLEM 5-25

Calculate the matrix elements $\langle \psi_n \mid x\psi_{n+1} \rangle$ and $\langle \psi_n \mid x\psi_{n-1} \rangle$ by means of the operators introduced in this section.

The formulation of the harmonic oscillator problem in terms of the dimensionless operators a and $a^\dagger$ has the important consequence that the results are independent of the specific dynamical variables of the original problem. In other words, the results are independent of the *representation*. Thus, whenever the Hamiltonian of a dynamical system can be written in the form

$$\mathscr{H} = q^2 + p^2 = aa^\dagger + a^\dagger a,$$

then the algebraic results of this section hold, regardless of the exact nature of these variables. An important application of this method is that of the creation-annihilation operator representation of field theory. In that formulation $a^\dagger$ creates a particle state, a annihilates a state, and $a^\dagger a$ counts or numbers the states. The latter is called the *number operator* and its function is readily seen as follows: $a^\dagger a\psi_n = a^\dagger \sqrt{n}\, \psi_{n-1} = \sqrt{n} \cdot \sqrt{n}\, \psi_n = n\psi_n$. Therefore, the eigenvalue of the operator $a^\dagger a$ is the integer which numbers the state of the eigenfunction upon which it operates. In physical problems involving the interactions of several different kinds of particles, it is customary to define a set of creation-annihilation operators for the states of each class of particle. Thus, in solids the many-body properties are conveniently treated as the collective normal-mode excitations of systems of harmonic oscillators. The quanta of excitation are called *phonons* for the set of oscillators associated with lattice vibrations. In a magnetic system the normal modes are called spinwaves and the quantized excitations are known as *magnons*. In a physical problem involving the interaction of electromagnetic waves with a magnetic solid, one might require pairs of creation-annihilation operators for photons, phonons, and magnons.

SUGGESTED REFERENCES

S. Borowitz, *Fundamentals of Quantum Mechanics*. W. A. Benjamin, Inc., 1967, particularly Chapter 10.

R. H. Dicke, and J. P. Wittke, *Introduction to Quantum Mechanics*. Addison-Wesley Publishing Co., Inc., Reading, Mass., 1960.

R. M. Eisberg, *Fundamentals of Modern Physics*. John Wiley and Sons, Inc., New York, 1961.

R. B. Leighton, *Principles of Modern Physics*. McGraw-Hill Book Co., Inc., New York, 1959.

P. T. Matthews, *Introduction to Quantum Mechanics*, 2nd ed. McGraw-Hill Book Co., New York, 1968.

E. Merzbacher, *Quantum Mechanics*. John Wiley and Sons, Inc., New York, 1961.

D. Park, *Introduction to the Quantum Theory*. McGraw-Hill Book Co., Inc., New York, 1964.

L. Pauling, and E. B. Wilson, *Introduction to Quantum Mechanics*. McGraw-Hill Book Co., Inc., New York, 1935.

J. L. Powell, and B. Crasemann, *Quantum Mechanics*. Addison-Wesley Publishing Co., Inc., Reading, Mass., 1961.

F. K. Richtmyer, E. H. Kennard, and J. N. Cooper, *Introduction to Modern Physics*, 6th ed. McGraw-Hill Book Co., New York, 1969.

D. S. Saxon, *Elementary Quantum Mechanics*. Holden-Day, Inc., San Francisco, 1968.

L. I. Schiff, *Quantum Mechanics*, 3rd ed. McGraw-Hill Book Co. Inc., New York, 1969.

R. L. White, *Basic Quantum Mechanics*. McGraw-Hill Book Co., Inc., New York, 1966.

K. Ziock, *Basic Quantum Mechanics*. John Wiley and Sons, Inc., New York, 1969.

CHAPTER 6

THE FORMAL STRUCTURE OF QUANTUM MECHANICS

Many of the concepts and some of the mathematical intricacies of quantum mechanics have been introduced in earlier chapters. Up to this point the emphasis has been upon making the development plausible to the reader. In the present chapter we will alter this approach and, instead, focus our attention on the logical structure of quantum mechanics and the formulation of matrix mechanics. The attractiveness of this formal approach is that once the postulates are accepted, the theory has a logical cohesiveness which has great intellectual appeal, perhaps more so than the heuristic arguments used earlier. Regardless of the plausibility or the logic of its postulates, the over-riding justification for quantum mechanics is its tremendous success in accounting for the observed behavior of atomic, nuclear, elementary particle, and quasi-particle systems.

I. THE POSTULATES OF QUANTUM MECHANICS

Although there is general agreement as to what ideas constitute the basic postulates of quantum mechanics, there is great diversity in the way these postulates are presented in different textbooks. Even the numbering of the postulates is quite arbitrary, since two or more postulates are sometimes combined into a single statement. In what follows, numerous theorems and definitions have been included in the discussions of the postulates, which are here presented as five in number. Wherever the term "particle" is used, the phrase "quantum mechanical system" could replace it, since "particle" is being used here in the sense of typifying the simplest quantum system.

Postulate I

The dynamical states of a particle can be described by a wave function which contains all that can be known about the particle.

A complex wave function can be used to represent a particle (see section 6 of Chapter 4), this wave function being the fundamental mathematical entity from which all observable quantities can be computed. The wave function may be denoted by $\Psi(\vec{r}, t)$ in the coordinate representation or by $\phi(\vec{k})$ in the momentum representation. Ψ and ϕ, which are related by the Fourier transforms, Equation 4.27, are equivalent representations of the particle. Not all functions make suitable wave functions. In order to be physically admissible, both the function and its derivative must be finite, continuous, and single-valued everywhere. In addition, the integral of the square of its modulus over all of space must be finite so that the wave function can be normalized. Suppose, for example, that

$$\int_\tau |\Psi(\vec{r}, t)|^2 \, dx \, dy \, dz = A,$$

where A is a finite number. Then a wave function which is normalized to unity may be constructed by writing the new function as $\Psi_N = (1/\sqrt{A})\Psi(\vec{r}, t)$. (See Problems 4-8, 4-9, 4-10, and 4-15.) A quantum mechanical state is defined by a specific set of values for the independent variables, such as $\Psi(\vec{r}_1, t_1)$ or $\phi(\vec{k}_1)$. Although neither $\Psi(\vec{r}_1, t_1)$ nor $\phi(\vec{k}_1)$ is directly observable, the square of its amplitude for a particular state is proportional to the probability that the particle will be observed in that state. We defined the probability density in Equation 4.18 as

$$\rho(\vec{r}_1, t_1) = \frac{|\Psi(\vec{r}_1, t_1)|^2}{\int_\tau |\Psi(\vec{r}_1, t_1)|^2 \, dx \, dy \, dz}.$$

For Ψ normalized, the probability of finding the particle in the volume element $d\tau$ centered around $\vec{r}_1$ is

$$dP = \rho(\vec{r}_1, t) \, d\tau = |\Psi(\vec{r}_1, t)|^2 \, d\tau. \tag{6.1}$$

It follows that the probability for finding the particle *somewhere* in space, that is, the total probability, is unity. Thus,

$$P = \int_\tau dP = \int_\tau |\Psi(\vec{r}, t)|^2 \, d\tau = 1,$$

when Ψ is properly normalized.

Definition

Two non-zero wave functions are said to be *orthogonal* if their scalar product is zero. Thus, the orthogonality condition is,

$$\int_\tau \Psi_i^* \Psi_j \, d\tau = 0.$$

Wave functions that are both normalized and orthogonal satisfy the combined orthonormality condition,

$$\int_\tau \Psi_i^* \Psi_j \, d\tau = \delta_{ij},\tag{6.2}$$

where δ_{ij} is unity for $i = j$, and zero for $i \neq j$.

Postulate 2

The superposition principle is valid for functions representing physically admissible states.

Wave functions can be superposed to form a new wave function which is, itself, a physically valid representation of a possible state of the particle. This is the fundamental assumption underlying the representation of a particle by a wave packet, where the Fourier integral provides the means for superposing a continuum of states. (See sections 7 and 8 of Chapter 4.) In the case of discrete states, the superposition postulate says that any linear combination of the functions describing these states is also a possible physical state. In general, then, such a state may be expressed as,

$$\Psi = \sum_i c_i \Psi_i,\tag{6.3}$$

where the c_i are the expansion coefficients and the Ψ_i are the known discrete states.

Postulate 3

The Schrödinger equation describes the behavior of a wave function in space and time.

If a wave function $\Psi(\vec{r}, t)$ is known at some particular time, then its development in time is given by the Schrödinger wave equation (see section 9 of Chapter 4),

$$\mathscr{H}\Psi(\vec{r}, t) = i\hbar \frac{\partial}{\partial t} \Psi(\vec{r}, t),\tag{6.4}$$

where $\mathscr{H}$ is the Hamiltonian operator for the system. In the particular case of a time-independent potential, Equation 6.4 is separable in space and time, and $\Psi(\vec{r}, t)$ may be written as the product of a spatial function and a time function. The spatial part is an energy eigenfunction; that is, it is one of the solutions of the energy eigenequation, Equation 4.57,

$$\mathscr{H}\psi_{E_n}(\vec{r}) = E_n \psi_{E_n}(\vec{r}),\tag{6.5}$$

where the eigenvalue E_n is the energy of the nth state. Since the energy E_n remains constant in time, the solutions are known as stationary states. The

n^{th} particular solution of Equation 6.4 is

$$\Psi_n(\vec{r}, t) = \psi_{E_n}(\vec{r})e^{-iE_n t/\hbar}. \tag{6.6}$$

Utilizing the superposition principle, a general solution of Equation 6.4 may then be expanded in terms of the solutions just obtained, namely,

$$\Psi(\vec{r}, t) = \sum_n c_n \psi_{E_n}(\vec{r})e^{-iE_n t/\hbar}. \tag{6.7}$$

The sufficient condition for Equation 6.7 to be a valid expansion for a general wave function Ψ is that the functions ψ_{E_n} form a complete, orthonormal set of functions. This matter will be discussed further in section 3 of this chapter.

Postulate 4

Each dynamical variable q can be directly associated with a linear, Hermitian operator Q. The only possible result of a measurement of the observable q is one of the eigenvalues of the operator Q.

Definition

The operator Q is *linear* if it commutes with constants and if it obeys the distributive law. Specifically, if

$$Q(c\psi) = cQ(\psi),$$

and if

$$Q(\psi_1 + \psi_2) = Q(\psi_1) + Q(\psi_2),$$

then Q is a linear operator.

Definition

An operator is *Hermitian* if it satisfies the following equality,

$$\int_\tau (Q\psi)^* \psi \, d\tau = \int_\tau \psi^* (Q\psi) \, d\tau.$$

In the Dirac notation of Equation 4.58, the Hermitian property is

$$\langle Q\psi \mid \psi \rangle = \langle Q\psi \mid \psi \rangle^* = \langle \psi \mid Q\psi \rangle. \tag{6.8}$$

A more general statement is obtained by letting $\psi = a\psi_1 + b\psi_2$. Then Equation 6.8 becomes

$$\langle Q(a\psi_1 + b\psi_2) \mid a\psi_1 + b\psi_2 \rangle = \langle a\psi_1 + b\psi_2 \mid Q(a\psi_1 + b\psi_2) \rangle.$$

Using the properties of linear operators this becomes

$$a^*a\langle Q\psi_1 \,|\, \psi_1\rangle + a^*b\langle Q\psi_1 \,|\, \psi_2\rangle + b^*a\langle Q\psi_2 \,|\, \psi_1\rangle + b^*b\langle Q\psi_2 \,|\, \psi_2\rangle$$
$$= a^*a\langle \psi_1 \,|\, Q\psi_1\rangle + a^*b\langle \psi_1 \,|\, Q\psi_2\rangle + b^*a\langle \psi_2 \,|\, Q\psi_1\rangle + b^*b\langle \psi_2 \,|\, Q\psi_2\rangle.$$

Then,

$$a^*a(\langle Q\psi_1 \,|\, \psi_1\rangle - \langle \psi_1 \,|\, Q\psi_1\rangle) + b^*b(\langle Q\psi_2 \,|\, \psi_2\rangle - \langle \psi_2 \,|\, Q\psi_2\rangle)$$
$$+ a^*b(\langle Q\psi_1 \,|\, \psi_2\rangle - \langle \psi_1 \,|\, Q\psi_2\rangle) + b^*a(\langle Q\psi_2 \,|\, \psi_1\rangle - \langle \psi_2 \,|\, Q\psi_1\rangle) = 0 \quad (6.9)$$

Since a and b are arbitrary, and their relative phase is arbitrary, the only way Equation 6.9 can be satisfied for all possible values of a and b is for each term to be identically zero. This immediately gives Equation 6.8 and the more general statement of the Hermitian property,

$$\langle \psi_1 \,|\, Q\psi_2\rangle = \langle Q\psi_1 \,|\, \psi_2\rangle. \tag{6.10}$$

A Hermitian operator is frequently called a *self-adjoint operator*. This terminology arises from the fact that an equation similar to Equation 6.10 can be written for any operator Q, provided that the *Hermitian adjoint** of Q, denoted by $Q^\dagger$, is used when it operates on the other function. Thus,

$$\langle \psi_1 \,|\, Q\psi_2\rangle = \langle Q^\dagger\psi_1 \,|\, \psi_2\rangle. \tag{6.11}$$

When Q is Hermitian, $Q^\dagger = Q$, which reduces Equation 6.11 to the result above. In the notation of Dirac, integrals involving Hermitian operators are frequently written as

$$\langle \psi| \, Q \, |\psi\rangle,$$

to indicate that the operator, being Hermitian, can operate equally well on the function to the left or to the right.

Now let us consider a measurement of the dynamical variable q which is represented by a linear, Hermitian operator Q. The fourth postulate says that the measured value must correspond to one of the eigenvalues of Q, but it does not tell us which one. In order to see what is implied here, let us suppose that we wish to measure the energy of a hypothetical state which can be represented by Equation 6.7, where the only non-zero expansion coefficients are c_1 and c_3, which are real numbers. That is,

$$\Psi(\vec{r}, t) = c_1\psi_{E_1}(\vec{r})e^{-iE_1t/\hbar} + c_3\psi_{E_3}(\vec{r})e^{-iE_3t/\hbar} = c_1\Psi_1 + c_3\Psi_3, \tag{6.12}$$

where normalization requires that $c_1^2 + c_3^2 = 1$. A measurement of the energy must yield a result which can be expressed mathematically by operating on Ψ with the Hamiltonian operator. Thus,

$$\mathcal{H}\Psi = c_1\mathcal{H}\Psi_1 + c_3\mathcal{H}\Psi_3 = c_1E_1\Psi_1 + c_3E_3\Psi_3. \tag{6.13}$$

---

* A precise definition of the Hermitian adjoint will be given in section 7.

The same result can be obtained by operating on Ψ with the energy operator $i\hbar(\partial/\partial t)$:

$$i\hbar \frac{\partial}{\partial t} \Psi(\vec{r}, t) = i\hbar \left[c_1 \psi_{E_1}(\vec{r}) \frac{\partial}{\partial t} e^{-iE_1 t/\hbar} + c_3 \psi_{E_3}(\vec{r}) \frac{\partial}{\partial t} e^{-iE_3 t/\hbar} \right]$$

$$= c_1 E_1 \Psi_1 + c_3 E_3 \Psi_3. \tag{6.14}$$

Equations 6.13 and 6.14 are evidently not eigenequations, since we cannot express the right hand side as the product of a scalar times the original function. In this example the only accessible eigenvalues of $\mathscr{H}$ are E_1 and E_3, so we conclude that a measurement of the energy will yield either the value E_1 or E_3, and not some intermediate or average value. The probability of measuring the value E_1 is $c_1^2/(c_1^2 + c_3^2)$ (or just c_1^2 if Ψ is normalized) and the probability of measuring E_3 is $c_3^2/(c_1^2 + c_3^2)$. The act of measurement, then, forces the system into one of its eigenstates, where it will remain unless it is disturbed. Once a system is known to be in an eigenstate, the result of a repeated measurement can be predicted with certainty.*

The previous discussion assumed, of course, that no time-dependent forces act upon the system. Should the Hamiltonian be time-dependent, an additional complication arises, since the Schrödinger equation is then no longer separable. We will assume that in such a case the wave function can still be expanded as in Equation 6.7 if we but add the assumption that the expansion coefficients, c_i, are also time-dependent (this will be done in Chapter 9). The act of measuring the energy will again yield one of the energy eigenvalues, say E_j, but in contrast with the previous case, the system will no longer remain in the jth state. The time dependence in the coefficients will cause some coefficients to grow at the expense of others, or even to oscillate. Therefore, successive measurements of the energy will not, in general, yield the same eigenvalue.

Theorem

The eigenvalues of a Hermitian operator are real.

PROOF. Given, $Q\psi_n = q_n\psi_n$, and $Q = Q^\dagger$. For convenience let $\langle \psi_n \mid \psi_n \rangle = 1$. The expectation value is $\langle \psi_n \mid Q\psi_n \rangle = \langle \psi_n \mid q_n\psi_n \rangle = q_n\langle \psi_n \mid \psi_n \rangle = q_n$. Also, $\langle Q\psi_n \mid \psi_n \rangle = \langle q_n\psi_n \mid \psi_n \rangle = q_n^*\langle \psi_n \mid \psi_n \rangle = q_n^*$. But the left hand sides are equal by the Hermitian property, so we have

$$q_n^* = q_n,$$

or q_n is real.

Theorem

The eigenfunctions of a Hermitian operator are orthogonal if they correspond to distinct eigenvalues.

---

* See the discussion of the Stern-Gerlach experiment in section 11 of Chapter 3.

PROOF. Given, $Q\psi_i = q_i\psi_i$ and $Q\psi_j = q_j\psi_j$, where $q_i \neq q_j$ and $Q = Q^\dagger$. Then, writing the integral,

$$\langle \psi_i \,|\, Q\psi_j \rangle = \langle \psi_i \,|\, q_j\psi_j \rangle = q_j\langle \psi_i \,|\, \psi_j \rangle.$$

Also,

$$\langle Q\psi_i \,|\, \psi_j \rangle = \langle q_i\psi_i \,|\, \psi_j \rangle = q_i\langle \psi_i \,|\, \psi_j \rangle,$$

where the fact that the eigenvalue is real for a Hermitian operator has been used. The left hand integrals of each line are equal by the Hermitian property, so by subtraction we obtain

$$0 = (q_j - q_i)\langle \psi_i \,|\, \psi_j \rangle.$$

Since $q_i \neq q_j$, we must have $\langle \psi_i \,|\, \psi_j \rangle = 0$.

Therefore the functions ψ_i and ψ_j are orthogonal (see the definition following Postulate 1). In section 4 it will be shown that if there are degenerate eigenfunctions (that is, two or more eigenfunctions have the same eigenvalue), these functions can be orthogonalized.

Postulate 5

The expectation value of a measurement of the variable q is given mathematically as

$$\langle q \rangle = \frac{\langle \Psi \,|\, Q\Psi \rangle}{\langle \Psi \,|\, \Psi \rangle}. \tag{6.15}$$

If Ψ is an eigenfunction of Q such that $Q\Psi = q\Psi$, then Ψ is said to be a *pure state*, and the expectation value of Q is simply the eigenvalue q each time the measurement is made. Suppose, however, that Ψ is a *mixed state* such as that given in Equation 6.12, namely, the superposition of two orthonormal eigenstates of Q. Then,

$$\Psi = c_1\Psi_1 + c_3\Psi_3,$$

$$Q\Psi_1 = q_1\Psi_1$$

$$Q\Psi_3 = q_3\Psi_3$$

and

$$\langle \Psi_i \,|\, \Psi_j \rangle = \delta_{ij}.$$

The expression for the expectation value of q becomes:

$$\langle q \rangle = \frac{\langle c_1\Psi_1 + c_3\Psi_3 \,|\, Q(c_1\Psi_1 + c_3\Psi_3) \rangle}{\langle c_1\Psi_1 + c_3\Psi_3 \,|\, c_1\Psi_1 + c_3\Psi_3 \rangle} = \frac{c_1^2 q_1 + c_3^2 q_3}{c_1^2 + c_3^2}.$$

Note that this number turns out to be a weighted average of the accessible eigenvalues of Q, although a single measurement must always correspond to just one of the eigenvalues of Q. Consequently, we interpret the expectation value as either the average value of many measurements of q on the same system, or the average value of the same measurement on many identical systems.

If we generalize these results to a mixed state that is a superposition of all of the eigenstates of Q, the wave function is given by Equation 6.3, and the expectation value becomes

$$\langle q \rangle = \frac{\sum_i q_i \, |c_i|^2}{\sum_i |c_i|^2} \, . \tag{6.16}$$

The probability that a single measurement will yield the eigenvalue q_i is

$$\frac{|c_i|^2}{\sum_i |c_i|^2} \, .$$

It has already been established that eigenvalues must be real—since they correspond to physical measurements—and we have seen that the use of Hermitian operators guarantees that this requirement is met. It is evident also from Equation 6.16 that the expectation value must be real, but this can be shown formally as follows.

Corollary

The necessary and sufficient condition for a real expectation value is that the dynamical variable be represented by a Hermitian operator.

PROOF. Given, $\langle Q \rangle = \langle Q \rangle^*$, that is, the expectation value of Q is real. Then, $\langle \psi \,|\, Q\psi \rangle = \langle \psi \,|\, Q\psi \rangle^* = \langle Q\psi \,|\, \psi \rangle$, which demonstrates that Q is Hermitian.

PROBLEM 6-1

An operator is said to be *skew Hermitian* or *anti-Hermitian* if $\langle Q\psi \,|\, \psi \rangle = -\langle \psi \,|\, Q\psi \rangle$. Show that the eigenvalues of a skew Hermitian operator are pure imaginary.

PROBLEM 6-2

Prove that the following operators are Hermitian for well-behaved wave functions:
(a) $p = -i\hbar \nabla$
(b) $E = i\hbar \dfrac{\partial}{\partial t}$
(c) x

PROBLEM 6-3

If A is a non-Hermitian operator, show that $(A + A^\dagger)$ and $i(A - A^\dagger)$ are both Hermitian. This is a useful result since the operator A can then be written as the following linear combination of two Hermitian operators:

$$A = \tfrac{1}{2}(A + A^\dagger) + \frac{1}{2i}[i(A - A^\dagger)].$$

PROBLEM 6-4

If A, B, and C are Hermitian operators, transform the following expressions so as to eliminate the daggers:
(a) $(ABC)^\dagger =$ (d) $(AB - BA)^\dagger =$
(b) $(A^n)^\dagger =$ (e) $(i[A, B])^\dagger =$
(c) $(AB + BA)^\dagger =$ (f) $(A + B)^\dagger =$
Which of the above are Hermitian?

PROBLEM 6-5

Show that the variance, defined by $\langle Q^2 \rangle - \langle Q \rangle^2$, is zero only when ψ is an eigenfunction of Q. (Let $\psi = \sum_i c_i \psi_i$ where the ψ_i satisfy $Q\psi_i = q_i\psi_i$.)

2. MEASUREMENTS OF COMPATIBLE OBSERVABLES: COMMUTING OPERATORS

Consider two operators P and Q associated with the observables p and q, respectively. We wish to know what to expect if a measurement of p is followed by a measurement of q and, conversely, if a measurement of q is followed by a measurement of p. These two expectation values are expressed mathematically by Equation 6.15 as

$$\langle QP \rangle = \langle \psi \,|\, QP\psi \rangle \quad \text{and} \quad \langle PQ \rangle = \langle \psi \,|\, PQ\psi \rangle,$$

provided that ψ is normalized. Now, if ψ is simultaneously an eigenfunction of each operator then we have the relations

$$P\psi = p_1\psi$$
$$Q\psi = q_1\psi,$$

and the two expectation values become:

$$\langle QP \rangle = p_1\langle \psi \,|\, Q\psi \rangle = p_1 q_1$$
$$\langle PQ \rangle = q_1\langle \psi \,|\, P\psi \rangle = q_1 p_1 = p_1 q_1.$$

Since p_1 and q_1 are simply numbers, they commute and the order of the measurements is immaterial. Physically, this means that a system that is in simultaneous eigenstates of two operators will be undisturbed by any sequence of measurements of the observables associated with these two operators. Mathematically, we see that for this to occur the operators must commute; that is,

$$[Q, P] \equiv QP - PQ = 0. \qquad (6.17)$$

The bracket on the left is defined as the *commutator bracket* or simply the *commutator* of Q and P. (See section 11 of Chapter 4.)

Corollary

If two operators P and Q commute and either P or Q has non-degenerate eigenvalues, its eigenfunctions are also eigenfunctions of the other operator.

PROOF. Given $[P, Q] = 0$ and $P\psi_i = p_i\psi_i$, where all p_i are distinct. Then,

$$QP\psi_i = Qp_i\psi_i = p_i(Q\psi_i).$$

Using the commuting property,

$$QP\psi_i = PQ\psi_i = P(Q\psi_i).$$

Equating the right members,

$$P(Q\psi_i) = p_i(Q\psi_i).$$

This says that $Q\psi_i$ is an eigenfunction of P with eigenvalue p_i, which will lead to a contradiction unless $Q\psi_i$ differs from ψ_i by a multiplicative constant. That is, we must have

$$Q\psi_i = q_i\psi_i.$$

Therefore, ψ_i is an eigenfunction of Q.

We will now examine the time dependence of an expectation value. From our previous definitions it is evident that an expectation value calculated for an eigenstate is a constant of the motion for the dynamical system described by that eigenstate. In the language of operators, a necessary condition for a dynamical variable to be a constant of the motion is that the operator associated with that variable commute with the Hamiltonian for the system. This may be shown as follows:

$$\frac{d}{dt}\langle Q \rangle = \frac{d}{dt}\langle \psi \mid Q\psi \rangle = \left\langle \frac{\partial \psi}{\partial t} \middle| Q\psi \right\rangle + \left\langle \psi \middle| \frac{\partial Q}{\partial t}\psi \right\rangle + \left\langle \psi \middle| Q\frac{\partial \psi}{\partial t} \right\rangle$$

$$= \frac{-1}{i\hbar}\langle \mathscr{H}\psi \mid Q\psi \rangle + \left\langle \frac{\partial Q}{\partial t} \right\rangle + \frac{1}{i\hbar}\langle \psi \mid Q\mathscr{H}\psi \rangle,$$

where Equation 6.4 has been used. Then,

$$\frac{d}{dt}\langle Q \rangle = \frac{1}{i\hbar}\langle \psi \,|\, (Q\mathcal{H} - \mathcal{H}Q)\psi \rangle + \left\langle \frac{\partial Q}{\partial t} \right\rangle,$$

where the Hermitian property has been used. Therefore,

$$\frac{d}{dt}\langle Q \rangle = \frac{1}{i\hbar}\langle [Q, \mathcal{H}] \rangle + \left\langle \frac{\partial Q}{\partial t} \right\rangle. \tag{6.18}$$

This tells us that the expectation value will be a constant of the motion if the operator has no explicit time-dependence, and if it commutes with the Hamiltonian.

PROBLEM 6-6

Use Equation 6.18 to verify the following expressions for the linear harmonic oscillator:

(a) $\dfrac{d}{dt}\langle p_x \rangle = -\left\langle \dfrac{\partial V}{\partial x} \right\rangle$

(b) $m\dfrac{d}{dt}\langle x \rangle = \langle p_x \rangle.$

The statement that the variables in the classical equations of motion can be replaced by quantum mechanical expectation values is known as Ehrenfest's theorem.[1]

PROBLEM 6-7

Evaluate the following commutators:

(a) $[p, x^2]$ (c) $[p, p^2]$

(b) $[p^2, x^2]$ (d) $[p^2, V(x)]$

PROBLEM 6-8

(a) Using mathematical induction, show that

$$[x^n, p] = i\hbar n x^{n-1} \quad \text{and} \quad [x, p^n] = i\hbar n p^{n-1}.$$

(b) If f is a polynomial function, show that

$$[f(x), p] = i\hbar \frac{\partial f}{\partial x} \quad \text{and} \quad [x, f(p)] = i\hbar \frac{\partial f}{\partial p}.$$

---

[1] P. Ehrenfest, *Z. Physik* **45,** 455 (1927).

PROBLEM 6-9

For arbitrary operators A, B, and C, show that

$$[A, BC] = [A, B]C + B[A, C].$$

3. LINEAR VECTOR SPACES

We will now construct an abstract, linear vector space in which each vector in the space is a wave function of the kind we have been using to represent a possible physical state of a quantum system (in either coordinate or momentum space). A linear function space of this type is called a *Hilbert space*. Any complete set of orthonormal eigenfunctions can serve as the *basis functions*— that is, the unit vectors—for such a space. The concept of the superposition of states is easily visualized by regarding the expansion coefficients, Equation 6.7, as the projections along the basis vectors made by the vector representing the state in question. When the state vector lies along one of the basis vectors, the system is in the eigenstate characterized by that basis vector. By way of illustration, the complete set of orthonormal harmonic oscillator energy eigenfunctions given in section 7 of Chapter 5 forms a basis for a Hilbert space of infinite dimension. Any superposition of harmonic oscillator states may be regarded as a vector in that space.

Now let us examine the properties of a linear vector space. (1) There is a null vector. (2) The associative and commutative laws hold for addition. Thus, if we regard the ψ's as vectors, $\psi_1 + \psi_2 + \psi_3 = \psi_1 + (\psi_2 + \psi_3) = (\psi_1 + \psi_2) + \psi_3$. Also, $\psi_1 + \psi_2 = \psi_2 + \psi_1$. (3) There is a scalar product defined by $\langle \psi_1 | \psi_2 \rangle$, which is linear (distributive) but not, in general, commutative. Thus, the scalar or inner product of ψ_1 and ψ_2 is:

$$\int_\tau \psi_1^* \psi_2 \, d\tau = \langle \psi_1 | \psi_2 \rangle = \langle \psi_2 | \psi_1 \rangle^*. \qquad (6.19)$$

In a real vector space (or Euclidean space) the scalar product is real and symmetric in the variables. In a complex vector space (or unitary space) the scalar product is complex and possesses symmetry in the variables under complex conjugation. The latter is known as *Hermitian symmetry*.

To illustrate the linearity of the scalar product, let us form the product of ψ_1 and $(\psi_2 + \psi_3)$:

$$\langle \psi_1 | \psi_2 + \psi_3 \rangle = \langle \psi_1 | \psi_2 \rangle + \langle \psi_1 | \psi_3 \rangle.$$

Definition

Two vectors are *orthogonal* if their scalar product is zero, provided that neither vector is the null vector.

Definition

A set of functions is *linearly independent* if the linear equation,

$$\sum_i c_i \psi_i = 0,$$

is satisfied *only when all the c_i's are zero.* For example, if the Euclidean vector $a\hat{i} + b\hat{j} + c\hat{k}$ is 0 only when $a = b = c = 0$, then $\hat{i}$, $\hat{j}$, and $\hat{k}$ are linearly independent. In other words, no one of these quantities can be written as a linear combination of the others.

A set of N linearly independent, orthonormal vectors, $\{\hat{\psi}_i\}$, can provide a basis for an N-dimensional space, since any vector in the space may then be written as a linear combination of the N vectors. The set of vectors is said to be *complete* if it *spans the space.** That is to say, when the space is fully spanned by the basis, no conceivable vector can have a projection which would require an additional basis vector. Another way of describing completeness is as follows. A set of functions is said to be complete if the addition of any other function makes the set linearly dependent.

Theorem

The representation of an arbitrary vector on a given basis is unique.

PROOF.   Given, $\Psi = \sum_{i=1}^{N} c_i \hat{\psi}_i$, and suppose that Ψ is also represented by $\Psi = \sum_{i=1}^{N} c_i' \hat{\psi}_i$, where the same basis is used in both cases. Subtracting these two expressions yields

$$0 = \sum_{i=1}^{N} (c_i - c_i') \hat{\psi}_i.$$

Since the basis vectors $\hat{\psi}_i$ are linearly independent, this result requires that $c_i = c_i'$. Therefore, the representation is unique.

Theorem

The inner product of a vector with itself is positive definite.

$$\langle \Psi \mid \Psi \rangle = \left\langle \sum_i^{N} c_i \hat{\psi}_i \mid \sum_j^{N} c_j \hat{\psi}_j \right\rangle = \sum_{i,j} c_i^* c_j \langle \hat{\psi}_i \mid \hat{\psi}_j \rangle = \sum_{i,j} c_i^* c_j \, \delta_{ij}$$

$$= \sum_i |c_i|^2.$$

Therefore, $\langle \Psi \mid \Psi \rangle \geq 0$, where the equal sign holds only for the null vector.

---

* The subject of completeness is discussed in R. Courant and D. Hilbert, op. cit., Vol. I, p. 369.

Definition

The norm of a vector is defined by

$$\text{norm } \Psi = \sqrt{\langle \Psi \,|\, \Psi \rangle} = \sqrt{\sum_i |c_i|^2}.$$

A vector with unit norm is said to be normalized. Any non-zero vector can be normalized and the result is unique except for an arbitrary phase factor.

Theorem

Two vectors Ψ_a and Ψ_b satisfy the Schwartz inequality,

$$\langle \Psi_a \,|\, \Psi_a \rangle \langle \Psi_b \,|\, \Psi_b \rangle \geq |\langle \Psi_a \,|\, \Psi_b \rangle|^2. \qquad (6.20)$$

PROOF.   Let $\Psi = \Psi_a + b\Psi_b$. Then, from the previous theorem, the inner product of Ψ with itself is positive definite. That is,

$$0 \leq \langle \Psi \,|\, \Psi \rangle = \langle \Psi_a + b\Psi_b \,|\, \Psi_a + b\Psi_b \rangle$$

$$= \langle \Psi_a \,|\, \Psi_a \rangle + b\langle \Psi_a \,|\, \Psi_b \rangle + b^*\langle \Psi_b \,|\, \Psi_a \rangle + |b|^2 \langle \Psi_b \,|\, \Psi_b \rangle.$$

Although the choice of b is arbitrary, it is convenient to give it the value that will minimize the right-hand side of the inequality. Setting $(\partial/\partial b)\langle \Psi \,|\, \Psi \rangle = 0$, we have

$$b = -\frac{\langle \Psi_a \,|\, \Psi_b \rangle}{\langle \Psi_b \,|\, \Psi_b \rangle}.$$

Then,

$$0 \leq \langle \Psi_a \,|\, \Psi_a \rangle \langle \Psi_b \,|\, \Psi_b \rangle - |\langle \Psi_a \,|\, \Psi_b \rangle|^2,$$

which proves the theorem. If either Ψ_a or Ψ_b is the null vector the equality is satisfied trivially. Otherwise, the equality is satisfied only if Ψ_a and Ψ_b are parallel, that is, if $\Psi_b = \lambda \Psi_a$, where λ is a scalar.

In ordinary 3-space the Schwartz inequality simply reduces to the statement

$$(\vec{X} \cdot \vec{X})(\vec{Y} \cdot \vec{Y}) \geq (\vec{X} \cdot \vec{Y})^2,$$

or

$$|X| \cdot |Y| \geq \vec{X} \cdot \vec{Y},$$

and

$$\cos(\vec{X}, \vec{Y}) = \frac{\vec{X} \cdot \vec{Y}}{|X| \cdot |Y|}.$$

By analogy we may define the "angle" between Ψ_a and Ψ_b as

$$\cos(\Psi_a, \Psi_b) = \frac{\langle \Psi_a \,|\, \Psi_b \rangle}{[\langle \Psi_a \,|\, \Psi_a \rangle \langle \Psi_b \,|\, \Psi_b \rangle]^{\frac{1}{2}}}.$$

4. THE SCHMIDT ORTHOGONALIZATION PROCEDURE

In section 1 (under Postulate 4) it was shown that non-degenerate eigenfunctions of a Hermitian operator are automatically orthogonal, but the case of degeneracy was not discussed at that time. We will now define what is meant by degenerate eigenfunctions and will show how to construct an orthogonalized set from a set of degenerate functions.

Definition

An eigenvalue q is *n-fold degenerate* if there are n linearly independent eigenfunctions corresponding to this eigenvalue. The n functions are called degenerate eigenfunctions.

Theorem

Any linear combination of degenerate, linearly independent eigenfunctions is also an eigenfunction having the same eigenvalue.

PROOF.   Given the set of n functions ψ_i, all having the same eigenvalue q associated with the operator Q. That is, $Q\psi_i = q\psi_i$, for $i = 1, 2, \ldots, n$. The most general linear combination of these n degenerate functions is $\Psi = \sum_{i=1}^{n} c_i \psi_i$. Then,

$$Q\Psi = \sum_{i=1}^{n} c_i Q\psi_i = \sum_{i=1}^{n} c_i q \psi_i = q \sum_{i=1}^{n} c_i \psi_i = q\Psi,$$

which proves the theorem.

Corollary

In the case of n-fold degeneracy, linear combinations of the n degenerate eigenfunctions may be found so as to form n linearly-independent, orthogonal eigenfunctions corresponding to the same eigenvalue.

A prescription for obtaining an orthogonal set of linearly independent functions is known as the Schmidt orthogonalization method. It will be illustrated in what follows. Let a degenerate set of eigenfunctions be given by $\psi_1, \psi_2, \ldots, \psi_N$. Normalize ψ_1 and let it be the first function of the new set, $\hat{\psi}_1$. Define the scalar product,

$$\langle \hat{\psi}_1 \mid \psi_2 \rangle = a_{12}.$$

Choose

$$\hat{\psi}_2 = \psi_2 - a_{12}\hat{\psi}_1.$$

Since we require that $\langle \hat{\psi}_1 \mid \hat{\psi}_2 \rangle = 0$, let us see if our choice for $\hat{\psi}_2$ satisfies this condition. Thus,

$$\langle \hat{\psi}_1 \mid \hat{\psi}_2 \rangle = \langle \hat{\psi}_1 \mid \psi_2 \rangle - a_{12} \langle \hat{\psi}_1 \mid \hat{\psi}_1 \rangle = a_{12} - a_{12} = 0, \quad \text{as required.}$$

Therefore, $\hat{\psi}_1$ and $\hat{\psi}_2$ are orthogonal. $\hat{\psi}_2$ is normalized by setting $\langle \hat{\psi}_2 \mid \hat{\psi}_2 \rangle = 1$. Now define the scalar products,

$$\langle \hat{\psi}_1 \mid \psi_3 \rangle = a_{13}$$

and

$$\langle \hat{\psi}_2 \mid \psi_3 \rangle = a_{23}.$$

Now take $\hat{\psi}_3 = \psi_3 - a_{13}\hat{\psi}_1 - a_{23}\hat{\psi}_2$ and require that

$$\langle \hat{\psi}_1 \mid \hat{\psi}_3 \rangle = 0,$$
$$\langle \hat{\psi}_2 \mid \hat{\psi}_3 \rangle = 0,$$

and

$$\langle \hat{\psi}_3 \mid \hat{\psi}_3 \rangle = 1.$$

Let us test the orthogonality of $\hat{\psi}_3$ with $\hat{\psi}_1$ and $\hat{\psi}_2$:

$$\langle \hat{\psi}_1 \mid \hat{\psi}_3 \rangle = \langle \hat{\psi}_1 \mid \psi_3 \rangle - a_{13}\langle \hat{\psi}_1 \mid \hat{\psi}_1 \rangle - a_{23}\langle \hat{\psi}_1 \mid \hat{\psi}_2 \rangle$$
$$= a_{13} - a_{13} = 0.$$
$$\langle \hat{\psi}_2 \mid \hat{\psi}_3 \rangle = \langle \hat{\psi}_2 \mid \psi_3 \rangle - a_{13}\langle \hat{\psi}_2 \mid \hat{\psi}_1 \rangle - a_{23}\langle \hat{\psi}_2 \mid \hat{\psi}_2 \rangle$$
$$= a_{23} - a_{23} = 0.$$

Therefore, $\hat{\psi}_1$, $\hat{\psi}_2$, and $\hat{\psi}_3$ form an orthonormal set of functions (vectors). This process may be continued for the entire set of degenerate functions. When the ψ_i are ordinary vectors instead of functions in a function space, the inner product $\langle \psi_i \mid \psi_j \rangle$ reduces to the familiar dot product.

PROBLEM 6-10

Suppose that an energy level E_0 is three-fold degenerate. If the orthonormal, degenerate eigenfunctions corresponding to this energy level are ψ_1, ψ_2, and ψ_3, show that

$$\psi = \frac{1}{\sqrt{5}} (\psi_1 + 2\psi_2 + 2\psi_3)$$

is also an eigenfunction of H corresponding to the eigenvalue E_0.

PROBLEM 6-11

Construct a set of orthonormal vectors which are linear combinations of the vectors whose components are $(1, 0, 2, 2)$,

$(1, 1, 0, 1)$, and $(1, 1, 0, 0)$, respectively.

$\Big($Ans.: Taking $\hat{\psi}_1 = \pm\frac{1}{3}(1, 0, 2, 2)$, then

$$\hat{\psi}_2 = \pm \frac{\sqrt{2}}{6} (2, 3, -2, 1) \quad \text{and} \quad \hat{\psi}_3 = \pm \frac{\sqrt{2}}{6} (2, 1, 2, -3).\Big)$$

PROBLEM 6-12

Consider a two-dimensional isotropic oscillator such that $\omega = \omega_x = \omega_y$. Construct normalized, linearly independent wave functions for the ground state and the first two excited levels. (Hint: the wave function for a degenerate level is a linear combination of all of the degenerate functions.)

5. LINEAR TRANSFORMATIONS

Once we have constructed a function space for representing any physical state of our quantum system, a question arises with regard to the interpretation of the role of the operators that we have been using. The operator equation,

$$\Psi_b = Q\Psi_a,$$

may now be thought of as a linear transformation in the vector space, such that the vector Ψ_a is transformed into the vector Ψ_b by the action of the linear operator Q. The eigenequation, Equation 4.56, thus appears as a special case in which the direction of Ψ_a is not changed under the action of the operator and Ψ_b is simply Ψ_a multiplied by a scalar, that is,

$$q_a\Psi_a = Q\Psi_a.$$

Two important mathematical operators which must be incorporated in the formalism are the *null operator* and the *identity operator*. Their properties are immediately evident from the following:

$$0\Psi_a = 0,$$
$$1\Psi_a = \Psi_a.$$

The product of two operators is defined in terms of the effect produced by performing successively the transformations associated with these operators. Thus, if $\Psi_c = P\Psi_b$ and $\Psi_b = Q\Psi_a$, then $\Psi_c = P(Q\Psi_a) = PQ\Psi_a = R\Psi_a$. This suggests that

$$R = PQ.$$

We define the product of two operators in this sense, that R operating on Ψ_a

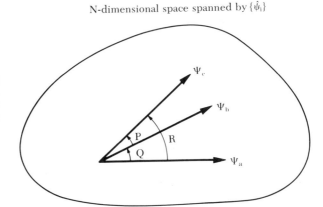

N-dimensional space spanned by $\{\hat{u}_i\}$

Figure 6-1 Schematic diagram of two successive transformations. R has the same effect upon the vector Ψ_a as the successive operations by Q and P.

is equivalent to operating first with Q and then with P. This is shown schematically in Figure 6-1.

If for every Ψ_a there is a unique vector Ψ_b given by $\Psi_b = Q\Psi_a$, and if for every Ψ_b there is a unique vector Ψ_a, given by $\Psi_a = P\Psi_b$, then it follows that

$$\Psi_a = P\Psi_b = PQ\Psi_a = 1\Psi_a.$$

Here we have the statement that

$$PQ = 1, \tag{6.21}$$

which means that if an operation by Q is followed by an operation with P, the net result is an operation with the identity operator. Equation 6.21 permits us to define the inverses of P and Q, provided that these operators are *non-singular* in the following sense:

Definition

An operator Q is non-singular if the equation $Q\Psi = 0$ is satisfied only by the null vector.

If P and Q are both non-singular, it follows from Equation 6.21 that

$$P = Q^{-1} \quad \text{and} \quad Q = P^{-1},$$

where P is the *left* inverse of Q and Q is the *right* inverse of P. If one can show that the operators P and Q commute, then the inverses defined above are valid for multiplication from either the left or the right.

6. DIRAC BRA-KET NOTATION

Dirac called the vectors of function space *ket vectors* or simply kets. In his notation the vector Ψ_a could be written as $|\Psi_a\rangle$ or as $|a\rangle$, where the label a

generally denotes one or more eigenvalues or quantum numbers. The basis functions are called *eigenkets* since they are the eigenfunctions of the older terminology. An arbitrary ket can be written as a linear combination of basis kets, such as

$$|\psi\rangle = c_1 |1\rangle + c_2 |2\rangle + \cdots = \sum_i c_i |i\rangle.$$

A linear operator acts on a ket from the left. Thus, the eigenequation takes the form,

$$Q |a\rangle = q_a |a\rangle. \tag{6.22}$$

The Hermitian conjugate of each ket is a *bra vector* which is written as $\langle a|$. The bras comprise a space having the same dimensionality as the ket space and which is dual to it. Since there is a bra for every ket, and vice versa, there is a one-to-one correspondence between these two spaces. Operators act upon bra vectors from the right, so that the equivalent form for the eigenbra equation is,

$$\langle a| Q^\dagger = q_a \langle a|, \tag{6.23}$$

where the Hermitian conjugate of Q is used. Note that multiplication from the left by the scalar q_a is a meaningful operation, whereas the statements $Q \langle a|$ or $Q^\dagger \langle a|$ are not defined.

The scalar product of a ket with itself is defined as the product of the ket with its corresponding bra. That is, the scalar product of $|a\rangle$ and $|a\rangle$ is

$$\langle a \mid a \rangle.$$

In general, the scalar product of $|b\rangle$ by $|a\rangle$ is

$$\langle a \mid b \rangle,$$

which is not the same as the product of $|a\rangle$ by $|b\rangle$. In agreement with Equation 6.19,

$$\langle a \mid b \rangle = \langle b \mid a \rangle^*.$$

The expectation value of the operator Q then becomes, using Equation 6.22,

$$\langle a\| Q |a\rangle = q_a \langle a \mid a \rangle = q_a,$$

for $|a\rangle$ normalized. From Equation 6.23 we have

$$\langle a| Q^\dagger \|a\rangle = q_a \langle a \mid a \rangle = q_a.$$

It is evident that if Q is Hermitian these results can be combined in the form

$$\langle a| Q |a\rangle = q_a,$$

which is in agreement with the discussion in section 13 of Chapter 4. Note that in the Dirac notation a closed pair of brackets $\langle\ \rangle$ is a scalar quantity (a number), while a half bracket, $|\ \rangle$ or $\langle\ |$, indicates a vector.

A new linear operator appears in this formalism, namely, the quantity $|a\rangle\langle a|$. Because of its structure it can operate on either a bra or a ket and in so doing it produces a transformed bra or ket, respectively. For example, let $|a\rangle\langle a|$ operate on the bra $\langle b|$ from the right:

$$\langle b|a\rangle\langle a| = \alpha\langle a|, \quad \text{where} \quad \alpha = \langle b\mid a\rangle.$$

We see that the vector $\langle b|$ is transformed to the direction of $\langle a|$ and that it is multiplied by a scalar factor which is numerically equal to the projection of $\langle b|$ along $\langle a|$. Now let this same operator act upon the ket $|c\rangle$:

$$|a\rangle\langle a\mid c\rangle = \beta|a\rangle, \quad \text{where} \quad \beta = \langle a\mid c\rangle.$$

Here $|c\rangle$ is transformed to the $|a\rangle$ direction and is multiplied by the scalar projection of $|c\rangle$ upon $|a\rangle$. As a result of these properties the operator $|a\rangle\langle a|$ is called the *projection operator*, $\mathscr{P}_a$. Multiple applications of this operator reduce to the equivalent of a single application, since,

$$\mathscr{P}_a^2 = \mathscr{P}_a\mathscr{P}_a = |a\rangle\langle a\mid a\rangle\langle a| = |a\rangle\, 1\, \langle a| = |a\rangle\langle a| = \mathscr{P}_a, \qquad (6.24)$$

and similarly for higher powers.

Suppose now that we have a complete, orthonormal set of basis kets, $|i\rangle$ (equivalent to the $\hat{\psi}_i$ of section 3). Define the projection operator $\mathscr{P}_i$, for projecting an arbitrary vector onto the ket $|i\rangle$. Let us consider the general vector Ψ_a which, when written in the earlier notation, has the expansion,

$$\Psi_a = \sum_{i=1}^{N} c_i\hat{\psi}_i.$$

Operating on Ψ_a with the projection operator yields the result

$$\mathscr{P}_i\Psi_a = |i\rangle\langle i\mid a\rangle = c_i|i\rangle,$$

since, by definition, the coefficient c_i is the projection of Ψ_a on $\hat{\psi}_i$. Now, if we sum over all possible projections we obtain,

$$\sum_{i=1}^{N}\mathscr{P}_i\Psi_a = \sum_{i=1}^{N}|i\rangle\langle i\mid a\rangle = \sum_{i=1}^{N}c_i|i\rangle \equiv |a\rangle.$$

This gives the important result,

$$\sum_{i=1}^{N}|i\rangle\langle i| = 1, \qquad (6.25)$$

which says that *the projection operator summed over any complete basis is the identity operator.*

Using the Dirac notation, show that (a) the eigenvalues of a Hermitian operator are real, and (b) eigenkets belonging to different eigenvalues of an operator are orthogonal.

7. MATRIX REPRESENTATIONS OF LINEAR OPERATORS

Let $\{\hat{\psi}_i\}$ be an orthonormal basis for the N-dimensional vector space such that

$$\langle \hat{\psi}_i \,|\, \hat{\psi}_j \rangle \equiv \langle i \,|\, j \rangle = \delta_{ij},$$

for all $i, j = 1, \ldots, N$.

Define $\Psi_a = \sum_{j=1}^{N} \alpha_j \hat{\psi}_j$, where the α_j are given by

$$\alpha_j = \langle \hat{\psi}_j \,|\, \Psi_a \rangle \equiv \langle j \,|\, a \rangle.$$

Then the transformation equation

$$\Psi_b = Q\Psi_a$$

becomes

$$\Psi_b = \sum_{k=1}^{N} \beta_k \hat{\psi}_k = Q \sum_{j=1}^{N} \alpha_j \hat{\psi}_j = \sum_{j=1}^{N} \alpha_j Q \hat{\psi}_j.$$

Taking the inner product with $\hat{\psi}_i$,

$$\sum_{k=1}^{N} \beta_k \langle i \,|\, k \rangle = \sum_{j=1}^{N} \alpha_j \, \langle i|\, Q \,|j \rangle = \sum_{j=1}^{N} Q_{ij} \alpha_j$$

Using the orthonormality property, we obtain

$$\beta_i = \sum_{j=1}^{N} Q_{ij} \alpha_j \quad \text{where} \quad Q_{ij} = \langle i|\, Q \,|j \rangle. \tag{6.26}$$

Thus we see that the components of the transformed vector can be characterized by the effect of the operator acting on the old components, where both sets of components are defined on the same basis.

The coefficients Q_{ij} can be written in a square array which we call the "matrix of Q" or the matrix (Q_{ij}). Then we can interpret the equation

$\Psi'_b = \mathbf{Q}\Psi'_a$ as a matrix equation in which Ψ'_b and Ψ'_a are column vectors with N components and $\mathbf{Q}$ is an $N \times N$ matrix. Thus:

$$\begin{pmatrix} \beta_1 \\ \beta_2 \\ \cdot \\ \cdot \\ \cdot \\ \beta_N \end{pmatrix} = \begin{pmatrix} Q_{11} & Q_{12} & \cdots & Q_{1N} \\ Q_{21} & Q_{22} & \cdots & Q_{2N} \\ \cdot & & & \cdot \\ \cdot & & & \cdot \\ \cdot & & & \cdot \\ Q_{N1} & Q_{N2} & \cdots & Q_{NN} \end{pmatrix} \begin{pmatrix} \alpha_1 \\ \alpha_2 \\ \cdot \\ \cdot \\ \cdot \\ \alpha_N \end{pmatrix}. \tag{6.27}$$

As an example of the correspondence between matrix multiplication and operator multiplication, consider the two successive transformations discussed above,

$$\Psi'_b = Q\Psi'_a$$

and $\hspace{12cm}$ (6.28)

$$\Psi'_c = P\Psi'_b = PQ\Psi'_a = R\Psi'_a.$$

Let $\Psi'_c = \sum_k^N \gamma_k \hat{\psi}_k$, where $\gamma_k = \langle k \,|\, c \rangle$. Then, analogous to Equation 6.26, $\gamma_k = \sum_i P_{ki}\beta_i$, and Equation 6.28 becomes:

$$\gamma_k = \sum_i P_{ki}\beta_i = \sum_i P_{ki} \sum_j Q_{ij}\alpha_j = \sum_j \left(\sum_i P_{ki}Q_{ij} \right) \alpha_j = \sum_j R_{kj}\alpha_j.$$

Therefore, the operator equation $R = PQ$ is equivalent to the matrix equation,

$$(R_{kj}) = \left(\sum_i P_{ki}Q_{ij} \right),$$

which will be recognized as the prescription for multiplying the matrix $\mathbf{Q}$ by the matrix $\mathbf{P}$.

In order to further establish the equivalence between the algebra of linear operators and matrix algebra, let us consider a few more important properties. The matrix equation $\Psi_b = \mathbf{Q}\Psi_a$ can be inverted to

$$\Psi_a = \mathbf{Q}^{-1}\Psi_b,$$

only if the inverse matrix $\mathbf{Q}^{-1}$ is defined and is non-singular.

Definition

A matrix is non-singular if its determinant,

$$|Q_{ij}| \neq 0.$$

Definition

The determinant $|Q_{ij}|$ may be written as an expansion in its cofactors,

$$\Delta = |Q_{ij}| = \sum_{i \text{ or } k} (-1)^{i+k} Q_{ik} |q_{ik}|,$$

where the summation is over only one index. Here, $|q_{ik}|$ is the minor of the element Q_{ik}. It is a scalar whose value is given by the $(N-1) \times (N-1)$ determinant remaining after striking out the ith row and the kth column of the original determinant. The cofactor of the element Q_{ij} is $(-1)^{i+j} |q_{ij}|$.

Definition

The *cofactor* matrix is the matrix of the cofactors of the elements of the original matrix.

$$\text{Cof} (Q_{ij}) = ((-1)^{i+j} |q_{ij}|).$$

Definition

The *transpose* of a matrix is the matrix formed by interchanging the rows and columns.

$$\text{The transpose of } (Q_{ij}) = (\tilde{Q}_{ij}) = (Q_{ji}).$$

Definition

The *adjoint* of a matrix is the transpose of its cofactor matrix.

$$\text{Adj} (Q_{ij}) = \text{Cof} (\tilde{Q}_{ij}) = \text{Cof} (Q_{ji}) = ((-1)^{i+j} |q_{ji}|).$$

Definition

The *inverse* of the matrix $\mathbf{Q}$ is determined by dividing the adjoint of $\mathbf{Q}$ by the determinant of $\mathbf{Q}$.

$$\mathbf{Q}^{-1} = (Q_{ij})^{-1} = \frac{\text{Adj} (Q_{ij})}{\Delta} = \frac{1}{\Delta} ((-1)^{i+j} |q_{ji}|).$$

Theorem

A matrix commutes with its inverse; that is,

$$\mathbf{QQ}^{-1} = \mathbf{Q}^{-1}\mathbf{Q} = \mathbf{1}.$$

PROOF.

$$(QQ^{-1})_{ij} = \sum_k Q_{ik}(Q^{-1})_{kj} = \frac{1}{\Delta} \sum_k (-1)^{j+k} Q_{ik} \, |q_{jk}| = \frac{\Delta \delta_{ij}}{\Delta} = \delta_{ij}$$

$$(Q^{-1}Q)_{ij} = \sum_k (Q^{-1})_{ik}Q_{kj} = \frac{1}{\Delta} \sum_k (-1)^{i+k} Q_{kj} \, |q_{ki}| = \frac{\Delta \delta_{ij}}{\Delta} = \delta_{ij}.$$

$$\text{Therefore, } \mathbf{QQ^{-1} = Q^{-1}Q} = \begin{pmatrix} 1 & 0 & 0 & \cdots \\ 0 & 1 & 0 & \cdots \\ 0 & 0 & 1 & \cdots \\ & & \ddots \end{pmatrix} = \mathbf{1}.$$

Theorem

The inverse of a product of non-singular matrices is equal to the product of the inverses in the reverse order.

PROOF.

Let

$$(\mathbf{ABC})^{-1} = \mathbf{D}$$

Then,

$$(\mathbf{ABC})(\mathbf{ABC})^{-1} = \mathbf{ABCD}$$

$$1 = \mathbf{ABCD}$$

$$\mathbf{A^{-1} = A^{-1}ABCD = BCD}$$

$$\mathbf{B^{-1}A^{-1} = B^{-1}BCD = CD}$$

$$\mathbf{C^{-1}B^{-1}A^{-1} = C^{-1}CD = D}$$

Therefore,

$$(\mathbf{ABC})^{-1} = \mathbf{C^{-1}B^{-1}A^{-1}}.$$

Following is a summary of useful definitions and properties of matrices.

Definition

Determinant: $\Delta = |Q_{ij}|$

Transpose: $(\tilde{Q}_{ij}) = (Q_{ji})$

Cofactor: $\text{Cof}\,(Q_{ij}) = ((-1)^{i+j} \, |q_{ij}|)$

Adjoint: $\text{Adj}\,(Q_{ij}) = \text{Cof}\,(\tilde{Q}_{ij})$

Inverse: $\mathbf{Q}^{-1} = \dfrac{1}{\Delta}\,\text{Adj}\,\mathbf{Q}$

Complex Conjugate: $(Q_{ij})^* = (Q_{ij}^*)$

Hermitian Adjoint: $(Q_{ij})^\dagger = (\tilde{Q}_{ij})^* = (Q_{ji}^*)$

A matrix is : *symmetric*   if $\mathbf{Q} = \tilde{\mathbf{Q}}$

                 skew-symmetric   if $\mathbf{Q} = -\tilde{\mathbf{Q}}$

                 real   if $\mathbf{Q} = \mathbf{Q}^*$

                 pure imaginary   if $\mathbf{Q} = -\mathbf{Q}^*$

                 Hermitian   if $\mathbf{Q} = \mathbf{Q}^\dagger$

                 skew Hermitian   if $\mathbf{Q} = -\mathbf{Q}^\dagger$

                 orthogonal   if $\mathbf{Q}^{-1} = \tilde{\mathbf{Q}}$

                 unitary   if $\mathbf{Q}^{-1} = \mathbf{Q}^\dagger$

Properties

$$\widetilde{(\mathbf{AB})} = \tilde{\mathbf{B}}\tilde{\mathbf{A}}$$
$$(\mathbf{AB})^\dagger = \mathbf{B}^\dagger\mathbf{A}^\dagger$$
$$(\mathbf{AB})^* = \mathbf{A}^*\mathbf{B}^*$$
$$(\mathbf{AB})^{-1} = \mathbf{B}^{-1}\mathbf{A}^{-1}$$

Properties of determinants

$$|\tilde{\mathbf{A}}| = |\mathbf{A}|$$
$$|\mathbf{A}^*| = |\mathbf{A}|^*$$
$$|\mathbf{A}^\dagger| = |\tilde{\mathbf{A}}^*| = |\mathbf{A}|^*$$

We will now show that a Hermitian operator Q is represented by a Hermitian matrix. Using the notation of Chapter 5,

$$Q_{ij} = \langle i \mid Qj \rangle,$$
$$Q_{ij}^* = \langle Qj \mid i \rangle = \langle j \mid Qi \rangle = Q_{ji},$$

where the Hermitian property of Q has been used here. Then,

$$(Q_{ij}) = (\tilde{Q}_{ij}^*) = (Q_{ij})^\dagger,$$

and the matrix is Hermitian. The Dirac notation of section 6 is even more transparent, as the following will show:

$$Q_{ij} = \langle i| Q |j \rangle,$$

and

$$Q_{ij}^* = \langle j| Q |i \rangle = Q_{ji}.$$

Therefore, $(Q_{ij}) = (Q_{ij})^\dagger$.

It is evident that the elements along the principal diagonal of a Hermitian matrix must all be real, that is,

$$\langle i| Q |i \rangle = \langle i| Q |i \rangle^*.$$

The matrix equivalent of this Hermitian form is

$$(\tilde{\psi}_i^* \mathbf{Q} \psi_i) = (\tilde{\psi}_i^* \mathbf{Q} \psi_i)^*,$$

where ψ_i is a column vector and its transpose, $\tilde{\psi}_i$, is a row vector. This expression may be readily verified by starting with the right hand side:

$$(\tilde{\psi}_i^* Q \psi_i)^* = (\tilde{\psi}_i Q^* \psi_i^*) = (\tilde{\psi}_i Q \psi_i^*) = ((\widetilde{Q\psi_i}) \psi_i^*) = (\tilde{\psi}_i^* Q \psi_i).$$

PROBLEM 6-14

Show that

$$\mathrm{Cof}\,(\tilde{Q}_{ij}) = \widetilde{\mathrm{Cof}}\,(Q_{ij}).$$

PROBLEM 6-15

Find the cofactor matrices of the following:

(a) $\begin{pmatrix} 1 & 0 & -1 \\ 1 & 2 & 1 \\ 2 & 2 & 3 \end{pmatrix}$ (b) $\begin{pmatrix} 1 & 2 & 2 \\ 0 & 2 & 1 \\ -1 & 2 & 2 \end{pmatrix}$

(c) $\begin{pmatrix} 2 & 2 & 1 \\ 1 & 3 & 1 \\ 1 & 2 & 2 \end{pmatrix}$ (d) $\begin{pmatrix} 2 & 0 & 0 \\ 0 & 3 & 0 \\ 0 & 0 & 5 \end{pmatrix}.$

Ans.:

(a) $\begin{pmatrix} 4 & -1 & -2 \\ -2 & 5 & -2 \\ 2 & -2 & 2 \end{pmatrix}$ (b) $\begin{pmatrix} 2 & -1 & 2 \\ 0 & 4 & -4 \\ -2 & -1 & 2 \end{pmatrix}$

(c) $\begin{pmatrix} 4 & -1 & -1 \\ -2 & 3 & -2 \\ -1 & -1 & 4 \end{pmatrix}$ (d) $\begin{pmatrix} 15 & 0 & 0 \\ 0 & 10 & 0 \\ 0 & 0 & 6 \end{pmatrix}$

PROBLEM 6-16

Evaluate the determinants of the four matrices in Problem 6-15.
(Ans.: 6, 4, 5, 30.)

PROBLEM 6-17

Find the inverses of the matrices given in Problem 6-15.
Ans.:

(a)
$$\begin{pmatrix} \frac{2}{3} & -\frac{1}{3} & \frac{1}{3} \\ -\frac{1}{6} & \frac{5}{6} & -\frac{1}{3} \\ -\frac{1}{3} & -\frac{1}{3} & \frac{1}{3} \end{pmatrix}$$

(b)
$$\begin{pmatrix} \frac{1}{2} & 0 & -\frac{1}{2} \\ -\frac{1}{4} & 1 & -\frac{1}{4} \\ \frac{1}{2} & -1 & \frac{1}{2} \end{pmatrix}$$

(c)
$$\begin{pmatrix} \frac{4}{5} & -\frac{2}{5} & -\frac{1}{5} \\ -\frac{1}{5} & \frac{3}{5} & -\frac{1}{5} \\ -\frac{1}{5} & -\frac{2}{5} & \frac{4}{5} \end{pmatrix}$$

(d)
$$\begin{pmatrix} \frac{1}{2} & 0 & 0 \\ 0 & \frac{1}{3} & 0 \\ 0 & 0 & \frac{1}{5} \end{pmatrix}$$

PROBLEM 6-18

By means of matrix multiplication show that the product of each matrix in Problem 6-15 with its inverse in Problem 6-17 equals the identity matrix.

8. THE MATRIX FORM OF THE EIGENVALUE PROBLEM

Consider the eigenequation

$$Q\,|a\rangle = \lambda\,|a\rangle,$$

which is equivalent to writing,

$$\sum_{j}^{N} Q_{ij}\alpha_j\hat{\psi}_j = \lambda \sum_{j}^{N} \alpha_j\hat{\psi}_j$$

or,

$$\sum_{j}^{N} \alpha_j(Q_{ij} - \lambda\,\delta_{ij})\hat{\psi}_j = 0.$$

The condition for non-trivial solutions of this set of linear equations is that the determinant of the coefficients is zero. That is,

$$|Q_{ij} - \lambda\,\delta_{ij}| = \begin{vmatrix} Q_{11} - \lambda & Q_{12} & Q_{13} & \cdot \\ Q_{21} & Q_{22} - \lambda & Q_{23} & \cdot \\ Q_{31} & Q_{32} & Q_{33} - \lambda & \cdot \\ \cdot & \cdot & \cdot & \cdot \\ \cdot & \cdot & \cdot & \cdot \\ \cdot & \cdot & \cdot & \cdot \end{vmatrix} = 0. \qquad (6.29)$$

This equation, which is called the *secular equation* or the *characteristic equation*, is a polynomial of order N in λ. Its N roots are the eigenvalues of the matrix **Q**.

It is evident that the problem of obtaining these eigenvalues would be greatly simplified if the matrix (Q_{ij}) were itself diagonal. In that case, Equation 6.29 would become

$$(Q_{11} - \lambda)(Q_{22} - \lambda)(Q_{33} - \lambda) \cdots (Q_{NN} - \lambda) = 0,$$

and the N eigenvalues would be the diagonal elements $Q_{11}, Q_{22}, \ldots, Q_{NN}$. We will show in the next section that *there is a transformation, called a unitary transformation, such that any Hermitian matrix may be written in diagonal form.* Such transformations are analogous to the principal axes theorems of classical mechanics and geometry, wherein cross products involving two different coordinates can be eliminated by choosing an appropriate transformation of coordinates. The off-diagonal terms of a matrix correspond to the cross terms in the geometrical problem. For example, a surface in 3-space with its principal axes along the coordinate axes may be described by the expression

$$\phi(\vec{r}) = \sum A_{ii}x_i^2 = \text{constant}.$$

But the vector $\vec{r}$ is in a principal direction when it is parallel to the normal to the surface; that is, when $(\vec{r} \cdot \hat{n})\hat{r} = \vec{r}$. But the direction cosines of the normal are given by $\partial\phi/\partial x_i$, so the component equations for the principal axes are

$$\frac{\partial\phi}{\partial x_i} = x_i$$

or,

$$A_{ii}x_i = \lambda x_i, \quad \text{where } \lambda \text{ is a constant.}$$

This is immediately rocognized as the eigenequation in matrix form.

PROBLEM 6-19

Show that:
(a) The eigenvalues of a Hermitian matrix are real.
(b) The eigenvectors of a Hermitian matrix corresponding to different eigenvalues are orthogonal.

We state without proof that Hermitian matrices have all of the properties of Hermitian operators; that is, the operators and matrices are isomorphic representations of the same abstract algebra.

9. CHANGE OF BASIS: UNITARY TRANSFORMATIONS

It was shown in the previous section that the operator equation, $\Psi_b = Q\Psi_a$, can be interpreted as a linear transformation of the vector Ψ_a into the vector Ψ_b by the linear operator Q in the N-dimensional vector space defined by the basis $\{\hat{\psi}_i\}$. Now we ask, is it posssible to describe this same transformation from a different coordinate system, that is, using a different set of basis functions?

We first define a new set of basis functions in terms of the old basis. Thus,

$$\hat{\psi}_i' = \sum_{j=1}^{N} u_{ij}^* \hat{\psi}_j,$$

where the basis $\{\hat{\psi}_j\}$ is a complete, orthonormal set of N functions. Since we require the new basis to be orthonormal, we have,

$$\langle \hat{\psi}_i' \mid \hat{\psi}_j' \rangle = \delta_{ij}$$

$$\left\langle \sum_k^N u_{ik}^* \hat{\psi}_k \,\middle|\, \sum_l^N u_{jl}^* \hat{\psi}_l \right\rangle = \delta_{ij}$$

$$\sum_{k,l} u_{ik} u_{jl}^* \langle \hat{\psi}_k \mid \hat{\psi}_l \rangle = \delta_{ij}$$

$$\sum_k u_{ik} u_{jk}^* = \delta_{ij}$$

$$\sum_k u_{ik} \tilde{u}_{kj}^* = \delta_{ij}. \tag{6.30}$$

Therefore, we conclude that

$$\mathbf{UU}^\dagger = \mathbf{1}$$

is the necessary and sufficient condition that the new basis be orthonormal. We conclude that: *A complete orthonormal basis for an N-dimensional space may be obtained by means of a unitary transformation on the old basis.*

Next, we wish to show that the transformations given by Equation 6.28 go into

$$\Psi_b' = Q'\Psi_a'$$

$$\Psi_c' = P'\Psi_b' = P'Q'\Psi_a' = R'\Psi_a',$$

under the unitary transformation which takes the basis $\{\hat{\psi}_i\}$ into the new basis $\{\hat{\psi}_i'\}$. It is desirable that Equation 6.28 preserve its form. Furthermore, we need to know if the operators P', Q', and R' differ from P, Q, and R.

Let us write $\Psi_b = Q\Psi_a$ in terms of the old and new bases. First, consider Ψ_a:

$$\Psi_a = \sum_{j=1}^{N} \alpha_j \hat{\psi}_j = \sum_i \alpha_i' \hat{\psi}_i' = \sum_i \alpha_i' u_{ij}^* \hat{\psi}_j.$$

Taking the scalar product with $\hat{\psi}_k$,

$$\alpha_k = \sum_i \alpha'_i u^*_{ij}\, \delta_{jk} = \sum_i \tilde{u}^*_{ki}\alpha'_i. \tag{6.31}$$

In matrix form this is $\Psi_a = U^\dagger \Psi'_a$ or

$$\begin{pmatrix} \alpha_1 \\ \alpha_2 \\ \cdot \\ \cdot \\ \cdot \\ \alpha_N \end{pmatrix} = (U^\dagger) \begin{pmatrix} \alpha'_1 \\ \alpha'_2 \\ \cdot \\ \cdot \\ \cdot \\ \alpha'_N \end{pmatrix}.$$

To obtain the inverse transformation multiply Equation 6.31 by u_{jk} and sum over k:

$$\sum_k u_{jk}\alpha_k = \sum_i \sum_k u_{jk}\tilde{u}^*_{ki}\alpha'_i = \sum_i \delta_{ji}\alpha'_i$$

where Equation 6.30 has been used.

$$\therefore \alpha'_j = \sum u_{jk}\alpha_k.$$

That is,

$$\Psi'_a = U\Psi_a$$

In like manner we have

$$\Psi_b = U^\dagger \Psi'_b \quad \text{and} \quad \Psi'_b = U\Psi_b.$$

Then, for Equation 6.28 we have:

$$\Psi_b = Q\Psi_a$$

$$U^\dagger \Psi'_b = QU^\dagger \Psi'_a \tag{6.32}$$

or

$$\Psi'_b = UQU^{-1}\Psi'_a, \tag{6.33}$$

since $U^\dagger = U^{-1}$. Thus we can preserve the form of Equation 6.28 if we define

$$Q' = UQU^{-1} \tag{6.34}$$

or, conversely,

$$Q = U^{-1}Q'U$$

We conclude that: *A given linear transformation can be expressed as a matrix equation in any representations obtained from the original representation by means of a unitary transformation.*

This transformation may be interpreted as

(1) a transformation of the basis vectors, Equation 6.32, or
(2) a transformation of the operator, Equation 6.33.

To complete our verification of the validity of the matrix formalism in the new representation, consider the equation $R = PQ$. Perform a unitary

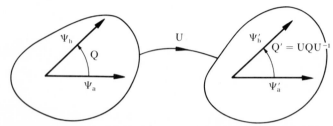

N-dimensional space spanned by $\{ \dot{\psi}_i \}$ N-dimensional space spanned by $\{ \dot{\psi}_i' \}$

Figure 6-2 Schematic diagram of a unitary transformation.

transformation on each member.

$$\mathbf{URU^{-1}} = \mathbf{UPQU^{-1}} = \mathbf{UP1QU^{-1}}$$
$$= \mathbf{UP(U^{-1}U)QU^{-1}}$$
$$= \mathbf{(UPU^{-1})(UQU^{-1})}$$
$$\therefore \mathbf{R'} = \mathbf{P'Q'}$$

We conclude that there is a one-to-one correspondence (an isomorphism) between operators in the vector space and their matrix representations. This is shown shematically in Figure 6-2:

Let us now examine what quantities are left invariant by a unitary transformation. First, consider the transformation of Equation 6.29:

$$|\mathbf{Q'} - \lambda'\mathbf{1}| = |\mathbf{UQU^{-1}} - \mathbf{U}\lambda\mathbf{1U^{-1}}| = |\mathbf{UQU^{-1}} - \lambda\mathbf{1}| = |\mathbf{U(Q} - \lambda\mathbf{1)U^{-1}}|,$$

since $\mathbf{U}$ and $\mathbf{U^{-1}}$ commute with both λ and 1. But the determinant of the product of several matrices equals the product of their determinants. Therefore,

$$|\mathbf{U(Q} - \lambda\mathbf{1)U^{-1}}| = |\mathbf{U}| \cdot |\mathbf{Q} - \lambda\mathbf{1}| \cdot |\mathbf{U^{-1}}|$$
$$= |\mathbf{U}| \cdot |\mathbf{U^{-1}}| \cdot |\mathbf{Q} - \lambda\mathbf{1}|$$
$$= |\mathbf{UU^{-1}}| \cdot |\mathbf{Q} - \lambda\mathbf{1}|$$
$$\therefore |\mathbf{Q'} - \lambda'\mathbf{1}| = |\mathbf{Q} - \lambda\mathbf{1}|.$$

Also, the *trace* of a matrix is invariant under a unitary transformation. The trace is defined as the sum of the elements on the principal diagonal. The proof of this is trivial since Q_{kk} is a number and thus it commutes with $\mathbf{U}$:

$$\sum_i Q'_{ii} = \sum_{i,k} U_{ik} Q_{kk} U_{ik}^{-1} = \sum_{i,k} Q_{kk} U_{ik} U_{ik}^{-1} = \sum_k Q_{kk}.$$

We now can list the following properties of a unitary transformation:

1. The *secular equation of the matrix* is *invariant*.
2. It follows that the *eigenvalues are invariant*.
3. The determinant—which is the *product of the eigenvalues*—is invariant.
4. The trace—which is the *sum* of the *eigenvalues*—is invariant.

The following properties of the trace are stated without proof:

1. The trace of a product is not changed by a cyclic permutation of the factors.

$$\mathrm{Tr}(ABCD) = \mathrm{Tr}(BCDA) = \mathrm{Tr}(CDAB), \text{ and so on.}$$

2. The trace of a commutator is always zero.

$$\mathrm{Tr}(AB - BA) = \mathrm{Tr}(AB) - \mathrm{Tr}(AB) = 0$$

3. The trace of a sum is equal to the sum of the traces.

$$\mathrm{Tr}(A + B) = \mathrm{Tr}\,A + \mathrm{Tr}\,B.$$

10. DIAGONALIZATION OF MATRICES

A unitary transformation is just one type of transformation among a larger class known as *similarity* transformations. If $\mathbf{U}^{-1} = \tilde{\mathbf{U}}$, then the similarity transformation $\mathbf{U}^{-1}\mathbf{Q}\mathbf{U}$ is called an *orthogonal* transformation. A unitary transformation, where $\mathbf{U}^{-1} = \tilde{\mathbf{U}}^* = \mathbf{U}^\dagger$, is an extension of the orthogonal transformation to complex matrices.

Although not all matrices can be diagonalized, any square matrix can be transformed by a similarity transformation such that its eigenvalues appear on the principal diagonal (although the off-diagonal elements are not necessarily all zero). More important for our purposes, the matrix representations of physical operators are usually Hermitian and these can be diagonalized.

The importance of a unitary transformation now becomes clear. For, if there is a transformation which will result in a diagonal representation of the operator Q, then the solutions to the eigenvalue problem are immediately obtained by the process of diagonalization. Defining $\mathbf{Q}$ as the diagonal matrix, $\mathbf{\Lambda} = \lambda\mathbf{1}$ (that is, $\Lambda_{ii} = \lambda_i\delta_{ij}$), and $\mathbf{Q}'$ as its non-diagonal form (now called $\mathbf{Q}$), we have from Equation 6.34,

$$\mathbf{QU} = \mathbf{U\Lambda}.$$

Writing out the matrices explicitly, this is

$$
\begin{pmatrix}
Q_{11} & Q_{12} & Q_{13} & \cdots \\
Q_{21} & Q_{22} & Q_{23} & \cdots \\
Q_{31} & Q_{32} & Q_{33} & \cdots \\
\cdot & \cdot & \cdot & \cdot \\
\cdot & \cdot & \cdot & \cdot
\end{pmatrix}
\begin{pmatrix}
u_1^1 & u_1^2 & u_1^3 & \cdots \\
u_2^1 & u_2^2 & u_2^3 & \cdots \\
u_3^1 & u_3^2 & u_3^3 & \cdots \\
\cdot & \cdot & \cdot & \cdot \\
\cdot & \cdot & \cdot & \cdot
\end{pmatrix}
=
$$

$$
\begin{pmatrix}
u_1^1 & u_1^2 & u_1^3 & \cdots \\
u_2^1 & u_2^2 & u_2^3 & \cdots \\
u_3^1 & u_3^2 & u_3^3 & \cdots \\
\cdot & \cdot & \cdot & \cdot \\
\cdot & \cdot & \cdot & \cdot
\end{pmatrix}
\begin{pmatrix}
\lambda_1 & 0 & 0 & \cdots \\
0 & \lambda_2 & 0 & \cdots \\
0 & 0 & \lambda_3 & \cdots \\
\cdot & \cdot & \cdot & \cdot \\
\cdot & \cdot & \cdot & \cdot
\end{pmatrix}.
$$

If we consider the columns of the **U** matrix as separate vectors, note that the above result is equivalent to the separate eigenequations:

$$\left.\begin{array}{c} \mathbf{Q}\mathscr{U}_1 = 1\lambda_1\mathscr{U}_1 \\ \mathbf{Q}\mathscr{U}_2 = 1\lambda_2\mathscr{U}_2 \\ \mathbf{Q}\mathscr{U}_3 = 1\lambda_3\mathscr{U}_3 \\ \cdots\cdots \end{array}\right\} \quad \text{or} \quad \left.\begin{array}{c} (\mathbf{Q} - \lambda_1\mathbf{1})\mathscr{U}_1 = 0 \\ (\mathbf{Q} - \lambda_2\mathbf{1})\mathscr{U}_2 = 0 \\ (\mathbf{Q} - \lambda_3\mathbf{1})\mathscr{U}_3 = 0 \\ \cdots\cdots\cdots \end{array}\right\}, \qquad (6.35)$$

where

$$\mathscr{U}_1 = \begin{pmatrix} u_1^1 \\ u_2^1 \\ u_3^1 \\ \cdot \\ \cdot \end{pmatrix}, \qquad \mathscr{U}_2 = \begin{pmatrix} u_1^2 \\ u_2^2 \\ u_3^2 \\ \cdot \\ \cdot \end{pmatrix}, \qquad \text{and so on.}$$

We now have the prescription for obtaining the matrix of the transformation **U**:

First find the eigenvalues of the operator matrix **Q**.
For each eigenvalue λ_i, obtain the eigenvector $\mathscr{U}_i$ from Equation 6.35. **U** is comprised of the $\mathscr{U}_i$ as columns.

EXAMPLE

Given,

$$\mathbf{Q} = \begin{pmatrix} 1 & 0 & -2 \\ 0 & 0 & 0 \\ -2 & 0 & 4 \end{pmatrix}.$$

Find the eigenvalues and eigenvectors of **Q**. Construct **U** and **U**$^{-1}$. Show that **U**$^{-1}$**QU** = Λ.

$$\begin{vmatrix} 1-\lambda & 0 & -2 \\ 0 & -\lambda & 0 \\ -2 & 0 & 4-\lambda \end{vmatrix} = (-\lambda)[(1-\lambda)(4-\lambda) - 4] = 0.$$

Then the eigenvalues are $\lambda = 0, 0, 5$. Taking $\lambda_1 = 0$, we have from Equation 6.35, $\mathbf{Q}\mathscr{U}_1 = \mathbf{0}$, or

$$\begin{pmatrix} 1 & 0 & -2 \\ 0 & 0 & 0 \\ -2 & 0 & 4 \end{pmatrix} \begin{pmatrix} u_1^1 \\ u_2^1 \\ u_3^1 \end{pmatrix} = 0,$$

which yields the results, $u_1^1 = 2u_3^1$.

Since u_2^1 is completely arbitrary, let it be zero. Then the simplest choice for a normalized $\mathcal{U}_1$ is:

$$\mathcal{U}_1 = \frac{1}{\sqrt{5}} \begin{pmatrix} 2 \\ 0 \\ 1 \end{pmatrix}.$$

Taking $\lambda_2 = 0$, we have as above, $u_1^2 = 2u_3^2$. Since $\mathcal{U}_2$ must be orthogonal to $\mathcal{U}_1$, choose $u_2^2 = 1$ and $u_1^2 = u_3^2 = 0$. Then,

$$\mathcal{U}_2 = \begin{pmatrix} 0 \\ 1 \\ 0 \end{pmatrix}.$$

Taking $\lambda_3 = 5$, the eigenequation is $(\mathbf{Q} - 5\mathbf{1})\mathcal{U}_3 = 0$, which becomes:

$$\begin{pmatrix} -4 & 0 & -2 \\ 0 & -5 & 0 \\ -2 & 0 & -1 \end{pmatrix} \begin{pmatrix} u_1^3 \\ u_2^3 \\ u_3^3 \end{pmatrix} = 0.$$

Then,

$$u_3^3 = -2u_1^3$$
$$u_2^3 = 0,$$

and we obtain

$$\mathcal{U}_3 = \frac{1}{\sqrt{5}} \begin{pmatrix} 1 \\ 0 \\ -2 \end{pmatrix}.$$

The transformation matrix is constructed by placing the $\mathcal{U}_i$ along the columns:

$$\mathbf{U} = \begin{pmatrix} \dfrac{2}{\sqrt{5}} & 0 & \dfrac{1}{\sqrt{5}} \\ 0 & 1 & 0 \\ \dfrac{1}{\sqrt{5}} & 0 & -\dfrac{2}{\sqrt{5}} \end{pmatrix} = \frac{1}{\sqrt{5}} \begin{pmatrix} 2 & 0 & 1 \\ 0 & \sqrt{5} & 0 \\ 1 & 0 & -2 \end{pmatrix}.$$

$$|\mathbf{U}| = -1, \; \mathrm{Cof}\,\mathbf{U} = \mathrm{Adj}\,\mathbf{U} = -\frac{1}{\sqrt{5}} \begin{pmatrix} 2 & 0 & 1 \\ 0 & \sqrt{5} & 0 \\ 1 & 1 & -2 \end{pmatrix},$$

and

$$\mathbf{U}^{-1} = \mathbf{U}.$$

$$\mathbf{U}^{-1}\mathbf{Q}\mathbf{U} = \tfrac{1}{5}\begin{pmatrix} 2 & 0 & 1 \\ 0 & \sqrt{5} & 0 \\ 1 & 0 & -2 \end{pmatrix}\begin{pmatrix} 1 & 0 & -2 \\ 0 & 0 & 0 \\ -2 & 0 & 4 \end{pmatrix}\begin{pmatrix} 2 & 0 & 1 \\ 0 & \sqrt{5} & 0 \\ 1 & 0 & -2 \end{pmatrix}$$

$$= \tfrac{1}{5}\begin{pmatrix} 2 & 0 & 1 \\ 0 & \sqrt{5} & 0 \\ 1 & 0 & -2 \end{pmatrix}\begin{pmatrix} 0 & 0 & 5 \\ 0 & 0 & 0 \\ 0 & 0 & -10 \end{pmatrix}$$

$$= \begin{pmatrix} 2 & 0 & 1 \\ 0 & \sqrt{5} & 0 \\ 1 & 0 & -2 \end{pmatrix}\begin{pmatrix} 0 & 0 & 1 \\ 0 & 0 & 0 \\ 0 & 0 & -2 \end{pmatrix} = \begin{pmatrix} 0 & 0 & 0 \\ 0 & 0 & 0 \\ 0 & 0 & 5 \end{pmatrix} \equiv \Lambda.$$

PROBLEM 6-20

Find the eigenvalues and eigenvectors for each of the matrices given in Problem 6-15.
(Ans.: (a) 1, 2, 3; (b) 1, 2, 2; (c) 5, 1, 1; (d) 2, 3, 5. The eigenvectors are the columns of the matrices given in Problem 6-21.)

PROBLEM 6-21

The matrix $\mathbf{U}$ of the transformation that will diagonalize each of the matrices of Problem 6-15 is given below.*

(a)
$$\begin{pmatrix} \dfrac{1}{\sqrt{2}} & \dfrac{2}{3} & \dfrac{1}{\sqrt{6}} \\[2ex] -\dfrac{1}{\sqrt{2}} & -\dfrac{1}{3} & -\dfrac{1}{\sqrt{6}} \\[2ex] 0 & -\dfrac{2}{3} & -\dfrac{2}{\sqrt{6}} \end{pmatrix}$$

* It should be pointed out that we use the symbol $\mathbf{U}$ for the transformation matrix even for those cases where $\mathbf{Q}$ is not Hermitian and $\mathbf{U}$ is not unitary. In general, the eigenvectors of $\mathbf{Q}$ will not be orthogonal unless $\mathbf{Q}$ is Hermitian. Also, $\mathbf{U}^{-1} \neq \mathbf{U}^{\dagger}$ unless $\mathbf{U}$ is unitary.

(b)   Cannot be diagonalized.

(c)
$$
\begin{pmatrix}
\dfrac{1}{\sqrt{3}} & \dfrac{1}{\sqrt{2}} & \dfrac{2}{\sqrt{5}} \\[2ex]
\dfrac{1}{\sqrt{3}} & 0 & -\dfrac{1}{\sqrt{5}} \\[2ex]
\dfrac{1}{\sqrt{3}} & -\dfrac{1}{\sqrt{2}} & 0
\end{pmatrix}
$$

(d)
$$
\begin{pmatrix}
1 & 0 & 0 \\
0 & 1 & 0 \\
0 & 0 & 1
\end{pmatrix}
$$

Find the inverse $\mathbf{U}^{-1}$ of each of these.

PROBLEM 6-22

Carry out the multiplication to show that the matrices of Problem 6-21 do indeed diagonalize the matrices given in Problem 6-15.

PROBLEM 6-23

For each of the matrices of Problem 6-15, verify that the trace is the sum of its eigenvalues and that its determinant is the product of its eigenvalues.

PROBLEM 6-24

Given the matrix,
$$
\mathbf{A} = \begin{pmatrix}
1 & i & 1 \\
-i & 0 & 0 \\
1 & 0 & 0
\end{pmatrix}
$$

(a)   Is $\mathbf{A}$ Hermitian?
(b)   Find the eigenvalues of $\mathbf{A}$.
(c)   Find the matrix $\mathbf{U}$ for the transformation that will diagonalize $\mathbf{A}$.
(d)   Show that the eigenvectors comprising $\mathbf{U}$ are orthonormal.

(e)   Show that **U** is unitary by obtaining $\mathbf{U}^\dagger$ and verifying that $\mathbf{UU}^\dagger = 1$ by means of matrix multiplication.

(f)   Carry out the matrix multiplication to verify that $\mathbf{U}^{-1}\mathbf{AU}$ is diagonal. (Remember that $\mathbf{U}^{-1} = \mathbf{U}^\dagger$ for a unitary matrix.)

Ans.: (b) 0, -1, 2;   (c) $\begin{pmatrix} 0 & \dfrac{1}{\sqrt{3}} & \dfrac{2}{\sqrt{6}} \\[2ex] \dfrac{1}{\sqrt{2}} & \dfrac{i}{\sqrt{3}} & -\dfrac{i}{\sqrt{6}} \\[2ex] -\dfrac{i}{\sqrt{2}} & -\dfrac{1}{\sqrt{3}} & \dfrac{1}{\sqrt{6}} \end{pmatrix}$.

PROBLEM 6-25

Given the Hermitian matrix,

$$\mathbf{\Gamma} = \begin{pmatrix} \gamma & 0 & i\beta\gamma \\ 0 & 1 & 0 \\ -i\beta\gamma & 0 & \gamma \end{pmatrix}.$$

(a)   Find the eigenvalues of $\mathbf{\Gamma}$.
(b)   Find the unitary matrix **U** for the transformation that will diagonalize $\mathbf{\Gamma}$.
(c)   Verify by matrix multiplication that $\mathbf{UU}^\dagger = 1$.
(d)   Verify that $\mathbf{U}^{-1}\mathbf{\Gamma U}$ is diagonal.
Ans.: (a) 1, $\gamma(1 \pm \beta)$.

(b) $\begin{pmatrix} \dfrac{1}{\sqrt{2}} & 0 & -\dfrac{1'}{\sqrt{2}} \\[2ex] 0 & 1 & 0 \\[2ex] \dfrac{i}{\sqrt{2}} & 0 & \dfrac{i}{\sqrt{2}} \end{pmatrix}$.

II.   APPLICATION OF MATRIX MECHANICS TO THE HARMONIC OSCILLATOR

In the previous sections it was shown that in a linear vector space an arbitrary vector may be denoted by giving its components along a complete, orthonormal set of basis vectors which span the space. Furthermore, each linear operator, to which these vectors are subject, can be represented by a matrix defined on the same basis vectors. The effect of applying a given operator to a vector, then, can be determined by multiplying the matrix representation of the operator times the column matrix of the vector components.

We have seen that the energy eigenfunctions of the harmonic oscillator form a complete, orthonormal set, the only complication being that the set is denumerably infinite. Thus, if we regard these eigenfunctions as our basis vectors we will be using a space of infinite dimension. For example, let us consider the energy matrix for the harmonic oscillator. This is, of course, the matrix of the Hamiltonian operator using the oscillator wave functions as the basis functions. Each matrix element is given by

$$E_{ij} = \langle i| \mathcal{H} |j\rangle = E_j \langle i | j \rangle = E_j \delta_{ij}.$$

Since both $\langle i|$ and $|j\rangle$ are eigenvectors of $\mathcal{H}$, the effect of the operator is merely to generate the eigenvalue E_{ij} which, being a number, can be moved outside of the integral. The integral is zero for all off-diagonal matrix elements because of the orthogonality of the oscillator wave functions. If we label the rows and columns in the order, $\psi_0, \psi_1, \psi_2, \ldots$, the matrix appears as follows:

$$(E) = \frac{\omega \hbar}{2} \begin{pmatrix} 1 & 0 & 0 & 0 & 0 & \cdot & \cdot \\ 0 & 3 & 0 & 0 & 0 & \cdot & \cdot \\ 0 & 0 & 5 & 0 & 0 & \cdot & \cdot \\ 0 & 0 & 0 & 7 & 0 & \cdot & \cdot \\ 0 & 0 & 0 & 0 & 9 & \cdot & \cdot \\ & & \cdot & \cdot & \cdot & \cdot & \cdot \\ & \cdot & \cdot & \cdot & \cdot & \cdot & \cdot \end{pmatrix}. \tag{6.36}$$

Equation 6.36 illustrates the important fact that an operator has a diagonal representation when its own eigenfunctions constitute the basis.

As a further example let us write the matrix of the operator x, whose matrix elements were given in Equation 5.41. It is as follows:

$$(x) = \frac{1}{\sqrt{2\alpha}} \begin{pmatrix} 0 & 1 & 0 & 0 & 0 & 0 & \cdot \\ 1 & 0 & \sqrt{2} & 0 & 0 & 0 & \cdot \\ 0 & \sqrt{2} & 0 & \sqrt{3} & 0 & 0 & \cdot \\ 0 & 0 & \sqrt{3} & 0 & 2 & 0 & \cdot \\ 0 & 0 & 0 & 2 & 0 & \sqrt{5} & \cdot \\ 0 & 0 & 0 & 0 & \sqrt{5} & 0 & \cdot \\ \cdot & \cdot & \cdot & \cdot & \cdot & \cdot & \cdot \end{pmatrix}. \tag{6.37}$$

Notice that the matrix of x is not diagonal. This was to be expected since the basis functions are not eigenfunctions of x.

For problems involving pure states, the value of the matrix formulation is not evident. A somewhat better example of its utility is that of Problem 5-22, in which an oscillator is in a superposition of two states. The time independent wave function as well as the function which includes the time dependence are given below:

$$\psi(x) = \begin{pmatrix} \dfrac{1}{\sqrt{2}} \\[2mm] \dfrac{1}{\sqrt{2}} \\[1mm] 0 \\ 0 \\ \cdot \\ \cdot \\ \cdot \end{pmatrix} \qquad \Psi(x,\,t) = \begin{pmatrix} \dfrac{1}{\sqrt{2}}\, e^{-iE_0 t/\hbar} \\[2mm] \dfrac{1}{\sqrt{2}}\, e^{-iE_1 t/\hbar} \\[1mm] 0 \\ 0 \\ \cdot \\ \cdot \\ \cdot \end{pmatrix}$$

The expectation value of the energy in the time varying state is:

$$\langle E \rangle = \left(\frac{1}{\sqrt{2}}\, e^{iE_0 t/\hbar},\ \frac{1}{\sqrt{2}}\, e^{iE_1 t/\hbar}\ 0 \cdots \right) \begin{pmatrix} \dfrac{\omega\hbar}{2} & 0 & 0 & \cdot \\[2mm] 0 & \dfrac{3\omega\hbar}{2} & 0 & \cdot \\[2mm] 0 & 0 & \dfrac{5\omega\hbar}{2} & \cdot \\[1mm] \cdot & \cdot & \cdot & \cdot \end{pmatrix} \begin{pmatrix} \dfrac{1}{\sqrt{2}}\, e^{-iE_0 t/\hbar} \\[2mm] \dfrac{1}{\sqrt{2}}\, e^{-iE_1 t/\hbar} \\[1mm] 0 \\ \cdot \\ \cdot \end{pmatrix}$$

$$= \left(\frac{1}{\sqrt{2}}\, e^{iE_0 t/\hbar},\ \frac{1}{\sqrt{2}}\, e^{iE_1 t/\hbar},\ 0,\ 0,\ \cdots \right) \begin{pmatrix} \dfrac{\omega\hbar}{2\sqrt{2}}\, e^{-iE_0 t/\hbar} \\[2mm] \dfrac{3\omega\hbar}{2\sqrt{2}}\, e^{-iE_1 t/\hbar} \\[1mm] 0 \\ 0 \\ \cdot \\ \cdot \\ \cdot \end{pmatrix}$$

$$= \frac{\omega\hbar}{4} + \frac{3\omega\hbar}{4} = \omega\hbar.$$

Although the last example is still quite elementary, one can easily visualize extending the method to more complicated situations. The real power of the matrix method will appear when we discuss perturbation theory in Chapter 9.

PROBLEM 6-26

   (a) Using Equations 5.56 and 5.57, obtain the matrices for the raising and lowering operators, $a^\dagger$ and a, with the harmonic oscillator wave functions as the basis.

   (b) With the help of Equation 5.58, obtain the matrix representations of the operators p and q.

   (c) In the same manner obtain the matrix representations for x, p_x, x^2, p_x^2, T, and V for the harmonic oscillator.

   (d) Which of the above matrices are Hermitian?

PROBLEM 6-27

Use the matrix method to calculate the expectation values $\langle x \rangle$, $\langle p_x \rangle$ and $\langle E \rangle$ for the following mixed state of the harmonic oscillator:

$$\Psi(x, t) = \frac{1}{\sqrt{10}} \Psi_0(x, t) + \frac{2}{\sqrt{10}} \Psi_1(x, t) + \frac{2}{\sqrt{10}} \Psi_2(x, t)$$

$$+ \frac{1}{\sqrt{10}} \Psi_3(x, t).$$

SUMMARY

A quantum system may be represented by a wave function which contains all of the information that can be known about the system. The wave function satisfies the Schrödinger wave equation, which provides a description of the development of the system as time progresses. The superposition principle is valid for wave functions. This means that any linear combination of wave functions which represent physical states of the system is itself a possible physical state. The connection between the mathematical wave function and physical measurements is made by associating a linear, Hermitian operator with each dynamical variable. A measurement of a particular variable can result in *only* one of the eigenvalues of the operator corresponding to that variable. If the system is known to be in an eigenstate of the operator, the result of a measurement of the corresponding variable can be predicted, with certainty, to be that eigenvalue. Otherwise, only a probabilistic prediction can be made. The average value of a large number of measurements on the same or identical systems is called the expectation value. This is the weighted mean of all of the accessible eigenvalues, where the weighting factor for each eigenvalue is interpreted as the probability of obtaining that eigenvalue. After a measurement of the variable q, the system is known to be in an eigenstate of the operator Q. If this is followed by a measurement of p, the act of measurement forces the system into an eigenstate of P. If the operators P and Q commute, the measurement of p will not disturb the eigenstate of Q, and, in fact, any sequence of measurements of p and q will repeatedly yield the same two eigenvalues of the operators P and Q. However, if P and Q do not commute, a measurement of p will destroy all previous knowledge concerning the state of Q, and vice versa.

A quantum mechanical state, which was just regarded as simply a wave function, can be visualized as a vector in an abstract function space (Hilbert space). Such a vector may be represented by a columnar matrix array of numbers which characterize its projections on a given set of basis vectors. In like manner, an operator Q may be represented by a matrix array composed of the elements $\langle \hat{\psi}_i | \, Q \, | \hat{\psi}_j \rangle$ calculated from a particular set of basis functions, $\{\hat{\psi}_i\}$. Just as a vector has an existence that is independent of the set of unit vectors used to define its components, an operator transcends the particular set of basis functions used to express its representation. The diagonal elements of the matrix $(i = j)$ are the expectation values of the operator Q, and the off-diagonal elements $(i \neq j)$ will later be identified as the *transition probabilities*. Although $\hat{\psi}_i$ and $\hat{\psi}_j$ are orthogonal, the operator itself might connect the two states. Physically, this means that the act of measurement produces an interaction which has a non-zero probability of causing a transition from one state to the other.

A coordinate system in 3-space can be rotated so that an arbitrary vector $\vec{A}$ lies parallel to one basis vector, say $\hat{k}$. Then the vector has a particularly simple form, namely,

$$\vec{A} = \begin{pmatrix} 0 \\ 0 \\ z \end{pmatrix}.$$

In like manner, a "rotation" in function space can be effected by applying a unitary transformation to the matrix $\mathbf{Q}$. There is one particular transformation that will enable $\mathbf{Q}$ to be written in diagonal form such that its elements are the eigenvalues of the operator Q and its basis functions are the eigenfunctions of Q. An equally valid procedure would be to construct the new basis by forming appropriate linear combinations of the old basis functions so as to make $\mathbf{Q}$ diagonal. In general, however, it is easier to obtain the eigenvalues directly by solving the secular equation. If the eigenfunctions are needed, they can then be obtained easily by means of the diagonal matrix $\mathbf{\Lambda}$.

SUGGESTED REFERENCES

David Bohm, *Quantum Theory*. Prentice-Hall, Inc., N.Y., 1951.
R. H. Dicke and J. P. Wittke, *Introduction to Quantum Mechanics*. Addison-Wesley Publishing Co., Inc., Reading, Mass., 1960.

P. A. M. Dirac, *The Principles of Quantum Mechanics, 3rd ed.* Oxford University Press, London, 1947.

R. B. Leighton, *Principles of Modern Physics.* McGraw-Hill Book Co., Inc., New York, 1959.

P. T. Matthews, *Introduction to Quantum Mechanics.* McGraw-Hill Book Co., New York, 1968.

E. Merzbacher, *Quantum Mechanics.* John Wiley and Sons, Inc., New York, 1961.

Albert Messiah, *Quantum Mechanics.* North-Holland Publishing Company, Amsterdam, 1958.

J. L. Powell and B. Crasemann, *Quantum Mechanics.* Addison-Wesley Publishing Co., Inc., Reading, Mass., 1961.

V. Rojansky, *Introductory Quantum Mechanics.* Prentice-Hall, Inc., Englewood Cliffs, N.J., 1938.

Leonard I. Schiff, *Quantum Mechanics, 3rd ed.* McGraw-Hill Book Co., New York, 1969.

Robert L. White, *Basic Quantum Mechanics.* McGraw-Hill Book Co., New York, 1966.

Klaus Ziock, *Basic Quantum Mechanics.* John Wiley & Sons, Inc., New York, 1969.

CHAPTER 7

THE WAVE EQUATION IN THREE DIMENSIONS

We will now apply quantum mechanics to three-dimensional systems. The reader has at his disposal the formal tools of both wave mechanics and matrix mechanics, as well as his experience in treating one-dimensional systems in Chapter 5. After a brief discussion of the separation of variables in rectangular coordinates we will turn to the two-body problem. This will be solved in spherical coordinates and will be applied to the hydrogen atom. The important role of the angular momentum operators and their eigenfunctions will be emphasized.

I. RECTANGULAR COORDINATES IN THREE DIMENSIONS

The time-independent Schrödinger equation in rectangular coordinates is

$$-\frac{\hbar^2}{2m}\nabla^2\psi_E(x, y, z) + V(x, y, z)\psi_E(x, y, z) = E\psi_E(x, y, z).$$

In the special case of the free particle $(V = 0)$, the solutions are plane waves of the form

$$\psi_E(x, y, z) = \left(\frac{1}{2\pi}\right)^{\frac{3}{2}}e^{i\vec{k}\cdot\vec{r}} = \left(\frac{1}{2\pi}\right)^{\frac{3}{2}}e^{ik_xx} \cdot e^{ik_yy} \cdot e^{ik_zz}. \tag{7.1}$$

Whenever the coordinates are independent, that is, they are not connected by interactions, the Schrödinger equation is separable and its solutions may be written as a product of one-dimensional solutions as in Equation 7.1. Furthermore, the total energy eigenvalue is equal to the sum of the eigenvalues of the one-dimensional problems, that is,

$$E = \frac{\hbar^2k^2}{2m} = \frac{\hbar^2}{2m}\left(k_x^2 + k_y^2 + k_z^2\right) = E_x + E_y + E_z.$$

Even when a non-zero potential exists, these same principles apply, provided that the potential can be expressed as

$$V(x, y, z) = V_a(x) + V_b(y) + V_c(z).$$

In such cases, the Schrödinger equation may be separated as follows:

$$\left[-\frac{\hbar^2}{2m} \frac{\partial^2}{\partial x^2} + V_a(x) \right] \psi_E + \left[-\frac{\hbar^2}{2m} \frac{\partial^2}{\partial y^2} + V_b(y) \right] \psi_E + \left[-\frac{\hbar^2}{2m} \frac{\partial^2}{\partial z^2} + V_c(z) \right] \psi_E$$
$$= E\psi_E.$$

Expressing the solution as a product function,

$$\psi_E = \psi_{E_x}(x) \cdot \psi_{E_y}(y) \cdot \psi_{E_z}(z),$$

we obtain for the x-equation,

$$\left[-\frac{\hbar^2}{2m} \frac{\partial^2}{\partial x^2} + V_a(x) \right] \psi_{E_x}(x) = E_x \psi_{E_x}(x).$$

Similar equations hold for the other coordinates with the result that

$$E = E_x + E_y + E_z.$$

Another example is that of a particle in a box. For simplicity, let the box be a cube of side L. Although there is no force acting upon the particle and its total energy is positive, it does *not* have a continuum of available energy states like a free particle. The presence of the rigid walls (infinite potential barriers) forces the particle to occupy one of a discrete set of energy states. We found these states for the one-dimensional case of an infinite square well in section 3 of Chapter 5. From Equation 5.18,

$$E_{n_x} = \frac{n_x^2 \pi^2 \hbar^2}{2mL^2},$$

and

$$\psi(x) = \sqrt{\frac{2}{L}} \sin \frac{n_x \pi x}{L}.$$

Then,

$$\psi_{n_x, n_y, n_z} = \left(\frac{2}{L} \right)^{\frac{3}{2}} \sin \frac{n_x \pi x}{L} \cdot \sin \frac{n_y \pi y}{L} \cdot \sin \frac{n_z \pi z}{L}, \qquad (7.2)$$

and

$$E_{n_x, n_y, n_z} = \frac{n^2 \pi^2 \hbar^2}{2mL^2}, \qquad (7.3)$$

where

$$n^2 = n_x^2 + n_y^2 + n_z^2.$$

PROBLEM 7-1

(a) Using Equation 7.3, what is the ground state energy of a particle in a box?

(b) How do the energies of the first and second excited levels compare with the ground state energy?

(c) What are the degeneracies of the three lowest levels?

(Ans.: (b) $2E_0$, $3E_0$; (c) 1, 3, 3.)

PROBLEM 7-2

Calculate the density of states (number of states per energy interval) for the particle in a box. (Hint: see the calculation of Jeans' number in Appendix A of Chapter 2.)

$$\left(\text{Ans.:} \quad \frac{L^3}{4\pi^2\hbar^3}(2m)^{\frac{3}{2}}E^{\frac{1}{2}}. \right)$$

PROBLEM 7-3

(a) Find the ground state energy of a three-dimensional harmonic oscillator by treating it as three one-dimensional oscillators having the same frequency.

(b) What are the energies and degeneracies of the three lowest levels?

(c) Show that the degeneracy of the nth level is given by $\frac{1}{2}(n+1)(n+2)$.

(Ans.: (b) $\frac{3}{2}\omega\hbar$, $\frac{5}{2}\omega\hbar$, $\frac{7}{2}\omega\hbar$; 1, 3, 6.)

2. SPHERICALLY SYMMETRIC POTENTIALS

A two-body problem can be reduced easily to a one-body problem in a conservative system in which the only force is one that is proportional to the distance between the two bodies. In such a case, the problem has spherical symmetry in a relative coordinate system whose origin is fixed at one of the bodies.

Consider, for example, a proton of mass M with position vector $\vec{r}_1$ and an electron of mass m and position vector $\vec{r}_2$ (see Figure 7-1). Assume that the only interaction between the particles is that given by the Coulomb potential,

$$V = \frac{-e^2}{|\vec{r}_2 - \vec{r}_1|} = \frac{-e^2}{r},$$

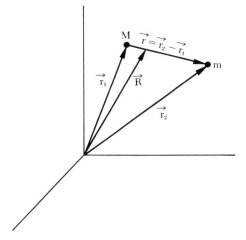

Figure 7-1 Coordinates for the two-body problem.

where r is the distance between the particles. The time-independent Schrödinger equation, Equation 4.54, becomes

$$\left[-\frac{\hbar^2}{2M}\nabla_1^2 - \frac{\hbar^2}{2m}\nabla_2^2 \right]\Psi_E - V(r)\Psi_E = E\Psi_E, \qquad (7.4)$$

where ∇_1^2 operates only on the coordinates of the proton and ∇_2^2 operates only on the coordinates of the electron. A difficulty arises with this expression, however, since the potential $V(r)$ is a function of the coordinates of *both* particles. This means that Equation 7.4 is not separable in the two sets of coordinates. For this reason it is convenient to transform the equation to center-of-mass and relative coordinates,

The position vector $\vec{R}$ for the center of mass of the two particles is given by the relation,

$$M\,|\vec{r}_1 - \vec{R}| = m\,|\vec{R} - \vec{r}_2|,$$

from which,

$$\vec{R} = \frac{M\vec{r}_1 + m\vec{r}_2}{M + m}. \qquad (7.5)$$

The kinetic energy of the center of mass is given by

$$T = \tfrac{1}{2}(M + m)\,\dot{R}^2,$$

while the kinetic energy of the two particles about the center of mass is

$$T' = \tfrac{1}{2}M\,|\dot{\vec{r}}_1 - \dot{\vec{R}}|^2 + \tfrac{1}{2}m\,|\dot{\vec{R}} - \dot{\vec{r}}_2|^2.$$

Using Equation 7.5, we can write

$$|\vec{r}_1 - \vec{R}| = \frac{m}{M + m}\,|\vec{r}_1 - \vec{r}_2| = \frac{m}{M + m}\,|\vec{r}|,$$

$$|\vec{R} - \vec{r}_2| = \frac{M}{M + m}\,|\vec{r}_1 - \vec{r}_2| = -\frac{M}{M + m}\,|\vec{r}|.$$

From the time derivatives of these expressions we readily obtain

$$T' = \frac{1}{2}\frac{Mm}{M+m}\dot{r}^2.$$

Therefore, the Hamiltonian for the system becomes simply

$$\mathcal{H} = T + T' + V = \frac{1}{2}(M+m)\dot{R}^2 + \frac{1}{2}\mu\dot{r}^2 + V(r)$$

$$= \frac{P^2}{2(M+m)} + \frac{p^2}{2\mu} + V(r) = -\frac{\hbar^2}{2(M+m)}\nabla^2_{cm}$$

$$+ \left[-\frac{\hbar^2}{2\mu}\nabla^2_{Rel} + V(r)\right] = \mathcal{H}_{cm} + \mathcal{H}_{Rel},$$

where

$$\vec{P} = (M+m)\dot{\vec{R}} = -i\hbar\vec{\nabla}_{cm},$$

$$\vec{p} = \mu\dot{\vec{r}} = -i\hbar\vec{\nabla}_{Rel},$$

and μ is the reduced mass defined in Equation 3.23. Since the Schrödinger equation is now separable in these variables, we may write its solution in the form of a product function,

$$\Psi_E = \psi_{cm}\cdot\psi_{Rel}.$$

For a conservative system the solution ψ_{cm} represents the continuum of free particle states for the system as a whole, and the function ψ_{Rel} represents one of the discrete eigenstates of the bound system. Thus the total energy eigen-equation may be written as

$$\mathcal{H}\Psi_E = (\mathcal{H}_{cm} + \mathcal{H}_{Rel})\psi_{cm}\cdot\psi_{Rel} = (E_{cm} + E_{Rel})\psi_{cm}\cdot\psi_{Rel} = E\Psi_E.$$

We will now focus our attention on the bound system and will disregard the energy of the system as a whole. Specifically, we will be concerned only with solutions of the problem

$$\mathcal{H}_{Rel}\psi_{Rel} = E_{Rel}\psi_{Rel}.$$

Hereafter we will drop the label "*Rel*" and will write the equation to be solved as

$$\left[-\frac{\hbar^2}{2\mu}\nabla^2 + V(r)\right]\psi_E = E\psi_E. \tag{7.6}$$

In Cartesian coordinates, Equation 7.6 would not be separable because $V(r)$ would involve the radical $(x^2 + y^2 + z^2)^{\frac{1}{2}}$. Therefore, it is simpler to transform ∇^2 to polar coordinates and to make use of the fact that the radial and angular parts of Equation 7.6 can be separated. It is convenient to denote the angular part of the Laplacian by $\mathcal{L}^2$ and the radial part by $\mathcal{R}$. That is, we let

$$\nabla^2 = \mathscr{R} + \frac{1}{r^2} \mathscr{L}^2,$$

where

$$\mathscr{R} = \left(\frac{\partial^2}{\partial r^2} + \frac{2}{r} \frac{\partial}{\partial r} \right) \qquad (7.7)$$

and

$$\mathscr{L}^2 = \frac{1}{\sin \theta} \frac{\partial}{\partial \theta} \left(\sin \theta \frac{\partial}{\partial \theta} \right) + \frac{1}{\sin^2 \theta} \frac{\partial^2}{\partial \phi^2}$$

PROBLEM 7-4

Show that the Laplacian operator in spherical coordinates is given by

$$\nabla^2 = \left(\frac{\partial^2}{\partial r^2} + \frac{2}{r} \frac{\partial}{\partial r} \right) + \frac{1}{r^2} \left[\frac{1}{\sin \theta} \frac{\partial}{\partial \theta} \left(\sin \theta \frac{\partial}{\partial \theta} \right) + \frac{1}{\sin^2 \theta} \frac{\partial^2}{\partial \phi^2} \right].$$

Since the potential term is independent of the angle variables, it is convenient to use a product wave function of the form

$$\psi_E(r, \theta, \phi) = R(r) \cdot Y(\theta, \phi).$$

Then Equation 7.6 becomes:

$$\left[-\frac{\hbar^2}{2\mu} \mathscr{R} R(r) \right] Y(\theta, \phi) - \frac{\hbar^2}{2\mu r^2} [\mathscr{L}^2 Y(\theta, \phi)] \cdot R(r)$$

$$+ V(r) \cdot R(r) \cdot Y(\theta, \phi) = E \cdot R(r) \cdot Y(\theta, \phi).$$

$Y(\theta, \phi)$ is written outside of the first bracket to indicate that it is not affected by the radial operator. Likewise, $R(r)$ is not affected by the angular operator in the second bracket. Multiplying each term by

$$\frac{2\mu r^2}{\hbar^2} \cdot \frac{1}{R(r) \cdot Y(\theta, \phi)},$$

we have the separated equations

$$\frac{r^2 \mathscr{R} R(r)}{R(r)} + \frac{2\mu r^2}{\hbar^2} [E - V(r)] = -\frac{\mathscr{L}^2 Y(\theta, \phi)}{Y(\theta, \phi)} \equiv \Lambda. \qquad (7.8)$$

Setting each member of Equation 7.8 equal to the separation constant, Λ, we have the two equations,

$$\mathscr{R}R(r) + \frac{2\mu}{\hbar^2}[E - V(r)]R(r) = \frac{\Lambda}{r^2}R(r) \tag{7.9}$$

and

$$\mathscr{L}^2 Y(\theta, \phi) = -\Lambda Y(\theta, \phi). \tag{7.10}$$

Since we know the differential operators $\mathscr{L}^2$ and $\mathscr{R}$, Equations 7.9 and 7.10 can be solved by the standard techniques for ordinary differential equations once the exact form of the potential is put into Equation 7.9. However, the reader should recognize Equation 7.10 as an eigenequation of the type that we have encountered before. Therefore, we prefer to solve Equation 7.10 by means of operator algebra in a manner similar to that used for the harmonic oscillator. But first we must digress to discuss angular momentum and its operators in three dimensions.

3. THE ANGULAR MOMENTUM OPERATORS

The classical definition of the angular momentum vector[1] is

$$\vec{L} = \vec{r} \times \vec{p},$$

having components given by

$$\left.\begin{array}{l} L_x = yp_z - zp_y \\ L_y = zp_x - xp_z \\ L_z = xp_y - yp_x \end{array}\right\}. \tag{7.11}$$

We use the same definitions for the components of the angular momentum operators, where it is understood that p_i is the momentum operator, $-i\hbar(\partial/\partial x_i)$. As we discovered earlier (Problem 4-22), a coordinate operator and its conjugate momentum operator do not commute. Hence we should not expect the components of the angular momentum operator to commute with each other.

PROBLEM 7-5

Verify the following commutation relations:

$$\left.\begin{array}{ll} \text{(a)} & [x_i, p_j] = i\hbar\delta_{ij} \\ \text{(b)} & [x_i, x_j] = 0 \\ \text{(c)} & [p_i, p_j] = 0 \end{array}\right\}. \tag{7.12}$$

---

[1] Although $\vec{L} = \vec{r} \times \vec{p}$ behaves like an ordinary vector (polar vector) under addition, it is actually a pseudovector (axial vector). Its different behavior appears under a reflection of all coordinates through the origin: $\vec{r}$ and $\vec{p}$ are reversed, but $\vec{L}$ is not.

In particular, consider the commutator $[L_x, L_y]$. The multiplication may be carried out explicitly, or it may be indicated as follows:

$$[L_x, L_y] = [(yp_z - zp_y), (zp_x - xp_z)]$$

$$= [yp_z, zp_x] - \overset{\overset{0}{\nearrow}}{[yp_z, xp_z]} - [zp_y, zp_x] + \overset{\overset{0}{\nearrow}}{[zp_y, xp_z]}$$

$$= yp_x[p_z, z] + xp_y[z, p_z]$$

$$= -i\hbar yp_x + i\hbar xp_y = i\hbar(xp_y - yp_x) = i\hbar L_z.$$

In the second line above, the two indicated commutators are zero because all of the factors in each bracket commute. In the third line, all commuting factors have been taken outside of the commutator brackets. The commutation relations for the other angular momentum components may be obtained easily from cyclic permutations of x, y, z in the expression

Thus,

and

$$\left.\begin{matrix} [L_x, L_y] = i\hbar L_z \\ \\ [L_y, L_z] = i\hbar L_x \\ \\ [L_z, L_x] = i\hbar L_y \end{matrix}\right\} . \qquad (7.13)$$

As in the case of vectors, the square of the total angular momentum operator may be defined by

$$L^2 = L_x^2 + L_y^2 + L_z^2.$$

PROBLEM 7-6

Verify that the commutation relations for the components of the angular momentum operator may be summarized by the mnemonic expression:

$$\vec{L} \times \vec{L} = i\hbar\vec{L}.$$

If $\vec{L}$ were an ordinary vector, $\vec{L} \times \vec{L}$ would be identically zero. Why is this not true here?

PROBLEM 7-7

Prove the following commutation relations:
(a) $[L^2, L_x] = [L^2, L_y] = [L^2, L_z] = 0.$
(b) $[L^2, L_x^2] = [L^2, L_y^2] = [L^2, L_z^2] = 0.$

The commutation relations derived in Problems 7-6 and 7-7 have important consequences with regard to measurements of the total angular momentum and its components in a physical system. The fact that the operators L_x, L_y, and L_z do not commute means that a measurement of any one component of the angular momentum introduces an uncertainty of the order of $\hbar$ in our knowledge of any other component of angular momentum. Thus, in a quantum system, a measurement of one component of the angular momentum essentially nullifies what might have been learned previously about some other component. In other words, a measurement of L_x forces the system into an eigenstate of L_x. This state cannot be an eigenstate of either L_y or L_z. However, the two operators L_x and L^2 commute, so these operators *can* have simultaneous eigenstates, and the knowledge of these states will be preserved throughout any sequence of measurements of L_x and L^2. An attempt to measure L_y or L_z will immediately destroy the eigenstate of L_x but will not disturb the eigenstate of L^2.

PROBLEM 7-8

Show that the components of the angular momentum in spherical coordinates are:

$$L_x = i\hbar\left(\sin\phi\,\frac{\partial}{\partial\theta} + \cot\theta\cos\phi\,\frac{\partial}{\partial\phi}\right)$$

$$L_y = -i\hbar\left(\cos\phi\,\frac{\partial}{\partial\theta} - \cot\theta\sin\phi\,\frac{\partial}{\partial\phi}\right)$$

$$L_z = -i\hbar\,\frac{\partial}{\partial\phi}$$

PROBLEM 7-9

Using the expressions in Problem 7-8 show that $L^2 = -\hbar^2\mathscr{L}^2$, where $\mathscr{L}^2$ is defined in Equation 7.7.

We see from the result of Problem 7-9 that the angular part of the Laplacian operator is identical with the square of the total angular momentum operator except for the factor $(-\hbar^2)$. Furthermore, if we rewrite Equation 7.10 as

$$L^2 Y(\theta,\phi) = \Lambda\hbar^2 Y(\theta,\phi), \tag{7.14}$$

it is evident that the eigenfunctions of the spatial operator $\mathscr{L}^2$ and the eigenfunctions of the angular momentum operator are the same set of functions. These will be shown later to be the *spherical harmonics*. This close relationship between geometry and momentum is real, and indeed, is summed up by the statements that *the conservation of linear momentum in the absence of external forces follows from the homogeneity of space* and *the conservation of angular momentum in the absence of external torques follows from the isotropy of space.*

In order to illustrate these remarks, consider an infinitesimal displacement of a particle from position $\vec{r}$ to $(\vec{r} + \delta\vec{r})$. This may be thought of as a transformation T_δ which transforms $\psi(\vec{r})$ into $\psi(\vec{r} + \delta\vec{r})$; that is,

$$T_\delta\psi(\vec{r}) = \psi(\vec{r} + \delta\vec{r}) \approx \psi(\vec{r}) + \delta\vec{r} \cdot \nabla\psi(\vec{r}), \tag{7.15}$$

where the last expression is obtained from the first two terms of the Taylor expansion of $\psi(\vec{r} + \delta\vec{r})$ about $\psi(\vec{r})$. If we make use of the expression $\vec{p} = -i\hbar\vec{\nabla}$, then we obtain

$$T_\delta\psi(\vec{r}) = \psi(\vec{r}) + \frac{i}{\hbar}\,\delta\vec{r}\cdot\vec{p}\psi(\vec{r}) = \left(1 + \frac{i}{\hbar}\,\delta\vec{r}\cdot\vec{p}\right)\psi(\vec{r}). \tag{7.16}$$

Therefore, the infinitesimal translation operator has the following explicit dependence on the linear momentum:

$$T_\delta = 1 + \frac{i}{\hbar}\,\delta\vec{r}\cdot\vec{p}. \tag{7.17}$$

A finite translation through a distance $\vec{s}$ can be regarded as the repeated application of the infinitesimal translation operator. Thus,

$$\begin{aligned}
T_{\vec{s}} &= \lim_{\delta\vec{r}\to 0}\left(1 + \frac{i}{\hbar}\,\delta\vec{r}\cdot\vec{p}\right)^{s/\delta r} \\
&= \sum_n^\infty \frac{1}{n!}\left(\frac{i}{\hbar}\,\vec{s}\cdot\vec{p}\right)^n \\
&= e^{(i/\hbar)\vec{s}\cdot\vec{p}}.
\end{aligned} \tag{7.18}$$

To show formally that translational invariance implies conservation of linear momentum, let us consider a system for which the total energy is independent of the choice of the origin of the coordinate system. That is, the Hamiltonian of the system possesses translational invariance. Mathematically, this implies that the Hamiltonian commutes with the translation operator, as the following will show. Since the Hamiltonian operator and the energy are unchanged by the translation we have

$$\mathscr{H}\psi_E = E\psi_E$$

and

$$\mathscr{H}T_{\vec{s}}\psi_E = ET_{\vec{s}}\psi_E. \tag{7.19}$$

Operating on the first of these equations with $T_{\vec{s}}$, we obtain

$$T_{\vec{s}}\mathscr{H}\psi_E = T_{\vec{s}}E\psi_E = ET_{\vec{s}}\psi_E. \tag{7.20}$$

Subtracting Equations 7.19 and 7.20,

$$\mathscr{H}T_{\vec{s}} - T_{\vec{s}}\mathscr{H} = 0,$$

or

$$[\mathscr{H}, T_{\vec{s}}] = 0, \tag{7.21}$$

Figure 7-2 The relationship between an infinitesimal displacement and an infinitesimal rotation. $\delta\vec{\phi}$ is perpendicular to the plane of $\vec{r}$ and $\delta\vec{r}$.

which is the required result. Since the operator $T_{\vec{s}}$ depends explicitly on the linear momentum $\vec{p}$, the latter must also commute with the Hamiltonian and is thus a conserved quantity.

Now let us consider an infinitesimal rotation through an angle $\delta\phi$ by regarding the displacement $\delta\vec{r}$ to be given by $\delta\vec{\phi} \times \vec{r}$. Then the counterpart of Equation 7.15 is:

$$R_\delta \psi(\vec{r}) = \psi(\vec{r}) + \delta\vec{\phi} \times \vec{r} \cdot \vec{\nabla}\psi(\vec{r})$$

and

$$R_\delta = 1 + \delta\vec{\phi} \times \vec{r} \cdot \vec{\nabla} = 1 + \delta\vec{\phi} \cdot \vec{r} \times \vec{\nabla}$$

$$= 1 + \frac{i}{\hbar}\delta\vec{\phi} \cdot \vec{L}. \tag{7.22}$$

As before, we can express a finite rotation about *the same axis* as the repeated application of the infinitesimal rotation operator:

$$R_\phi = \lim_{\delta\vec{\phi} \to 0} \left(1 + \frac{i}{\hbar}\delta\vec{\phi} \cdot \vec{L}\right)^{\phi/\delta\phi}$$

$$= \sum_n^\infty \frac{1}{n!}\left(\frac{1}{\hbar}\vec{\phi} \cdot \vec{L}\right)^n$$

$$= e^{(i/\hbar)\vec{\phi}\cdot\vec{L}}$$

Therefore, when the Hamiltonian has rotational symmetry about an axis, the component of angular momentum along that axis commutes with the Hamiltonian and is conserved. Further, if the Hamiltonian has spherical symmetry the total angular momentum and its component along any single axis are conserved.

PROBLEM 7-10

Using the approach of Equation 7.19 to 7.21, show that angular momentum is conserved when $\mathscr{H}$ is invariant under any rotation.

One important point should be emphasized before closing this discussion, namely, that *finite rotations about different axes do not commute.* It is now easy to understand why this is so, since we know that rotations are intimately connected with angular momenta and that the components of the angular momentum do not commute with one another. Suppose we now consider two infinitesimal rotations about the x– and y–axes to see what effect the order of the operations will have on the result. This is given by the commutator,

$$[R_{\delta x}, R_{\delta y}] = -\frac{1}{\hbar^2} [L_x, L_y] \, \delta\phi_x \, \delta\phi_y \qquad (7.23)$$

$$\approx 0,$$

provided the second-order infinitesimal $\delta\phi_x \, \delta\phi_y$ is negligibly small. To the extent that the latter is true, we can construct any finite rotation from a large number of infinitesimal rotations. This difficulty does not occur in the case of translations along different axes, since the coordinate operators commute for both finite and infinitesimal translations.

4. EIGENVALUES OF THE ANGULAR MOMENTUM OPERATORS

As in the case of the harmonic oscillator, the algebraic method for obtaining the eigenvalues of the angular momentum operators requires the use of raising and lowering operators which we will denote by the symbols L_+ and L_-, respectively. They are defined as follows:

$$\left. \begin{array}{l} L_+ = L_x + iL_y = \hbar e^{i\phi} \left(\dfrac{\partial}{\partial\theta} + i \cot\theta \, \dfrac{\partial}{\partial\phi} \right) \\[4mm] L_- = L_x - iL_y = -\hbar e^{-i\phi} \left(\dfrac{\partial}{\partial\theta} - i \cot\theta \, \dfrac{\partial}{\partial\phi} \right) \end{array} \right\} . \qquad (7.24)$$

Let us first ascertain the commutation properties of these new operators. Thus,

$$L_+L_- = (L_x + iL_y)(L_x - iL_y) = L_x^2 + L_y^2 - i[L_x, L_y]$$
$$L_-L_+ = (L_x - iL_y)(L_x + iL_y) = L_x^2 + L_y^2 + i[L_x, L_y].$$

Therefore,

$$[L_+, L_-] = L_+L_- - L_-L_+ = -2i[L_x, L_y] = 2\hbar L_z . \qquad (7.25)$$

PROBLEM 7-11

Verify the following commutation relations:

(a) $[L_z, L_+] = \hbar L_+$

(b) $[L_z, L_-] = -\hbar L_-$ $\left. \phantom{\begin{array}{c}a\\b\\c\end{array}} \right\}$ $\qquad (7.26)$

(c) $[L^2, L_+] = [L^2, L_-] = 0$

It is also useful to note that

$$L_+L_- + L_-L_+ = 2(L_x^2 + L_y^2).$$

This permits us to write the operator for the square of the total angular momentum in any of the four following forms:

$$\left.\begin{aligned}
L^2 &= L_x^2 + L_y^2 + L_z^2 \\
&= \tfrac{1}{2}(L_+L_- + L_-L_+) + L_z^2 \\
&= L_+L_- - \hbar L_z + L_z^2 \\
&= L_-L_+ + \hbar L_z + L_z^2
\end{aligned}\right\}. \qquad (7.27)$$

The advantages of using L_+ and L_- and the expressions derived above will become clear shortly.

The problem we have before us is that of solving Equation 7.14, which is derived from Equation 7.10, for both Λ and Y. That is, for each allowed value of Λ (eigenvalue) there is a function Y (eigenfunction) that satisfies Equation 7.14. Our goal is to obtain the whole set of these functions. We also have another important bit of information at our disposal: the operator L^2 commutes with the operator for *one* component of angular momentum, which we will elect to be L_z. This means that the functions Y are also eigenfunctions of L_z (see section 2 of Chapter 6). It was stated in section 11 of Chapter 3 that the z component of the orbital angular momentum is an integer times $\hbar$, say $m\hbar$. This will be derived in what follows below. Let us then write:

$$L^2 Y^m = \Lambda \hbar^2 Y^m \qquad (7.28)$$

and

$$L_z Y^m = m\hbar Y^m, \qquad (7.29)$$

where we have labeled a particular eigenfunction with its eigenvalue of L_z (without the $\hbar$).

Now let us operate on Equation 7.29 with the operator L_+:

$$L_+ L_z Y^m = m\hbar L_+ Y^m. \qquad (7.30)$$

But we may use the commutation relation (Equation 7.26) to rewrite the left member of Equation 7.30 as:

$$L_+ L_z Y^m = (L_z L_+ - \hbar L_+) Y^m = L_z L_+ Y^m - \hbar L_+ Y^m. \qquad (7.31)$$

Equating the right members of Equations 7.30 and 7.31, we obtain:

$$L_z(L_+ Y^m) = (m + 1)\hbar(L_+ Y^m). \qquad (7.32)$$

Equation 7.32 states that the quantity $L_+ Y^m$ may be regarded as a new eigenfunction of L_z having a new eigenvalue $(m + 1)\hbar$. Now let us see if this new function satisfies Equation 7.28:

$$L^2(L_+Y^m) = L_+(L^2Y^m) = L_+(\Lambda\hbar^2 Y^m) = \Lambda\hbar^2(L_+Y^m). \qquad (7.33)$$

Thus, we see that L_+ is a *raising operator*. If Y^m is an eigenfunction of L_z and L^2, operation on Y^m with L_+ will generate a new eigenfunction associated with the same eigenvalue of L^2 but with an eigenvalue of L_z which is greater by one unit of $\hbar$.

Operating now on Equation 7.32 with L_+ we have

$$L_+L_z(L_+Y^m) = (m+1)\hbar(L_+^2 Y^m), \qquad (7.34)$$

and from the commutation relation the left member becomes

$$L_+L_z(L_+Y^m) = L_zL_+^2 Y^m - \hbar L_+^2 Y^m. \qquad (7.35)$$

Equating the right members of Equation 7.34 and 7.35, we have

$$L_z(L_+^2 Y^m) = (m+2)\hbar(L_+^2 Y^m).$$

Also,

$$L^2(L_+^2 Y^m) = L_+^2 L^2 Y^m = \Lambda\hbar^2(L_+^2 Y^m).$$

Thus, once again we have obtained a new simultaneous eigenfunction of L_z and L^2, where the eigenvalue of L_z has been raised another unit of $\hbar$ but the eigenvalue of L^2 is unchanged. Generalizing these results, one may write

$$L_z(L_+^r Y^m) = (m+r)\hbar(L_+^r Y^m)$$
$$L^2(L_+^r Y^m) = \Lambda\hbar^2(L_+^r Y^m).$$

PROBLEM 7-12

If $L_zY^m = m\hbar Y^m$ and $L^2Y^m = \Lambda\hbar^2 Y^m$, show that $L_zL_-^3 Y^m = (m-3)\hbar L_-^3 Y^m$ and $L^2L_-^3 Y^m = \Lambda\hbar^2 L_-^3 Y^m$.

We conclude from the above that a whole set of simultaneous eigenfunctions of L_z and L^2 can be generated by operating with L_+ or L_-, and it is quite natural to ask what limitations there are on the number of these functions. Recall that in the treatment of the harmonic oscillator the raising operator can be applied any number of times, because there is no limit to the number of higher states. However, there *is* a finite limit to the number of lowering operations in the harmonic oscillator, because eventually the ground state will be reached. In fact, it is this limit which allows us to obtain the ground state eigenfunction (see Equation 5.51). In order to determine the limitations in this present case of the angular momentum eigenfunctions, we write:

$$(L_x^2 + L_y^2)Y^m = (L^2 - L_z^2)Y^m = (\Lambda - m^2)\hbar^2 Y^m.$$

Since the angular momentum operators L_x, L_y, L_z, and L^2 are Hermitian operators with real expectation values, the squares of these expectation values must be *positive* numbers. Thus the operator $(L_x^2 + L_y^2)$ must have a positive eigenvalue and $\Lambda \geq m^2$. This means that there is a limit on the number of times that either the lowering or the raising operator may be applied to a given function Y^m, and thus the set of eigenfunctions must terminate with a minimum and a maximum eigenvalue of L_z. For convenience let ℓ_1 be the maximum value of m and $-\ell_2$ be the minimum value. It follows that

$$L_+ Y^{\ell_1} = 0$$

and

$$L_- Y^{-\ell_2} = 0$$

(7.36)

since, by definition, Y^{ℓ_1} is the function associated with the top of the ladder of states and $Y^{-\ell_2}$ is the bottom function. Letting n be the integer representing the number of steps between Y^{ℓ_1} and $Y^{-\ell_2}$, we have

$$n = \ell_1 - (-\ell_2) = \ell_1 + \ell_2.$$

(7.37)

Using Equations 7.27, 7.28, 7.29, and 7.36, we obtain:

$$L^2 Y^{\ell_1} = (L_- L_+ + \hbar L_z + L_z^2) Y^{\ell_1} = (\hbar^2 \ell_1 + \hbar^2 \ell_1^2) Y^{\ell_1} = \Lambda \hbar^2 Y^{\ell_1},$$

$$L^2 Y^{-\ell_2} = (L_+ L_- - \hbar L_z + L_z^2) Y^{-\ell_2} = (\hbar^2 \ell_2 + \hbar^2 \ell_2^2) Y^{-\ell_2} = \Lambda \hbar^2 Y^{-\ell_2}$$

or

$$\Lambda = \ell_1(\ell_1 + 1) = \ell_2(\ell_2 + 1).$$

(7.38)

This implies either that $\ell_2 = \ell_1$ or that $\ell_2 = -(\ell_1 + 1)$. The latter choice is rejected, however, since it contradicts the assumption that both ℓ_1 and ℓ_2 are positive (Equation 7.36). From Equations 7.37 and 7.38 we find that:

$$\ell = \ell_1 = \ell_2 = \frac{n}{2}$$

(7.39)

and

$$\Lambda = \ell(\ell + 1).$$

(7.40)

From Equation 7.39 we learn that the maximum value of m is ℓ and the minimum value of m is $-\ell$, where ℓ can be an integer *or a half-integer*. For the classical orbital angular momentum which we are considering here, only integral values of ℓ are physically meaningful, so the integer n must be even. However, the reader can probably guess that the existence of the half-integral states for odd n will permit this same formalism to be used for treating half-integral spin angular momentum. From Equation 7.40 we see that the eigenvalue of the square of the total angular momentum is $\ell^2 + \ell$, where ℓ is the maximum observable component of the total angular momentum. Thus, one can never measure a component of angular momentum equal in magnitude to the total angular momentum. A measurement of, say, the z-component will

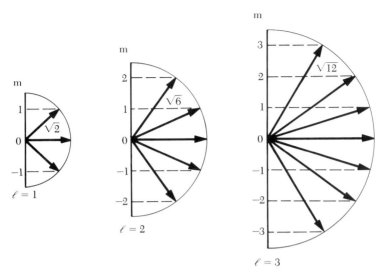

Figure 7-3   The allowed projections of the angular momentum for the cases of $\ell = 1, 2,$ and 3.

always yield one of the $2\ell + 1$ allowed m values given by $\ell, \ell - 1, \ell - 2, \ldots,$ $0, -1, \ldots, -\ell$. This is illustrated for a few ℓ-values in Figure 7-3.

5.   THE ANGULAR MOMENTUM EIGENFUNCTIONS

Using the results of the last section, we will now denote the angular momentum eigenfunctions by Y_ℓ^m, where m, actually $m\hbar$, is the eigenvalue of L_z and ℓ is called the orbital quantum number. The latter is related to the eigenvalue of L^2 by the expression $\ell(\ell + 1)\hbar^2$. If any one eigenfunction is known, then the other 2ℓ functions corresponding to the same ℓ value can be obtained by repeated applications of the raising or lowering operators. In particular, if the known function is either Y_ℓ^ℓ or $Y_\ell^{-\ell}$, corresponding to the top or bottom of the ladder, then repeated applications of the same operator will generate the whole set of functions.

The clue to obtaining either of these functions is contained in Equation 7.36. Using Equation 7.24, we may write the following equation for Y_ℓ^ℓ:

$$L_+ Y_\ell^\ell = \hbar e^{i\phi} \left(\frac{\partial}{\partial\theta} + i \cot\theta \, \frac{\partial}{\partial\phi} \right) Y_\ell^\ell (\theta, \phi) = 0,$$

or   (7.41)

$$\left(\frac{\partial}{\partial\theta} + i \cot\theta \, \frac{\partial}{\partial\phi} \right) Y_\ell^\ell(\theta, \phi) = 0.$$

Since this is a partial differential equation in two variables, we will employ the same separation procedure as before.

Writing

$$Y_\ell^\ell(\theta, \phi) = P(\theta) \cdot \Phi(\phi),$$

Equation 7.41 becomes

$$\frac{1}{P(\theta)} \frac{\partial}{\partial \theta} P(\theta) + \frac{i \cot \theta}{\Phi(\phi)} \frac{\partial}{\partial \phi} \Phi(\phi) = 0.$$

If we let the new separation constant be k, we obtain the two ordinary differential equations

$$\frac{d\Phi(\phi)}{\Phi(\phi)} = ik \, d\phi \tag{7.42}$$

$$\frac{dP(\theta)}{P(\theta)} = k \cot \theta \, d\theta, \tag{7.43}$$

whose solutions may be readily written in the form

$$\Phi(\phi) = e^{ik\phi}$$

and

$$P(\theta) = (\sin \theta)^k,$$

where the constants of integration have been neglected for the time being. Then the product function Y_ℓ^ℓ is:

$$Y_\ell^\ell = (C_\ell^\ell)(\sin \theta)^k e^{ik\phi},$$

where C_ℓ^ℓ is the constant of integration, which will be determined later from the normalization of the eigenfunctions.

PROBLEM 7-13 ▬▬▬▬▬▬▬▬▬▬▬▬▬▬▬▬▬▬▬▬▬▬▬▬▬▬

Use the eigenequations

$$L_z Y_\ell^\ell = \ell \hbar Y_\ell^\ell$$

and

$$L^2 Y_\ell^\ell = \ell(\ell + 1)\hbar^2 Y_\ell^\ell$$

to show that the separation constant $k = \ell$, and that the product function is

$$Y_\ell^\ell = (C_\ell^\ell)(\sin \theta)^\ell e^{i\ell\phi}. \tag{7.44}$$

▬▬▬▬▬▬▬▬▬▬▬▬▬▬▬▬▬▬▬▬▬▬▬▬▬▬▬▬▬▬▬▬▬▬

The eigenfunctions corresponding to the other 2ℓ m-values can be obtained by operating on Equation 7.44 with L_-. It simplifies the bookkeeping if the old normalization constant is replaced by the new one each time L_+ or L_-

operates. Thus,

$$Y_\ell^{\ell-1} = \frac{C_\ell^{\ell-1}}{\hbar\, C_\ell^\ell}\, L_- Y_\ell^\ell \tag{7.45}$$

$$= -C_\ell^{\ell-1} e^{-i\phi}\left(\frac{\partial}{\partial\theta} - i\cot\theta\,\frac{\partial}{\partial\phi}\right)(\sin\theta)^\ell e^{i\ell\phi}$$

$$= -C_\ell^{\ell-1} e^{i(\ell-1)\phi}\left(\frac{d}{d\theta} + \ell\cot\theta\right)(\sin\theta)^\ell.$$

But the operator

$$\frac{d}{d\theta} + \ell\cot\theta \equiv \frac{1}{(\sin\theta)^\ell}\frac{d}{d\theta}(\sin\theta)^\ell. \tag{7.46}$$

PROBLEM 7-14

Verify the identity given in Equation 7.46 by operating on a function of θ with each member.

Using Equation 7.46 we may then write:

$$Y_\ell^{\ell-1} = -C_\ell^{\ell-1} e^{i(\ell-1)\phi}\frac{1}{(\sin\theta)^\ell}\frac{d}{d\theta}(\sin\theta)^{2\ell}.$$

For the sth application of L_-,

$$Y_\ell^{\ell-s} = \left(\frac{1}{\hbar}\right)^s \frac{C_\ell^{\ell-s}}{C_\ell^\ell}\, L_-^s Y_\ell^\ell = (-1)^s C_\ell^{\ell-s} e^{i(\ell-s)\phi}\frac{1}{(\sin\theta)^\ell}\left(\frac{d}{d\theta}\right)^s(\sin\theta)^{2\ell}$$

$$\tag{7.47}$$

$$= (-1)^s C_\ell^{\ell-s} e^{i(\ell-s)\phi}\frac{1}{(\sin\theta)^{\ell-s}}\left(\frac{1}{\sin\theta}\frac{d}{d\theta}\right)^s(\sin\theta)^{2\ell}.$$

Now let us make the following substitutions:

$$\cos\theta = \eta$$

$$\frac{d}{d\theta} = -\sin\theta\,\frac{d}{d\eta}\,.$$

Then,

$$Y_\ell^{\ell-s}(\eta) = C_\ell^{\ell-s} e^{i(\ell-s)\phi}\left(\frac{1}{1-\eta^2}\right)^{(\ell-s)/2}\left(\frac{d}{d\eta}\right)^s(1-\eta^2)^\ell,$$

and if we set $\ell - s = m$,

$$Y_\ell^m(\eta) = C_\ell^m e^{im\phi}\left(\frac{1}{1-\eta^2}\right)^{m/2}\left(\frac{d}{d\eta}\right)^{\ell-m}(1-\eta^2)^\ell. \tag{7.48}$$

Equation 7.48 may be simplified by using the definition of the associated Legendre functions, namely,[2]

$$P_\ell^m(\eta) = (-1)^m \left(\frac{1}{1-\eta^2}\right)^{m/2} \left(\frac{d}{d\eta}\right)^{\ell-m} \cdot (\eta^2 - 1)^\ell. \qquad (7.49)$$

Then,

$$Y_\ell^m(\eta) = (-1)^m C_\ell^m \, P_\ell^m(\eta) e^{im\phi}. \qquad (7.50)$$

For convenience, a few of the associated Legendre functions are tabulated below. The argument here is η, and the functions have not been normalized. It will next be necessary to determine the coefficient C_ℓ^m in Equation 7.50.

$$P_0^0 = 1$$

$$P_1^0 = 2 \cos \theta$$

$$P_1^1 = \sin \theta$$

$$P_2^0 = 4(3 \cos^2 \theta - 1)$$

$$P_2^1 = 4 \sin \theta \cos \theta$$

$$P_2^2 = \sin^2 \theta$$

$$P_3^0 = 24(5 \cos^3 \theta - 3 \cos \theta)$$

$$P_3^1 = 6 \sin \theta (5 \cos^2 \theta - 1)$$

$$P_3^2 = 6 \sin^2 \theta \cos \theta$$

$$P_3^3 = \sin^3 \theta$$

PROBLEM 7-15

(a) Show that the associated Legendre functions given by Equation 7.49 are orthogonal.

(b) Evaluate the integral

$$\langle P_\ell^m(\eta) \mid P_\ell^m(\eta) \rangle = (2^\ell \ell!)^2 \cdot \frac{4\pi}{2\ell + 1} \frac{(\ell - m)!}{(\ell + m)!}.$$

6. NORMALIZATION OF THE ANGULAR MOMENTUM EIGENFUNCTIONS

The normalization procedure which will now be used to determine the constants C_ℓ^m is straightforward but tedious. First it will be necessary to obtain the normalization factor C_ℓ^ℓ and then we will obtain the dependence

[2] E. Merzbacher, *Quantum Mechanics*. John Wiley and Sons, Inc., New York, 1961, p. 179.

upon m which enters when we operate with L_-. Our starting point is the requirement that

$$\langle Y_\ell^\ell \mid Y_\ell^\ell \rangle = 1.$$

Then,

$$1 = |C_\ell^\ell|^2 \int_\Omega [(\sin \theta)^\ell e^{i\ell\phi}]^* \cdot [(\sin \theta)^\ell e^{i\ell\phi}] \, d\Omega$$

$$= 2\pi |C_\ell^\ell|^2 \int_0^\pi (\sin \theta)^{2\ell} \sin \theta \, d\theta.$$

Using the substitution $x = \cos \theta$, this becomes

$$1 = 2\pi |C_\ell^\ell|^2 \int_{-1}^{1} (1 - x)^\ell (1 + x)^\ell \, dx. \tag{7.51}$$

This integral may be evaluated by means of the properties of beta and gamma functions:

$$B(m, n) = \int_0^1 y^{m-1}(1 - y)^{n-1} \, dy = \frac{\Gamma(m)\,\Gamma(n)}{\Gamma(m + n)},$$

for m and n positive. In order to transform our integral to that of the beta function, we next make the substitution $t = \frac{1}{2}(1 + x)$. Then we have

$$1 + x = 2t$$
$$1 - x = 2(1 - t)$$
$$dx = 2 \, dt,$$

and Equation 7.51 becomes:

$$1 = 2\pi |C_\ell^\ell|^2 \cdot 2^{2\ell+1} \int_0^1 t^\ell (1 - t)^\ell \, dt$$

$$= 2\pi |C_\ell^\ell|^2 \cdot 2^{2\ell+1} B(\ell + 1, \ell + 1)$$

$$= 2\pi |C_\ell^\ell|^2 \cdot 2^{2\ell+1} \frac{\Gamma(\ell + 1)\Gamma(\ell + 1)}{\Gamma(2\ell + 2)}$$

$$= 2\pi |C_\ell^\ell|^2 \cdot 2^{2\ell+1} \frac{\ell!\,\ell!}{(2\ell + 1)!},$$

since $\Gamma(\ell + 1) = \ell!$ Therefore,

$$C_\ell^\ell = \frac{1}{2^\ell \ell!} \sqrt{\frac{(2\ell + 1)!}{4\pi}} \tag{7.52}$$

and

$$Y_\ell^\ell = \frac{1}{2^\ell \ell!} \sqrt{\frac{(2 + 1)!}{\pi}} (\sin \theta)^\ell e^{i\ell\phi}. \tag{7.53}$$

In order to introduce the m-dependence which arises when the lowering operator is used, we write Equation 7.45 in the form

$$Y_\ell^m = \frac{1}{\hbar} \frac{C_\ell^m}{C_\ell^{m+1}} L_- Y_\ell^{m+1}. \tag{7.54}$$

The square of the norm of Y_ℓ^m is:

$$\langle Y_\ell^m \mid Y_\ell^m \rangle = \frac{1}{\hbar^2} \left| \frac{C_\ell^m}{C_\ell^{m+1}} \right|^2 \langle L_- Y_\ell^{m+1} \mid L_- Y_\ell^{m+1} \rangle$$

$$= \frac{1}{\hbar^2} \left| \frac{C_\ell^m}{C_\ell^{m+1}} \right|^2 \langle Y_\ell^{m+1} \mid L_+ L_- Y_\ell^{m+1} \rangle,$$

where we use the fact that L_+ and L_- are Hermitian conjugates. Applying the commutation relations, the ket on the right becomes (see Equation 7.27):

$$|L_+ L_- Y_\ell^{m+1} \rangle = |(L^2 + \hbar L_z - L_z^2) Y_\ell^{m+1} \rangle$$

$$= \hbar^2 [\ell(\ell + 1) + (m + 1) - (m + 1)^2] \, | Y_\ell^{m+1} \rangle$$

$$= \hbar^2 (\ell - m)(\ell + m + 1) \, | Y_\ell^{m+1} \rangle.$$

Then

$$\langle Y_\ell^m \mid Y_\ell^m \rangle = \left| \frac{C_\ell^m}{C_\ell^{m+1}} \right|^2 \cdot (\ell - m)(\ell + m + 1) \langle Y_\ell^{m+1} \mid Y_\ell^{m+1} \rangle.$$

Using the same procedure, one readily obtains:

$$\langle Y_\ell^{m+1} \mid Y_\ell^{m+1} \rangle = \left| \frac{C_\ell^{m+1}}{C_\ell^{m+2}} \right|^2 \cdot (\ell - m - 1)(\ell + m + 2) \langle Y_\ell^{m+2} \mid Y_\ell^{m+2} \rangle.$$

It follows that

$$\langle Y_\ell^m \mid Y_\ell^m \rangle = \left| \frac{C_\ell^m}{C_\ell^{m+2}} \right|^2 \cdot (\ell - m)(\ell - m - 1)(\ell + m + 1)(\ell + m + 2)$$

$$\times \langle Y_\ell^{m+2} \mid Y_\ell^{m+2} \rangle.$$

Ultimately,

$$\langle Y_\ell^m \mid Y_\ell^m \rangle = \left| \frac{C_\ell^m}{C_\ell^\ell} \right|^2 \cdot (\ell - m)(\ell - m - 1) \cdots (1)(\ell + m + 1)(\ell + m + 2)$$

$$\times \cdots (2\ell) \langle Y_\ell^\ell \mid Y_\ell^\ell \rangle.$$

Thus,

$$\langle Y_\ell^m \mid Y_\ell^m \rangle = \left| \frac{C_\ell^m}{C_\ell^\ell} \right|^2 \cdot (\ell - m)! \cdot \frac{(2\ell)!}{(\ell + m)!} \langle Y_\ell^\ell \mid Y_\ell^\ell \rangle,$$

and setting both norms equal to unity, we obtain

$$C_\ell^m = C_\ell^\ell \sqrt{\frac{(\ell+m)!}{(2\ell)!(\ell-m)!}} = \frac{1}{2^\ell \ell!} \sqrt{\frac{2\ell+1}{4\pi}} \sqrt{\frac{(\ell+m)!}{(\ell-m)!}}, \qquad (7.55)$$

where Equation 7.52 has been used in the last expression. The value of this coefficient may now be put into our earlier expression (Equation 7.47) to obtain:

$$Y_\ell^m = \left(\frac{1}{\hbar}\right)^{\ell-m} \sqrt{\frac{(\ell+m)!}{(2\ell)!(\ell-m)!}} L_-^{\ell-m} Y_\ell^\ell, \qquad (7.56)$$

$$Y_\ell^m(\eta, \phi) = \frac{(-1)^m}{2^\ell \ell!} \sqrt{\frac{2\ell+1}{4\pi}} \sqrt{\frac{(\ell+m)!}{(\ell-m)!}} P_\ell^m(\eta) e^{im\phi}, \qquad (7.57)$$

and

$$Y_\ell^m(\eta, \phi) = \frac{1}{2^\ell \ell!} \sqrt{\frac{2\ell+1}{4\pi}} \sqrt{\frac{(\ell+m)!}{(\ell-m)!}} e^{im\phi} \left(\frac{1}{1-\eta^2}\right)^{m/2} \left(\frac{d}{d\eta}\right)^{\ell-m} (\eta^2-1)^\ell$$

$$(7.58)$$

where $\eta = \cos\theta$.

The above results were obtained for m positive. In order to obtain the functions for negative m, it is most convenient to use the relations

$$P_\ell^{-m}(\eta) = P_\ell^m(\eta) \qquad (7.59)$$

and

$$Y_\ell^{-m}(\eta) = (-1)^m (Y_\ell^m)*(\eta). \qquad (7.60)$$

Of course, these expressions could have been obtained by following the procedure above, using $Y_\ell^{-\ell}$ as the starting point and operating with L_+. In this case the counterparts of Equations 7.54 and 7.56 are:

$$Y_\ell^m = \frac{1}{\hbar} \frac{C_\ell^m}{C_\ell^{m-1}} L_+ Y_\ell^{m-1}, \qquad (7.61)$$

and

$$Y_\ell^m = \left(\frac{1}{\hbar}\right)^{\ell+m} \sqrt{\frac{(\ell-m)!}{(2\ell)!(\ell+m)!}} L_+^{\ell+m} Y_\ell^{-\ell}. \qquad (7.62)$$

The angular momentum eigenfunctions, as given in Equations 7.56 and 7.58, are known as the *spherical harmonics*. They comprise a complete, ortho-normal set of functions on the unit sphere. A few of the normalized spherical harmonics are tabulated below because of their frequent use:

$$Y_0^0 = \frac{1}{2\sqrt{\pi}}$$

$$Y_1^0 = \frac{1}{2}\sqrt{\frac{3}{\pi}}\cos\theta$$

$$Y_1^{\pm 1} = \mp\frac{1}{2}\sqrt{\frac{3}{2\pi}}\sin\theta \cdot e^{\pm i\phi}$$

$$Y_2^0 = \frac{1}{4}\sqrt{\frac{5}{\pi}}\cdot(3\cos^2\theta - 1)$$

$$Y_2^{\pm 1} = \mp\frac{1}{2}\sqrt{\frac{15}{2\pi}}\cdot\sin\theta\cdot\cos\theta\cdot e^{\pm i\phi}$$

$$Y_2^{\pm 2} = \frac{1}{4}\sqrt{\frac{15}{2\pi}}\cdot\sin^2\theta\cdot e^{\pm 2i\phi}$$

$$Y_3^0 = \frac{1}{4}\sqrt{\frac{7}{\pi}}\cdot(5\cos^3\theta - 3\cos\theta)$$

$$Y_3^{\pm 1} = \mp\frac{1}{8}\sqrt{\frac{21}{\pi}}\cdot\sin\theta\cdot(5\cos^2\theta - 1)\cdot e^{\pm i\phi}$$

$$Y_3^{\pm 2} = \frac{1}{4}\sqrt{\frac{105}{2\pi}}\cdot\sin^2\theta\cdot\cos\theta\cdot e^{\pm 2i\phi}$$

$$Y_3^{\pm 3} = \mp\frac{1}{8}\sqrt{\frac{35}{\pi}}\cdot\sin^3\theta\cdot e^{\pm 3i\phi}$$

PROBLEM 7-16

Use the results of Problem 7-15 to prove the orthonormality of the Y_l^m as expressed by Equation 7.57.

7. THE ANGULAR MOMENTUM MATRICES

Since the spherical harmonics are simultaneous eigenfunctions of the operators L^2 and L_z, they constitute a basis for a matrix representation of these operators in diagonal form. The matrix elements are given by:

$$\langle Y_{\ell'}^{m'}| L^2 |Y_\ell^m\rangle = \hbar^2\ell(\ell + 1)\,\delta_{\ell',\ell}\,\delta_{m',m}$$

and

$$\langle Y_{\ell'}^{m'}| L_z |Y_\ell^m\rangle = m\hbar\,\delta_{\ell',\ell}\,\delta_{m',m}.$$

Since the off-diagonal elements are characterized by either $\ell' \neq \ell$ or $m' \neq m$ (or both), the Kronecker deltas ensure that all off-diagonal elements are zero. If we order the basis functions as follows, $Y_0^0, Y_1^1, Y_1^0, Y_1^{-1}, Y_2^2, Y_2^1, Y_2^0, \ldots$, then we can write the matrix of L^2 as:

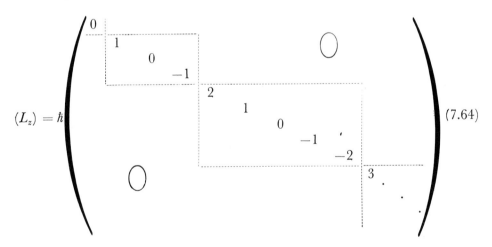

$$(L^2) = \hbar^2 \begin{pmatrix} 0 & & & & & & \\ & 2 & & & & & \\ & & 2 & & & & \\ & & & 2 & & & \\ & & & & 6 & & \\ & & & & & \ddots & \\ & & & & & & 12 \\ & & & & & & & \ddots \end{pmatrix}, \qquad (7.63)$$

where all of the off-diagonal elements are zero. The boxes correspond to the $2\ell + 1$ functions associated with the same eigenvalue of L^2. That is, the one-by-one submatrix corresponds to $\ell = 0$, the three-by-three to $\ell = 1$, the five-by-five to $\ell = 2$, and so forth. In like manner, the matrix representation for the operator L_z is:

$$(L_z) = \hbar \begin{pmatrix} 0 & & & & & & & \\ & 1 & & & & & & \\ & & 0 & & & & & \\ & & & -1 & & & & \\ & & & & 2 & & & \\ & & & & & 1 & & \\ & & & & & & 0 & \\ & & & & & & & -1 \\ & & & & & & & & -2 \\ & & & & & & & & & 3 \\ & & & & & & & & & & \ddots \end{pmatrix} \qquad (7.64)$$

Although the matrices shown in Equations 7.63 and 7.64 can be of infinite dimension, each set of $2\ell + 1$ functions associated with a given angular momentum forms a *subspace* of dimension $2\ell + 1$. Furthermore, *the $2\ell + 1$ spherical harmonics associated with that ℓ-value constitute a complete, orthonormal basis in that subspace.*

The operators L_+, L_-, L_x, and L_y can also be represented by matrices, although their representations are more difficult to obtain. One approach is to begin with Equations 7.54 and 7.55 and to write:

$$L_- Y_\ell^m = \hbar \sqrt{(\ell + m)(\ell - m + 1)} \, Y_\ell^{m-1}. \qquad (7.65)$$

Then, operating on Equation 7.65 with L_+, we obtain:

$$L_+L_-Y_\ell^m = \hbar\sqrt{(\ell + m)(\ell - m + 1)}\, L_+Y_\ell^{m-1}.$$

The left member may be simplified by using Equation 7.27.

$$(L^2 + \hbar L_z - L_z^2)Y_\ell^m = \hbar\sqrt{(\ell + m)(\ell - m + 1)}\, L_+Y_\ell^{m-1}$$

$$L_+Y_\ell^{m-1} = \frac{\hbar[\ell(\ell + 1) + m - m^2]}{\sqrt{(\ell + m)(\ell - m + 1)}}\, Y_\ell^m$$

$$= \hbar\sqrt{(\ell + m)(\ell - m + 1)}\, Y_\ell^m$$

or

$$L_+Y_\ell^m = \hbar\sqrt{(\ell - m)(\ell + m + 1)}\, Y_\ell^{m+1} \tag{7.66}$$

Equations 7.65 and 7.66 may be combined in the form

$$L_\pm Y_\ell^m = \hbar\sqrt{(\ell \mp m)(\ell \pm m + 1)}\, Y_\ell^{m\pm 1}. \tag{7.67}$$

Now it is possible to find the matrix elements of $L_\pm$. Thus:

$$\langle Y_{\ell'}^{m'} \mid L_\pm Y_\ell^m \rangle = \hbar\sqrt{(\ell \mp m)(\ell \pm m + 1)}\, \delta_{\ell',\ell}\, \delta_{m',m\pm 1}, \tag{7.68}$$

because of the orthonormality of the spherical harmonics. Hence, the orthog-onality property requires that all matrix elements of $(L_\pm)$ be zero except for those along one diagonal adjacent to the principal diagonal. By way of illustration, the matrices for the case of $\ell = 2$ are shown below:

$$(L_+) = \hbar \begin{pmatrix} 0 & 2 & 0 & 0 & 0 \\ 0 & 0 & \sqrt{6} & 0 & 0 \\ 0 & 0 & 0 & \sqrt{6} & 0 \\ 0 & 0 & 0 & 0 & 2 \\ 0 & 0 & 0 & 0 & 0 \end{pmatrix},$$

and

$$(L_-) = \hbar \begin{pmatrix} 0 & 0 & 0 & 0 & 0 \\ 2 & 0 & 0 & 0 & 0 \\ 0 & \sqrt{6} & 0 & 0 & 0 \\ 0 & 0 & \sqrt{6} & 0 & 0 \\ 0 & 0 & 0 & 2 & 0 \end{pmatrix},$$

where all other elements are zero. Then, from the definitions of Equation

7.24, for $\ell = 2$ we see that

$$(L_x) = \tfrac{1}{2}(L_+) + \tfrac{1}{2}(L_-) = \frac{\hbar}{2} \begin{pmatrix} 0 & 2 & 0 & 0 & 0 \\ 2 & 0 & \sqrt{6} & 0 & 0 \\ 0 & \sqrt{6} & 0 & \sqrt{6} & 0 \\ 0 & 0 & \sqrt{6} & 0 & 2 \\ 0 & 0 & 0 & 2 & 0 \end{pmatrix}$$

and

$$(L_y) = \frac{i\hbar}{2} \begin{pmatrix} 0 & -2 & 0 & 0 & 0 \\ 2 & 0 & -\sqrt{6} & 0 & 0 \\ 0 & \sqrt{6} & 0 & -\sqrt{6} & 0 \\ 0 & 0 & \sqrt{6} & 0 & -2 \\ 0 & 0 & 0 & 2 & 0 \end{pmatrix}.$$

PROBLEM 7-17

Derive Equation 7.65.

PROBLEM 7-18

(a) Write the matrix representations of L_x, L_y, and L_z for $\ell = 3$.

(b) Using matrix algebra, verify that the matrices obtained in (a) satisfy the equation: $(L^2) = (L_x^2) + (L_y^2) + (L_z^2)$.

PROBLEM 7-19

(a) Operate on Y_2^2 with L_- to obtain Y_2^1 and Y_2^0.

(b) Operate on Y_2^{-2} with L_+ to obtain Y_2^{-1} and Y_2^0. Hint: Make use of Equations 7.24 and 7.67.

PROBLEM 7-20

(a) Find the matrix (U) that will diagonalize (L_x). (Do this for the 3 × 3 matrix).

(b) What does this transformation do to (L_y) and (L_z)?

(c) Show that the transformed matrices still satisfy $(L_x)^2 + (L_y)^2 + (L_z)^2 = 2\hbar^2(1)$. Physically, this is equivalent to changing the axis of quantization from the z-axis to the x-axis by, for example, rotating a magnetic field from the z-direction to the x-direction.

8. HYDROGENIC ATOMS

As in section 4 of Chapter 3, we will adopt the model of an atom consisting of a fixed nucleus of charge Z and a single electron of reduced mass μ moving in a spherically symmetric Coulomb potential. However, we have refined our earlier model in two important ways. First, the electron is no longer viewed as a hard sphere of charge $-e$ whose precise position can be specified in space and time. Instead, we describe it by means of a wave function $\Psi(r, \theta, \phi, t)$, whose square is a measure of the probability that the electron can be found at position (r, θ, ϕ) at time t. In the second place, we now understand the intimate relationship between coordinate space and angular momentum. Therefore, instead of artificially imposing Bohr's quantum condition on circular orbits, we now know that the allowed wave functions for the electron must be eigenfunctions of the square of the total angular momentum operator.

Our task is to solve the time-independent Schrödinger equation for the allowed energies and wave functions of the electron. Thus, we begin by setting $V(r) = -(Ze^2/r)$ in Equation 7.6. Since the Coulomb potential is spherically symmetric we can immediately write:

$$\psi_{\ell,m}(r, \theta, \phi) = R(r) \cdot Y_\ell^m(\theta, \phi),$$

where the wave function has been labelled with the angular momentum quantum numbers corresponding to those of the angular function. We already know that the angular functions are the spherical harmonics given by Equation 7.57. It now remains to determine the radial functions. Rewriting Equation 7.9, we obtain:

$$\left(\frac{\partial^2}{\partial r^2} + \frac{2}{r}\frac{\partial}{\partial r}\right)R(r) - \frac{2\mu}{\hbar^2}\left[\left(\frac{\ell(\ell+1)\hbar^2}{2\mu r^2} - \frac{Ze^2}{r}\right) - E\right]R(r) = 0. \quad (7.69)$$

Before proceeding with the solution of Equation 7.69, it should be pointed out that the quantity in parentheses in the second term is often called the *effective* potential seen by the electron. A non-zero angular momentum results in a so-called "centrifugal barrier," which must be overcome as the electron approaches the origin. Hence the effective potential is the algebraic sum of the Coulomb potential and the so-called "centrifugal potential." (See Figure 7-4.)

Equation 7.69 can be simplified by means of the substitution

$$R(r) = \frac{g(r)}{r},$$

which, upon differentiation with respect to r, has the derivatives:

$$R'(r) = \frac{g'(r)}{r} - \frac{g(r)}{r^2}$$

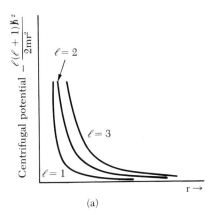

$\ell = 2$

$\ell = 3$

$\ell = 1$

$r \rightarrow$

(a)

Figure 7-4 (a) The centrifugal potential for several values of ℓ. (b) The effective hydrogenic potential due to both the Coulomb and centrifugal terms.

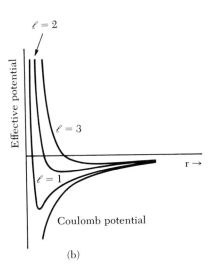

$\ell = 2$

$\ell = 3$

$\ell = 1$

$r \rightarrow$

Coulomb potential

(b)

and

$$R''(r) = \frac{g''(r)}{r} - \frac{2g'(r)}{r^2} + \frac{2g(r)}{r^3}.$$

Then Equation 7.69 becomes:

$$g''(r) - \frac{2\mu}{\hbar^2}\left[\frac{\ell(\ell+1)\hbar^2}{2\mu r^2} - \frac{Ze^2}{r} - E\right]g(r) = 0. \qquad (7.70)$$

Since this cannot be solved in closed form, we must employ either the power series method[3] or an analytical approach which utilizes a knowledge of the asymptotic solutions in the limits of very small and very large r. Choosing

---

[3] See L. Pauling and E. B. Wilson, *Introduction to Quantum Mechanics*. McGraw-Hill Book Co., Inc., New York, 1935, p. 121.

the latter approach, we note that for very large r, Equation 7.70 may be approximated by

$$g''(r) + \frac{2\mu E}{\hbar^2} g(r) = 0,$$

whose solutions are

$$g(r) \sim \exp\left(\pm\frac{i}{\hbar}\sqrt{2\mu E}\,r\right).$$

Now, for bound solutions the total energy E must be negative. Therefore, we write

$$\pm\frac{i}{\hbar}\sqrt{-2\mu\,|E|} = \pm\frac{1}{\hbar}\sqrt{2\mu\,|E|} \equiv \pm\lambda,$$

and then

$$g(r) \sim e^{\pm\lambda r}.$$

Since the plus sign leads to a function which diverges for large r, we cannot use it for a physical system. Hence our only acceptable solution is

$$g(r) \sim e^{-\lambda r}. \tag{7.71}$$

The oscillatory solutions which result from positive total energies (which were excluded above) correspond, physically, to scattering states in the energy continuum for free particles. Each continuum state is infinitely degenerate since it can be assigned an infinite set of ℓ-values. A detailed discussion of these states will be deferred until the section on the quantum mechanical treatment of scattering.

In the limit of very small r, the dominant term in the bracket of Equation 7.70 is the $1/r^2$ term. Thus, in the asymptotic limit for small r, the solutions of Equation 7.70 will be dominated by the solutions of

$$g''(r) - \frac{\ell(\ell+1)}{r^2} g(r) = 0.$$

Trying a solution of the form

$$g(r) \sim r^p$$

leads to the relation:

$$p(p-1) = \ell(\ell+1).$$

The latter is satisfied only for $p = -\ell$ or $p = \ell+1$, which correspond to the asymptotic solutions (for small r), $r^{-\ell}$ or $r^{\ell+1}$. Since the first of these diverges at the origin, whereas the second is well-behaved, we choose for small r the solution

$$g(r) \sim r^{\ell+1}. \tag{7.72}$$

Our present knowledge of the final solution is shown schematically in Figure 7-5. We will now seek a polynomial solution for the intermediate region.

It is convenient at this point to change to a dimensionless variable by

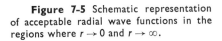

Figure 7-5 Schematic representation of acceptable radial wave functions in the regions where $r \to 0$ and $r \to \infty$.

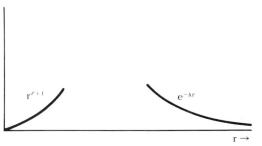

means of the substitution

$$\rho = 2\lambda r.$$

Then, $g''(r) = 4\lambda^2 g''(\rho)$, and Equation 7.70 becomes:

$$g''(\rho) - \left[\frac{\ell(\ell+1)}{\rho^2} - \frac{n}{\rho} + \frac{1}{4}\right]g(\rho) = 0. \tag{7.73}$$

Here n is an unspecified number defined as

$$n = \frac{\mu Z e^2}{\hbar^2 \lambda} = \frac{Z}{a_0 \lambda}. \tag{7.74}$$

We construct a product solution from our asymptotic solutions and a power series in ρ:

$$g(\rho) = e^{-\rho/2} \cdot \rho^{l+1} \cdot \sum_{j=0}^{\infty} a_j \rho^j = e^{-\rho/2} \cdot \sum_{j=0}^{\infty} a_j \rho^{l+1+j}.$$

Differentiating and substituting into Equation 7.73, we obtain the following recursion relation on the coefficients a_j:

$$a_{k+1} = a_k \cdot \frac{\ell+k+1-n}{(k+1)(2\ell+k+2)}.$$

In order to obtain a satisfactory wave function, the infinite power series must be truncated to a polynomial so that it can be normalized. If the polynomial is to terminate with the kth term, it is necessary that $a_{k+1} = 0$. That is, we require that

$$n = \ell + k + 1,$$

where $k = 0, 1, 2, \ldots$. Hence, n takes on integral values given by $\ell + 1$, $\ell + 2$, $\ell + 3, \ldots$. It is customary to call n the *principal* or *total quantum number;* it is allowed to take on integral values but not zero. For a given value of n, the angular momentum quantum number ℓ can have any one of the set of values $0, 1, 2, \ldots, n-1$.

The polynomial solutions of Equation 7.73 may be obtained by means of series solutions in a manner analogous to the treatment of the harmonic oscillator in Chapter 5. For a detailed treatment the reader is referred to one of the references listed at the end of this chapter. The polynomials are members of a class of well-known functions called the *associated Laguerre polynomials*,[4]

$$L_{n+\ell}^{2\ell+1}(\rho) = \left(\frac{d}{d\rho}\right)^{2\ell+1}\left[e^{\rho}\left(\frac{d}{d\rho}\right)^{n+\ell}(\rho^{n+\ell}e^{-\rho})\right]. \qquad (7.75)$$

To obtain the complete radial solution for the hydrogenic atom we must retrace our steps through all of the substitutions and obtain

$$R_{n\ell}(\rho) = \frac{g(\rho)}{\rho} = e^{-\rho/2}\rho^{\ell}L_{n+\ell}^{2\ell+1}(\rho),$$

where $\rho = 2\lambda r = (2Z/na_0)r$. Normalization of the radial functions over ρ may be achieved by evaluating the integral

$$\int_0^{\infty} e^{-\rho}\rho^{2\ell}[L_{n+\ell}^{2\ell+1}(\rho)]^2\rho^2\,d\rho = \frac{2n[(n+\ell)!]^3}{(n-\ell-1)!}.$$

Normalization over r is obtained as follows:

$$\int_0^{\infty}[R_{n\ell}(r)]^2 r^2\,dr = \left(\frac{1}{2\lambda}\right)^3\int_0^{\infty}[R_{n\ell}(\rho)]^2\rho^2\,d\rho = \left(\frac{na_0}{2Z}\right)^3\int_0^{\infty}[R_{n\ell}(\rho)]^2\rho^2\,d\rho.$$

Then the normalized radial wave functions are given by:

$$R_{n\ell}(r) = -2\left(\frac{Z}{na_0}\right)^{\frac{3}{2}}\sqrt{\frac{(n-\ell-1)!}{n[(n+\ell)!]^3}}\left(\frac{2Zr}{na_0}\right)^{\ell}e^{-Zr/na_0}L_{n+\ell}^{2\ell+1}\left(\frac{2Zr}{na_0}\right). \qquad (7.76)$$

Figure 7-6 shows the general behavior of some of the radial functions. A few of the normalized radial functions are given below for convenience.

$$R_{10}(r) = \left(\frac{Z}{a_0}\right)^{\frac{3}{2}}\cdot 2\exp\left(-\frac{Zr}{a_0}\right)$$

$$R_{20}(r) = \left(\frac{Z}{2a_0}\right)^{\frac{3}{2}}\cdot\left(2-\frac{Zr}{a_0}\right)\exp\left(-\frac{Zr}{2a_0}\right)$$

$$R_{21}(r) = \left(\frac{Z}{2a_0}\right)^{\frac{3}{2}}\cdot\frac{1}{\sqrt{3}}\cdot\frac{Zr}{a_0}\cdot\exp\left(-\frac{Zr}{2a_0}\right)$$

$$R_{30}(r) = \left(\frac{Z}{3a_0}\right)^{\frac{3}{2}}\cdot 2\left[1-\frac{2}{3}\cdot\frac{Zr}{a_0}+\frac{2}{27}\left(\frac{Zr}{a_0}\right)^2\right]\cdot\exp\left(-\frac{Zr}{3a_0}\right)$$

$$R_{31}(r) = \left(\frac{Z}{3a_0}\right)^{\frac{3}{2}}\cdot\frac{4\sqrt{2}}{3}\cdot\frac{Zr}{a_0}\left(1-\frac{1}{6}\frac{Zr}{a_0}\right)\cdot\exp\left(-\frac{Zr}{3a_0}\right)$$

$$R_{32}(r) = \left(\frac{Z}{3a_0}\right)^{\frac{3}{2}}\cdot\frac{2\sqrt{2}}{27\sqrt{5}}\cdot\left(\frac{Zr}{a_0}\right)^2\exp\left(-\frac{Zr}{3a_0}\right).$$

---

[4] See, for example, M. R. Spiegel, *Mathematical Handbook of Formulas and Tables.* McGraw-Hill Book Co., New York, 1968, p. 155.

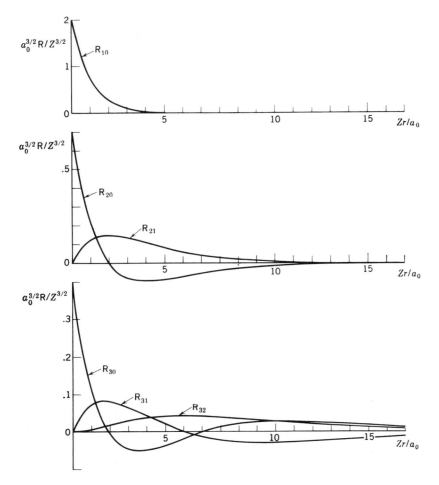

Figure 7-6 Graphs of the radial wave functions $R_{n\ell}(r)$ for $n = 1, 2,$ and 3 and $\ell = 0, 1, 2$. (From *Principles of Modern Physics* by R. B. Leighton. Copyright 1959 by McGraw-Hill Book Company. Used with permission of McGraw-Hill Book Company.)

The radial solutions obtained above should be recognized as the energy eigenfunctions for the hydrogenic atom. This result could have been anticipated from Equations 7.9 and 7.10, where the energy is seen to appear only in the radial equation. The energy eigenvalues corresponding to the radial solutions are readily obtained from Equation 7.74. Thus:

$$\lambda = \frac{1}{\hbar} \sqrt{2\mu |E_n|} = \frac{Z}{na_0},$$

or,

$$E_n = -\frac{Z^2 e^2}{2a_0 n^2} = -\frac{w_0 Z^2}{n^2}.$$

The allowed energies are precisely the same as those obtained for circular Bohr orbits and for Sommerfeld's non-relativistic elliptical orbits. Such good agreement between classical and quantum mechanical calculations is fortuitous in the case of the Coulomb potential, and should not be expected in general.

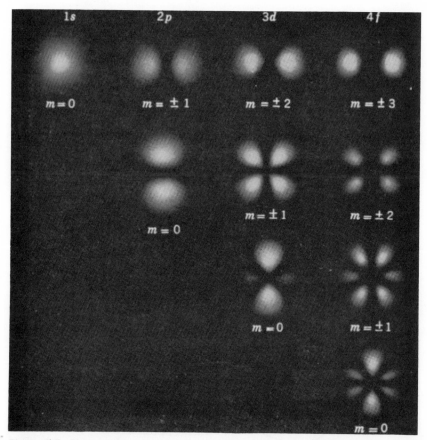

Figure 7-7 Photographic representation of the electron probability density distribution $\psi^*\psi$ for several energy eigenstates. These may be regarded as sectional views of the distributions in a plane containing the polar axis, which is vertical and in the plane of the paper. The scale varies from figure to figure. (From *Principles of Modern Physics* by R. B. Leighton. Copyright 1959 by McGraw-Hill Book Company. Used with permission of McGraw-Hill Book Company.)

Since the energy levels are independent of the quantum numbers ℓ and m, we can express the degeneracy of each level by the sum

$$\sum_{\ell=0}^{n-1} (2\ell + 1) = 1 + 3 + 5 + \cdots + 2n - 1 = n^2.$$

The $(2\ell + 1)$-fold degeneracy of the m-values is characteristic of all spherically symmetric potentials, while the n-fold degeneracy of the ℓ-values occurs only for a potential such as the Coulomb potential, which satisfies a specific invariance condition.[5] Historical usage has coined the unfortunate phrase "accidental degeneracy" to describe the latter. The reader can anticipate some of the results of later sections by speculating that any alteration of the physical system which destroys the spherical symmetry or the pure Coulombic potential will lead to the removal of some or all of the degeneracies.

---

[5] The n-fold degeneracy will occur for any potential that is invariant under the group O(4). See L. I. Schiff, *Quantum Mechanics*, 3rd. ed. McGraw-Hill Book Co., New York, 1969, p. 237.

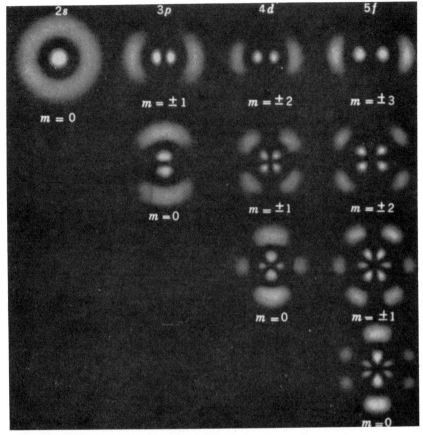

Figure 7-7 Continued.

The total hydrogenic wave functions can now be written as

$$\Psi_{n\ell m}(r,\,\theta,\,\phi,\,t) = \psi_{n\ell m}(r,\,\theta,\,\phi) \cdot e^{-iE_n t/\hbar} = R_{n\ell}(r)\,Y_\ell^m(\eta,\,\phi)e^{-iE_n t/\hbar},$$

where n can be any integer, ℓ can take on the values $0, 1, 2, \ldots, (n-1)$, and m can have any of the values $\ell, \ell-1, \ldots, 0, -1, \ldots, -\ell$. Although the radial and angular functions have been given above, a few of the normalized spatial, hydrogenic wave functions are tabulated below to facilitate their use in the problems:

$$\psi_{1s} = \psi_{100} = \frac{1}{\sqrt{\pi}}\left(\frac{Z}{a_0}\right)^{\frac{3}{2}} \cdot e^{-Zr/a_0}$$

$$\psi_{2s} = \psi_{200} = \frac{1}{4\sqrt{2\pi}}\left(\frac{Z}{a_0}\right)^{\frac{3}{2}} \cdot \left(2 - \frac{Zr}{a_0}\right) \cdot e^{-Zr/2a_0}$$

$$\psi_{2p} = \psi_{210} = \frac{1}{4\sqrt{2\pi}}\left(\frac{Z}{a_0}\right)^{\frac{3}{2}} \cdot \frac{Zr}{a_0} \cdot e^{-Zr/2a_0}\cos\theta$$

$$\psi_{2p} = \psi_{21\pm1} = \frac{1}{8\sqrt{\pi}}\left(\frac{Z}{a_0}\right)^{\frac{3}{2}} \cdot \frac{Zr}{a_0} \cdot e^{-Zr/2a_0}\sin\theta \cdot e^{\pm i\phi}$$

$$\psi_{3s} = \psi_{300} = \frac{1}{81\sqrt{3\pi}}\left(\frac{Z}{a_0}\right)^{\frac{3}{2}} \cdot \left(27 - 18\frac{Zr}{a_0} + \frac{2Z^2 r^2}{a_0^2}\right)e^{-Zr/3a_0}$$

$$\psi_{3p} = \psi_{310} = \frac{1}{81}\sqrt{\frac{2}{\pi}}\left(\frac{Z}{a_0}\right)^{\frac{3}{2}}\left(6 - \frac{Zr}{a_0}\right) \cdot \frac{Zr}{a_0} e^{-Zr/3a_0} \cos\theta$$

$$\psi_{3p} = \psi_{31\pm1} = \frac{1}{81\sqrt{\pi}}\left(\frac{Z}{a_0}\right)^{\frac{3}{2}}\left(6 - \frac{Zr}{a_0}\right) \cdot \frac{Zr}{a_0} \cdot e^{-Zr/3a_0} \sin\theta \cdot e^{\pm i\phi}$$

$$\psi_{3d} = \psi_{320} = \frac{1}{81\sqrt{6\pi}}\left(\frac{Z}{a_0}\right)^{\frac{3}{2}} \cdot \frac{Z^2 r^2}{a_0^2} \cdot e^{-Zr/3a_0} \cdot (3\cos^2\theta - 1)$$

$$\psi_{3d} = \psi_{32\pm1} = \frac{1}{81\sqrt{\pi}}\left(\frac{Z}{a_0}\right)^{\frac{3}{2}} \cdot \frac{Z^2 r^2}{a_0^2} \cdot e^{-Zr/3a_0} \cdot \sin\theta \cdot \cos\theta \cdot e^{\pm i\phi}$$

$$\psi_{3d} = \psi_{32\pm2} = \frac{1}{162\sqrt{\pi}}\left(\frac{Z}{a_0}\right)^{\frac{3}{2}} \cdot \frac{Z^2 r^2}{a_0^2} \cdot e^{-Zr/3a_0} \cdot \sin^2\theta \cdot e^{\pm 2i\phi}$$

The states associated with wave functions for which the angular momentum is zero (that is, $\ell = 0$) are called *s-states*. Note that all *s*-states are spherically symmetric. States associated with $\ell = 1, 2, 3, 4, \ldots$, are called *p-*, *d-*, *f-*, *g-*, $\ldots$ states, respectively, and their degeneracies are 3-, 5-, 7-, 9-, $\ldots$ fold. This terminology had its origin in optical spectroscopy where certain sets of spectral lines were characterized as *s*harp, *p*rincipal, *d*iffuse, and *f*ine. The ground state of hydrogen is the ψ_{1s} state.

For a hydrogenic system, the probability that the single electron will be found in the volume element $d\tau$ is

$$|\psi_{n\ell m}|^2 \, d\tau = |\psi_{n\ell m}|^2 r^2 \, dr \, d\Omega.$$

Pictorial representations of a few of these probability densities are shown in Figure 7-7. The probability of finding the electron in a spherical shell of radius r and thickness dr is obtained by integrating the above expression over solid angle; thus,

$$\int_\Omega |\psi_{n\ell m}|^2 r^2 \, dr \, d\Omega = |R_{n\ell}(r)|^2 r^2 \, dr \int_\Omega |Y_\ell^m(\Omega)|^2 \, d\Omega = r^2 \, |R_{n\ell}(r)|^2 \, dr. \quad (7.77)$$

The quantity $r^2 \, |R_{n\ell}(r)|^2$ is called the *radial probability density* or the radial distribution function. Figure 7-8 shows sketches of this quantity for a few hydrogenic states.

It is often convenient to express the angular parts of the hydrogenic functions in rectangular coordinates. By taking appropriate linear combinations of a set of degenerate functions, it is possible to obtain a new set of functions having easily-visualized symmetries with respect to the rectangular axes. For example, the three degenerate *p*-functions for any value of n may be written in the form,

$$\psi_{n1,0} = f(r) \cdot r \cos\theta = f(r) \cdot z$$

and

$$\sqrt{2} \, \psi_{n1,\pm1} = f(r) \cdot r \sin\theta \cdot e^{\pm i\phi} = f(r)(x \pm iy).$$

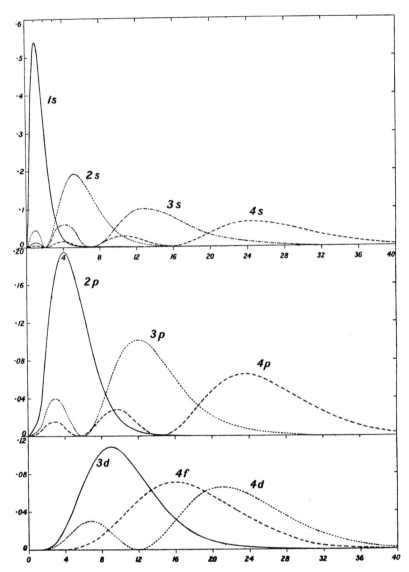

Figure 7-8 The radial probability distribution function $|rR_{n\ell}|^2$ for several values of the quantum numbers n, ℓ. (From E. U. Condon and G. H. Shortley, *The Theory of Atomic Spectra*, Cambridge University Press, Cambridge, 1953. Used with permission.)

Then we may define:

$$\psi_{p_x} = f(r) \cdot x = \frac{1}{\sqrt{2}} \left(\psi_{n11} + \psi_{n1,-1} \right)$$

$$\psi_{p_y} = f(r) \cdot y = \frac{1}{i\sqrt{2}} \left(\psi_{n11} - \psi_{n1,-1} \right)$$

$$\psi_{p_z} = f(r) \cdot z = \psi_{n10}. \tag{7.78}$$

Each of these wave functions resembles a figure 8 having rotational symmetry about one of the coordinate axes. Such wave functions have applications in crystal lattices and certain molecular structures when rectangular symmetry exists.

PROBLEM 7-21

Find the appropriate linear combinations of the d-functions which will lead to the following d-orbitals:

$$d_{xy}, d_{xz}, d_{yz}, d_{x^2-y^2}, d_{3z^2-r^2}.$$

PROBLEM 7-22

Calculate the expectation value $\langle r \rangle$ for the ψ_{100}, the ψ_{210} and the ψ_{320} states of hydrogen. Compare with the Bohr theory results.

PROBLEM 7-23

Find the values of r for which the radial probability density is a maximum for each of the states ψ_{100}, ψ_{210}, and ψ_{320} of hydrogen. Compare with the preceding problem and the Bohr theory result.
(Ans.: $a_0, 4a_0, 9a_0$.)

PROBLEM 7-24

(a) For the state ψ_{210} of hydrogen calculate the expectation values $\langle 1/r \rangle$ and $\langle p^2 \rangle$.
(b) From the results of part (a), find $\langle T \rangle$ and $\langle V \rangle$ and show that $\langle T \rangle = -\frac{1}{2}\langle V \rangle$. Compare with the Bohr theory. (This is an example of the virial theorem.)

PROBLEM 7-25

Show that ψ_{100} and ψ_{200} are orthogonal.

9. MAGNETIC MOMENTS OF HYDROGENIC ELECTRONS

It is instructive to examine the probability density currents associated with electronic motion in hydrogenic states. By way of illustration, let us consider the $2s$ and $2p$ states of hydrogen. From Equation 4.47 we have for the probability density current,

$$\vec{S} = -\frac{\hbar}{m_0} \text{Imag.} [\psi(\nabla\psi^*)],$$

where

$$\nabla(r,\ \theta,\ \phi) = \frac{\partial}{\partial r}\hat{r} + \frac{1}{r}\frac{\partial}{\partial \theta}\hat{\theta} + \frac{1}{r\sin\theta}\frac{\partial}{\partial \phi}\hat{\phi}.$$

Then, for the 2s state of hydrogen, $\psi(\nabla\psi^*)$ is real, and $\vec{S} = 0$. It should be evident that this will be true for all s-states. For the 2p states, however, we write:

$$\psi_{2p} = R_{21}(r)\,Y_1^m(\theta,\ \phi),$$

and

$$\psi_{2p}^* = R_{21}(r)\,Y_1^{-m}(\theta,\ \phi).$$

Then,

$$\psi_{2p}(\nabla\psi_{2p}^*) = R_{21}'(r)\cdot R_{21}(r)\,|Y_1^m|^2\,\hat{r}$$

$$+ \frac{1}{r}|R_{21}(r)|^2 Y_1^m\frac{\partial}{\partial\theta}(Y_1^{-m})\,\hat{\theta}$$

$$+ \frac{1}{r\sin\theta}|R_{21}(r)|^2\,Y_1^m\frac{\partial}{\partial\phi}(Y_1^{-m})\,\hat{\phi}.$$

Since the radial component has no imaginary part, it can make no contribution to the current, as in the case of s-states. Let us now examine the θ term. Its angular factors may be written as the product $Y_1^m = P_1(\theta)e^{im\phi}$. Then,

$$Y_1^m\frac{\partial}{\partial\theta}Y_1^{-m} = P_1(\theta)\cdot P_1'(\theta),$$

which is also real. Therefore, the θ term does not contribute to a current. It remains to examine the $\hat{\phi}$ term, which may be written in the form:

$$\frac{-im}{r\sin\theta}\cdot|R_{21}(r)|^2\cdot|P_1(\theta)|^2.$$

Since this is an imaginary term it *does* contribute to a probability current density in the $\hat{\phi}$ direction, given by

$$\vec{S}_\phi = \frac{\hbar m}{m_0}\frac{|R_{21}(r)P_1(\theta)|^2}{r\sin\theta}\,\hat{\phi}.$$

This can be converted to an electrical current density by multiplying by the charge of the electron, that is,

$$\vec{j}_\phi = -e\vec{S}_\phi.$$

Now, the magnetic moment due to this current density is

$$d\vec{M} = \tfrac{1}{2}(\vec{r}\times\vec{j})\,d\tau$$

or,

$$dM_z \doteq \frac{-e\hbar m}{2m_0 c} \frac{|R_{21}(r)P_1(\theta)|^2}{\sin \theta} \, d\tau$$

$$= -m\mu_B \cdot \frac{|R_{21}(r)P_1(\theta)|^2}{\sin \theta} \, d\tau,$$

and

$$M_z = -m\mu_B \int_0^\infty |R_{21}(r)P_1(\theta)|^2 \, 4\pi r^2 \, dr. \qquad (7.79)$$

PROBLEM 7-26

   (a)   From Equation 7.79 show that the magnetic moments associated with the states $\psi_{21\pm1}$ are $\mp\mu_B$, respectively.

   (b)   What are the magnetic moments for the $3d$ states?

PROBLEM 7-27

The magnetic moment operator may be written in the compact form,

$$\boldsymbol{\mu} = -\mu_B \mathbf{L}.$$

Show that this operator provides a direct means of obtaining magnetic moments from the angular momentum matrices.

10. THE PARITY OPERATOR

In Chapter 5 we noted, for both the square well and the harmonic oscillator, that each eigenfunction can be either symmetric or antisymmetric in the coordinates. In quantum mechanics the term *parity* is used to characterize the symmetry properties of wave functions with respect to inversion (that is, reflection through the origin), which is equivalent to changing the sign of each coordinate. Thus, *even* and *odd* parity refer to the symmetric and antisymmetric cases, respectively.

It is convenient to define a parity operator Π which changes the signs of all coordinates. For example,

$$\Pi\psi(x, y, z) = \psi(-x, -y, -z).$$

Then,

$$\Pi^2\psi(x, y, z) = \Pi\psi(-x, -y, -z) = 1\psi(x, y, z).$$

Since the eigenvalue of Π^2 is $+1$, the eigenvalues of the parity operator, for states which have definite parity, are ±1, where $+1$ corresponds to even parity and -1 corresponds to odd parity. This leads to the equivalent statements:

plus parity = even parity = symmetry under inversion

minus parity = odd parity = antisymmetry under inversion

Let us now examine the parity of the hydrogenic wave functions. The parity operator transforms the variables as follows:

$$r \to r, \ \theta \to \pi - \theta, \quad \text{and} \quad \phi \to \phi + \pi.$$

Then, from Equation 7.49,

$$\Pi P_\ell^m(\eta) = (-1)^{\ell-m} P_\ell^m(\eta),$$

which results from converting the derivatives $d/d(\cos \theta)$ to $d/d[\cos (-\theta)]$. Also,

$$\Pi(e^{im\phi}) = e^{im\phi}e^{im\pi} = (-1)^m e^{im\phi}.$$

Using the above and Equation 7.57, we obtain:

$$\Pi Y_\ell^m(\eta, \phi) = (-1)^\ell Y_\ell^m(\eta, \phi).$$

It follows that

$$\Pi \psi_{n\ell m} = \Pi R_{n\ell} Y_\ell^m = R_{n\ell} \Pi Y_\ell^m = (-1)^\ell \psi_{n\ell m},$$

which says that the parity of the hydrogenic functions is the same as the parity of the spherical harmonics, namely, $(-1)^\ell$. Using the same terminology, the parity of a harmonic oscillator function is $(-1)^n$ and the parity of a square well function is $(-1)^{n+1}$.

When states of different parity have the same energy, the degeneracy is called *accidental* degeneracy. This occurs for the different ℓ-values in a hydrogenic potential when all non-Coulombic interactions are neglected.* If these interactions are included in the Hamiltonian, the ℓ-degeneracy is removed and all of the hydrogenic states become states of definite parity.

One might ask whether the parity of a state is a constant of the motion. To determine the answer we examine the commutator,

$$[\mathcal{H}, \Pi] = \mathcal{H}\Pi - \Pi\mathcal{H} = \mathcal{H}(x, y, z) - \mathcal{H}(-x, -y, -z).$$

This commutator is zero whenever the Hamiltonian is even in the coordinates. Therefore, from Equation 6.18, we conclude that *the parity of a state is constant in time if the Hamiltonian is even in each coordinate.* There is, of course, the important example of the weak nuclear interactions in which parity is not conserved. The

---

* See footnote 5.

Nobel prize was awarded to Lee and Yang in 1957 for their theoretical prediction of this phenomenon.[6]

Frequent use of the concept of parity is made in calculating matrix elements between states having definite parity. Since an integral of an odd integrand over symmetric limits is zero, one immediately knows that all matrix elements of an odd operator (antisymmetric in the coordinates) between states of the same parity are zero. Likewise, matrix elements of a symmetric operator between states of opposite parity are zero. This will be very useful in later discussions of the selection rules governing transitions under the influence of perturbations. Since the parity of an isolated system is conserved for all but the weak nuclear interactions, it is often useful for obtaining information about a state where the exact wave function is unknown.[7]

SUMMARY

The two-body problem in three dimensions is shown to be separable in the coordinates of the center of mass of the system and in the relative coordinates which give the position of one body with respect to the other. The system can be then treated as a one-body problem by using the reduced mass in the relative coordinate system. For a spherically symmetric potential, the Schrödinger equation can be separated into a radial and an angular equation. The differential operator of the angular equation (that is, the angular part of the Laplacian operator), is shown to be the square of the total angular momentum operator. Hence, the angular momentum eigenfunctions (the spherical harmonics) are the solutions of the angular part of the Schrödinger equation when space is spherically symmetric (isotropic). In other words, when the Hamiltonian is invariant under a rotation, it follows that: space is isotropic, angular momentum is conserved, the spherical harmonics are the eigenfunctions, and ℓ and m_ℓ are good quantum numbers. This close connection between space and momentum is also apparent in translational motion. For if the Hamiltonian is invariant under a translation, space is homogeneous, linear momentum is conserved, the momentum wave functions are the eigenfunctions, and p is a good quantum number.

As in the case of the harmonic oscillator, there are raising and lowering operators for the angular momentum. Using these operators, as well as the operators for the square of the total angular momentum and its z-component, simultaneous eigenfunctions of L^2 and L_z are obtained by means of operator algebra. These functions, which are the spherical harmonics, are the angular solutions of all physical problems having spherical symmetry.

The solutions of the radial equation for a hydrogenic atom are shown to be the associated Laguerre polynomials. Hence, the total wave function for a

---

[6] T. D. Lee and C. N. Yang, *Phys. Rev.* **104,** 254 (1956); C. S. Wu, E. Ambler, R. W. Hayward, D. D. Hoppe, and R. P. Hudson, *Phys. Rev.* **105,** 1413 (1957); R. L. Garwin, L. M. Lederman, and M. Weinrich, *Phys. Rev.* **105,** 1415 (1957).

[7] See, for example, J. M. Blatt and V. F. Weisskopf, *Theoretical Nuclear Physics*, John Wiley and Sons, New York, 1952, or R. D. Evans, *The Atomic Nucleus*, McGraw-Hill Book Co., Inc., New York, 1955.

hydrogenic electron (neglecting spin) may be written as the product of a Laguerre polynomial, a spherical harmonic, and a time factor depending on the energy of the state. Each wave function is characterized by three spatial quantum numbers, n, ℓ, and m. The principal quantum number n takes on integral values (not zero) and denotes the "orbit" or "shell" in which the electron is moving. The angular momentum quantum number ℓ can have any one of the values $0, 1, 2, \ldots, n - 1$, for a given value of n. The states associated with the different ℓ-values are sometimes called "sub-shells," and their wave functions are frequently referred to as *orbitals*. The spectroscopic designations of these orbital states are s, p, d, f, g, $\ldots$, respectively, for $\ell = 0, 1, 2, 3, 4, \ldots$. When solving molecular or solid state problems it is often convenient to form linear combinations of the spherical harmonics so as to remove all of the angular variables. They may then be written in terms of r and the rectangular coordinates. The magnetic quantum number m (or m_ℓ) can have $2\ell + 1$ values for each value of ℓ, ranging through ℓ, $\ell - 1, \ldots, 0$, $-1, \ldots, -\ell$. Each value of m_ℓ prescribes an allowed orientation for an orbital plane, since m_ℓ represents the projection on the quantization axis of the normal to the plane. The name "magnetic" quantum number derives from the fact that each orbital state of a charged particle possesses an effective magnetic moment proportional to its m-value. Thus, for an electronic state in a constant magnetic field, the component of the orbital magnetic moment parallel to the field is $\mu_\ell = m_\ell \mu_B$, where μ_B is the Bohr magneton. The torque produced by the magnetic field causes the orbital plane to precess about the field direction at a fixed polar angle. The energy of each state is altered by an amount given by $\pm m_\ell \mu_B \mathscr{B}$. The removal of the orbital degeneracy in this fashion is what gives rise to the normal Zeeman effect.

The absolute square of a hydrogenic wave function integrated over a solid angle is called the radial probability density of the electron. For the case of hydrogen, the radii corresponding to the maxima of the radial probability density function are in good agreement with the values given by the Bohr theory.

The parity of a hydrogenic state is given by $(-1)^\ell$, where ± 1 are the eigenfunctions of the parity operator. Since the parity operation is a reflection through the origin (inversion), the eigenvalue $+1$ corresponds to an even or symmetric function of the coordinates, whereas the eigenvalue -1 corresponds to an odd or antisymmetric function. The terms "plus" and "minus" are often used for even and odd, respectively. The efficacy of the concept of parity derives from the fact that parity is conserved when the parity operator commutes with the Hamiltonian, that is, when the latter is symmetric in the coordinates.

CHAPTER 7   APPENDIX A

Average Values[8] of Powers of r for a Hydrogenic Atom in the State $\psi_{n\ell m}$.

$$\langle r \rangle = \frac{a_0 n^2}{Z}\left[1 + \frac{1}{2}\left\{ 1 - \frac{\ell(\ell+1)}{n^2}\right\} \right]$$

$$\langle r^2 \rangle = \frac{a_0^2 n^4}{Z^2}\left[1 + \frac{3}{2}\left\{ 1 - \frac{\ell(\ell+1) - \frac{1}{3}}{n^2}\right\} \right]$$

$$\left\langle \frac{1}{r} \right\rangle = \frac{Z}{a_0 n^2}$$

$$\left\langle \frac{1}{r^2} \right\rangle = \frac{Z^2}{a_0^2 n^3 (\ell + \frac{1}{2})}$$

$$\left\langle \frac{1}{r^3} \right\rangle = \frac{Z^3}{a_0^3 n^3 \ell(\ell + \frac{1}{2})(\ell + 1)}$$

$$\left\langle \frac{1}{r^4} \right\rangle = \frac{\frac{3}{2}Z^4\left\{ 1 - \frac{\ell(\ell+1)}{3n^2}\right\}}{a_0^4 n^3 (\ell + \frac{3}{2})(\ell + 1)(\ell + \frac{1}{2})\ell(\ell - \frac{1}{2})}$$

SUGGESTED REFERENCES

David Bohm, *Quantum Theory*, Prentice-Hall, Inc., New York, 1951.
Sidney Borowitz, *Fundamentals of Quantum Mechanics*. W. A. Benjamin, Inc., New York, 1967.
R. H. Dicke and J. P. Wittke, *Introduction to Quantum Mechanics*. Addison-Wesley Publishing Co., Inc., Reading, Mass., 1960.
P. T. Matthews, *Introduction to Quantum Mechanics*. McGraw-Hill Book Co., New York, 1968.
E. Merzbacher, *Quantum Mechanics*. John Wiley and Sons, Inc., New York, 1961.
Albert Messiah, *Quantum Mechanics*. North-Holland Publishing Co., Amsterdam, 1958.
D. Park, *Introduction to the Quantum Theory*. McGraw-Hill Book Co., Inc., New York, 1964.
L. Pauling and E. B. Wilson, *Introduction to Quantum Mechanics*. McGraw-Hill Book Co., Inc., New York, 1935.
J. L. Powell and B. Crasemann, *Quantum Mechanics*. Addison-Wesley Publishing Co., Inc., Reading, Mass., 1961.
F. K. Richtmyer, E. H. Kennard, and J. N. Cooper, *Introduction to Modern Physics*, 6th ed. McGraw-Hill Book Co., New York, 1969.
D. S. Saxon, *Elementary Quantum Mechanics*. Holden-Day, Inc., San Francisco, 1968.
Leonard I. Schiff, *Quantum Mechanics*, 3rd ed. McGraw-Hill Book Co., New York, 1969.
Robert L. White, *Basic Quantum Mechanics*. McGraw-Hill Book Co., New York, 1966.
Klaus Ziock, *Basic Quantum Mechanics*. John Wiley & Sons, Inc., New York, 1969.

[8] L. Pauling and E. B. Wilson, *op. cit.*, section 21c.

CHAPTER 8

SPIN, ADDITION OF ANGULAR MOMENTA, AND IDENTICAL PARTICLES

We have freely alluded to the fact that the elementary particles possess an intrinsic angular momentum which we call *spin*. Thus, we noted in Chapter 3 that a magnetic moment is associated with the spin of a nucleon, as well as that of an electron, and that this moment can be accurately measured by beam deflection or magnetic resonance experiments. It was from the number of components in the beam experiments that the half-integral value of $\frac{1}{2}\hbar$ for the spin of the electron became apparent (see section 11 of Chapter 3). The existence of an electronic spin angular momentum of $\frac{1}{2}\hbar$ was postulated by Uhlenbeck and Goudsmit in 1928 in order to explain the fine structure splitting of hydrogen.[1] It was soon observed that this assumption also explained the "anomalous" Zeeman effect, the doublet structure of the spectra of the alkali metals, and the anomalous magnetic moment reported by Compton,[2] provided that one also assumed the "anomalous" gyromagnetic ratio of roughly twice the classical value.[3]

The Schrödinger formulation of quantum mechanics is built, as we have seen, on the classical Hamiltonian function. Thus, the Schrödinger equation does not include spin as an intrinsic variable, since spin is a relativistic phenomenon having no classical counterpart. For example, there are no additional generalized coordinates that will permit us to write the spin in the classical form,

$$\vec{S} = \vec{r} \times \vec{p}_s.$$

This conceptual disadvantage turns out to be an advantage, however, since the uniqueness of the spin permits S^2 to commute with all other dynamical variables. Hence, the spin is a constant of the motion and s is a good quantum

---

[1] G. E. Uhlenbeck and S. A. Goudsmit, *Naturwiss.* **13**, 593 (1925).
[2] A. H. Compton, *J. Franklin Inst.* **192**, 144 (1921).
[3] W. Pauli, *Z. Physik* **43**, 601 (1927).

number. (S_z may or may not be a constant of the motion, depending upon the nature of the spin-dependent interactions, if any, that are included in the Hamiltonian.)

Although it is an *ad hoc* assumption in the Schrödinger treatment of quantum mechanics, the half-integral spin of the electron arises quite naturally out of the relativistic theory of Dirac.[4] Thus, its validity as an additional coordinate has been firmly established. Having this assurance we can, in many instances, treat spin phenomenologically by including it in the Schrödinger wave functions and in the Hamiltonian in such a way as to account for the physical effect being considered. The success of this *ad hoc* method is attested by the frequent use of a "spin Hamiltonian" in the Schrödinger equation for solving problems.[5]

I. SPIN IN THE SCHRÖDINGER FORMULATION

The spin hypothesis may be incorporated in the Schrödinger formalism by regarding the spin angular momentum as an additional degree of freedom. It is assumed that the components of the spin obey the same commutation relations as the components of the orbital angular momentum. However, in the case of a single electron, the spin quantum number s has a fixed value of $\frac{1}{2}$, unlike the value of ℓ, which can be zero or any integer up to $n - 1$. The projection of the magnitude, $\sqrt{s(s + 1)}$, of the spin vector on an axis of quantization behaves in exactly the same manner as the projection of the length, $\sqrt{\ell(\ell + 1)}$, of the angular momentum vector. That is, successive values of m_s must differ by unity. For $s = \frac{1}{2}$ there are only $2s + 1 = 2$ allowed values of m_s, namely, $\pm\frac{1}{2}$. (See Figure 8-1.)

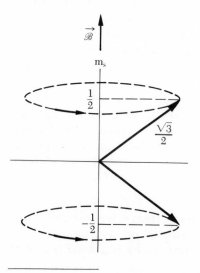

Figure 8-1 The two spin orientations for the electron. If the axis of quantization is defined by a magnetic field, the spin precesses as shown.

---

[4] P. A. M. Dirac, *Proc. Royal Society* (*London*) **117,** 610 (1928) and **118,** 351 (1928).
[5] For examples in the field of magnetism see K. W. H. Stevens, "Spin Hamiltonians," in G. T. Rado and H. Suhl, *Magnetism*, Vol I, Academic Press, New York, 1963.

In order to include spin in the Schrödinger wave function we must use the fact that the spin coordinates are independent of the coordinates of configuration space. This enables us to write the total wave function as a product function,

$$\Psi_{total} = \psi_{n\ell m}(r, \theta, \phi) \cdot \chi(\text{spin}) \cdot e^{-iE_{nt}t/\hbar} = |R_{n\ell} \cdot e^{-iE_{nt}t/\hbar}\rangle \, |\ell, m_\ell\rangle \, |s, m_s\rangle. \quad (8.1)$$

Here the radial and time functions have been lumped together in the same eigenket, since our main concern in this chapter will be with the last two kets, representing the angular momentum and spin eigenfunctions.

For spin $\frac{1}{2}$ we know that there are two basis functions for the two-dimensional spin space. If we adopt the convention that an "up" spin has a positive projection on the axis of quantization (usually the z-axis), these spin functions may be written in the following equivalent ways:

$$\chi_+ = \chi(\text{up}) \quad = \begin{pmatrix} 1 \\ 0 \end{pmatrix} = |\tfrac{1}{2}, \quad \tfrac{1}{2}\rangle$$

$$\chi_- = \chi(\text{down}) = \begin{pmatrix} 0 \\ 1 \end{pmatrix} = |\tfrac{1}{2}, -\tfrac{1}{2}\rangle. \quad (8.2)$$

Since either of these functions may be used in Equation 8.1, we see that *each hydrogenic spatial state is doubly degenerate in spin*. The spin wave functions, or *spinors*, given in Equation 8.2 are simultaneous eigenfunctions of the spin operators S^2 and S_z. Thus, using ket notation:

$$S^2 |\tfrac{1}{2}, \tfrac{1}{2}\rangle = \tfrac{1}{2}(\tfrac{1}{2} + 1)\hbar^2 |\tfrac{1}{2}, \tfrac{1}{2}\rangle = \frac{3\hbar^2}{4} |\tfrac{1}{2}, \tfrac{1}{2}\rangle$$

$$S^2 |\tfrac{1}{2}, -\tfrac{1}{2}\rangle = \tfrac{1}{2}(\tfrac{1}{2} + 1)\hbar^2 |\tfrac{1}{2}, -\tfrac{1}{2}\rangle = \frac{3\hbar^2}{4} |\tfrac{1}{2}, -\tfrac{1}{2}\rangle$$

$$S_z |\tfrac{1}{2}, \tfrac{1}{2}\rangle = \frac{\hbar}{2} |\tfrac{1}{2}, \tfrac{1}{2}\rangle$$

$$S_z |\tfrac{1}{2}, -\tfrac{1}{2}\rangle = -\frac{\hbar}{2} |\tfrac{1}{2}, -\tfrac{1}{2}\rangle.$$

In spite of the non-classical origin of spin, it is assumed that the algebra of the angular momentum operators can be applied directly to the spin operators. These expressions will not be repeated here since they will be summarized for a general angular momentum in the next section.

The matrix elements for the spin $\frac{1}{2}$ operators are:

$$\langle \tfrac{1}{2}, m_{s'}| \, S^2 \, |\tfrac{1}{2}, m_s\rangle = \hbar^2[\tfrac{1}{2}(\tfrac{1}{2} + 1)]\delta_{m_{s'}, m_s} = \frac{3\hbar^2}{4} \delta_{m_{s'}, m_s}$$

$$\langle \tfrac{1}{2}, m_{s'}| \, S_z \, |\tfrac{1}{2}, m_s\rangle = m_s\hbar \, \delta_{m_{s'}, m_s}$$

$$\langle \tfrac{1}{2}, m_{s'}| \, S_\pm \, |\tfrac{1}{2}, m_s\rangle = \hbar\sqrt{(\tfrac{1}{2} \mp m_s)(\tfrac{3}{2} \pm m_s)} \, \delta_{m_{s'}, m_s \pm 1},$$

for which the full matrices are

$$\mathbf{S}^2 = \frac{3\hbar^2}{4}\begin{pmatrix} 1 & 0 \\ 0 & 1 \end{pmatrix} \qquad\qquad \mathbf{S}_z = \frac{\hbar}{2}\begin{pmatrix} 1 & 0 \\ 0 & -1 \end{pmatrix}$$

$$\mathbf{S}_+ = \hbar\begin{pmatrix} 0 & 1 \\ 0 & 0 \end{pmatrix} \qquad\qquad \mathbf{S}_- = \hbar\begin{pmatrix} 0 & 0 \\ 1 & 0 \end{pmatrix}$$

$$\mathbf{S}_x = \tfrac{1}{2}(\mathbf{S}_+ + \mathbf{S}_-) = \frac{\hbar}{2}\begin{pmatrix} 0 & 1 \\ 1 & 0 \end{pmatrix} \qquad \mathbf{S}_y = \frac{i}{2}(\mathbf{S}_- - \mathbf{S}_+) = \frac{i\hbar}{2}\begin{pmatrix} 0 & -1 \\ 1 & 0 \end{pmatrix}.$$

It is convenient to define the *Pauli spin matrices*:

$$\sigma_x = \frac{2}{\hbar}\mathbf{S}_x = \begin{pmatrix} 0 & 1 \\ 1 & 0 \end{pmatrix}$$

$$\sigma_y = \frac{2}{\hbar}\mathbf{S}_y = \begin{pmatrix} 0 & -i \\ i & 0 \end{pmatrix} \qquad\qquad (8.3)$$

$$\sigma_z = \frac{2}{\hbar}\mathbf{S}_z = \begin{pmatrix} 1 & 0 \\ 0 & -1 \end{pmatrix}$$

PROBLEM 8-1

(a) Show that the Pauli matrices anti-commute, that is, that any pair of them satisfies

$$\{\sigma_x, \sigma_y\} = \sigma_x\sigma_y + \sigma_y\sigma_x = 0.$$

(b) Show that $\sigma_x\sigma_y = i\sigma_z$, $\sigma_y\sigma_z = i\sigma_x$, and $\sigma_z\sigma_x = i\sigma_y$.
(c) Show that $\sigma_x\sigma_y\sigma_z = i\mathbf{1}$.

PROBLEM 8-2

(a) Show that the Pauli matrices are unitary.
(b) Operate on χ_+ and χ_- with each of the following matrix operators:

$$\sigma_x, \sigma_y, \sigma_z, \mathbf{S}_+ \text{ and } \mathbf{S}_-.$$

(c) Write the eigenvectors of σ_x, σ_y, σ_z as linear combinations of the basis vectors χ_+ and χ_-.

PROBLEM 8-3

(a) Show that $\sigma_x^2 = \sigma_y^2 = \sigma_z^2 = \mathbf{1}$.
(b) Show that the commutation relations for the Pauli

matrices can be summarized by the expression

$$\vec{\sigma} \times \vec{\sigma} = 2i\vec{\sigma},$$

where the cross product has the same significance here as it does in Problem 7-6.

2. THE SPIN-ORBIT INTERACTION

The total wave function given in Equation 8.1 for an electron possessing both orbital momentum and spin assumes that there is no interaction between $\vec{L}$ and $\vec{S}$, that is, that $[L, S] = 0$. Hence, Equation 8.1 is an eigenfunction of both L_z and S_z, which means that both m_ℓ and m_s are good quantum numbers; in other words, the projections of $\vec{L}$ and $\vec{S}$ are constants of the motion. In reality, however, there *is* an interaction between $\vec{L}$ and $\vec{S}$—called the *spin-orbit interaction*—which removes the spin degeneracy, since it produces an energy difference between the states for which the orbital and spin magnetic moments are parallel and antiparallel. We will see later that this interaction can be expressed in terms of the quantity $\vec{L} \cdot \vec{S}$. Since $\vec{L} \cdot \vec{S}$ does not commute with either $\vec{L}$ or $\vec{S}$, Equation 8.1 is no longer the correct wave function, and m_s and m_ℓ are no longer independently quantized; hence, they cease to be good quantum numbers.

An elementary way to picture the spin-orbit interaction is to regard the stationary spin magnetic moment as interacting with the magnetic field produced by the orbiting *nucleus* (see Problem 3-18). In the rest frame of the electron there is an electric field, $\vec{\mathscr{E}} = Ze\hat{r}/r^2$, and a magnetic field

$$\vec{H}_e = \frac{\vec{j} \times \hat{r}}{r^2},$$

where $\hat{r}$ is directed from the nucleus toward the electron. Assuming that $\vec{v}$ is the velocity of the electron in the rest frame of the nucleus, the current produced by the nuclear motion is $\vec{j} = -(Ze/c)\vec{v}$ in the rest frame of the electron. Then

$$\vec{H}_e = -\frac{Ze}{c}\frac{\vec{v} \times \hat{r}}{r^2} = -\frac{1}{c}\vec{v} \times \vec{\mathscr{E}}.$$

The spin moment of the electron precesses in this field at the Larmor frequency (see section 10 of Chapter 3),

$$\vec{\omega}_e = \gamma \vec{H}_e = -\frac{e}{m_0 c^2}\vec{v} \times \vec{\mathscr{E}}, \tag{8.4}$$

with potential energy,

$$E_e = -\vec{\mu}_s \cdot \vec{H}_e = -\vec{\omega}_e \cdot \vec{S}. \tag{8.5}$$

Equations 8.4 and 8.5 hold in the rest frame of the electron. Since the rest frame of the nucleus is of more importance experimentally, we must make a transformation to that frame. This introduces a factor of $\frac{1}{2}$ which is called the *Thomas factor*.[6] This factor results from a relativistic effect known as the *Thomas precession*. Hence, an observer in the rest frame of the nucleus would observe the electron to precess with an angular velocity of

$$\vec{\omega}_L = -\frac{e}{2m_0c^2}\vec{v} \times \vec{\mathscr{E}}, \tag{8.6}$$

and an additional energy given by

$$\Delta E = -\tfrac{1}{2}\vec{\omega}_e \cdot \vec{S}. \tag{8.7}$$

Equations 8.6 and 8.7 can be put in a more general form if we restrict V to be any central potential with spherical symmetry. Then,

$$\vec{F} = -\hat{r}\frac{\partial V}{\partial r} = -e\vec{\mathscr{E}},$$

and

$$\vec{v} \times \vec{\mathscr{E}} = \frac{1}{e}\frac{\partial V}{\partial r}\vec{v} \times \hat{r} = \frac{1}{e}\frac{1}{r}\frac{\partial V}{\partial r}\vec{v} \times \vec{r} = \frac{1}{em_0}\frac{1}{r}\frac{\partial V}{\partial r}\vec{L}.$$

Using Equation 8.6,

$$\vec{\omega}_L = -\frac{1}{2m_0^2c^2}\frac{1}{r}\frac{\partial V}{\partial r}\vec{L}$$

and

$$\Delta E = -\frac{1}{2m_0^2c^2}\frac{1}{r}\frac{\partial V}{\partial r}\vec{L} \cdot \vec{S}. \tag{8.8}$$

Equation 8.8 illustrates the explicit dependence of the spin-orbit energy on the relative orientations of $\vec{L}$ and $\vec{S}$. For spin $= \frac{1}{2}$ the energy splitting is:

$$|\Delta E| = \frac{\hbar^2 m_\ell}{4m_0^2c^2}\frac{1}{r}\frac{\partial V}{\partial r}.$$

For the Coulomb potential this can be approximated by $\Delta E = \lambda_c^2 m_\ell \cdot Ze^2/r^3$. Since the average value of $1/r^3$ is[7]

$$\frac{Z^3}{a_0^3 n^3 \ell(\ell + \frac{1}{2})(\ell + 1)} \qquad \text{(for } \ell \neq 0\text{)}, \tag{8.9}$$

the spin-orbit splitting of the $2p$ state of hydrogen is approximately

$$\Delta E_{2p} \sim \frac{\lambda_c^2 e^2}{24a_0^3} \sim \left(\frac{\lambda_c}{a_0}\right)^2 \cdot \frac{w_0}{12} \sim 10^{-4}\text{ eV}.$$

---

[6] L. H. Thomas, *Nature* **117**, 514 (1926).
[7] See Appendix A of Chapter 7.

This is the same order of magnitude as the fine structure correction which goes as $\alpha^2 \sim 10^{-4}$ (see section 2 of Chapter 4).

PROBLEM 8-4

Assuming a circular orbit, estimate the angular velocity of precession for the electron in the ground state of hydrogen. How does this compare with the electron's orbital angular velocity?

PROBLEM 8-5

Use the relativistic velocity transformation to derive the Thomas factor by considering three inertial frames, one at the nucleus and one each for two successive positions of the electron. Assume circular motion.

PROBLEM 8-6

Derive the Thomas factor by calculating the time dilation between the rest frame of the nucleus and that of the electron.

PROBLEM 8-7

(a) Using the classical expression for the energy of two interacting dipoles, estimate the spin-orbit energy splitting for the $2p$ state of hydrogen.

(b) If a nucleus has a non-zero spin I, there will be a spin-spin interaction between the magnetic moments associated with the nuclear and electronic spins. Using the classical dipole expression, estimate the energy of this interaction (known as the *hyperfine* interaction) for the proton and a $2p$ electron, and compare it with the spin-orbit interaction.

(Ans.: (a) $\sim 10^{-5}$ eV; (b) $\sim 10^{-8}$ eV.)

3. THE VECTOR MODEL FOR COMBINING ANGULAR MOMENTA

As a result of the spin-orbit interaction, $\vec{L}$ and $\vec{S}$ exert torques on each other via their magnetic moments, and neither can be treated independently as a constant of the motion. However, if there is no external torque acting on the system the *total* angular momentum must be a constant of the motion. Using the so-called *vector model* we define the total angular momentum $\vec{J}$ to be just the

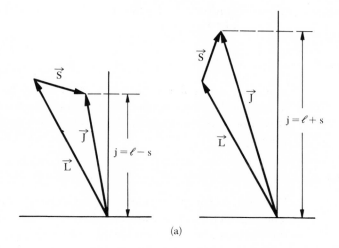

(a)

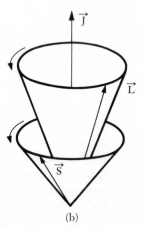

(b)

Figure 8-2   (a) The two possible orientations of the spin and orbital moments for a single electron.   (b) Spin-orbit coupling causes the vectors $\vec{L}$ and $\vec{S}$ to precess about $\vec{J}$ at the same angular
velocity.

vector sum,

$$\vec{J} = \vec{L} + \vec{S}.$$

The vectors $\vec{L}$ and $\vec{S}$ precess about the vector $\vec{J}$ at the same angular velocity
(see Figure 8-2).   The quantum conditions on angular momentum now apply
to $\mathbf{J}^2$ and $\mathbf{J}_z$, instead of to $\mathbf{L}^2$, $\mathbf{S}^2$, $\mathbf{L}_z$, and $\mathbf{S}_z$ separately.   That is, we can define
the eigenvectors $|j, m_j\rangle$ such that

$$\mathbf{J}^2 |j, m_j\rangle = j(j + 1)\hbar^2 |j, m_j\rangle,$$

and

$$\mathbf{J}_z |j, m_j\rangle = m_j\hbar |j, m_j\rangle.$$

The good or conserved quantum numbers are j and m_j, where j can have the
values $|\ell + s|$, $|\ell + s - 1|$, ..., $|\ell - s|$.   For each value of j there are $2j + 1$

possible values of m_j. In the case of a single electron there are always just two values of j, namely, $j = \ell \pm \frac{1}{2}$. For an s-electron the only possible value of j is $\frac{1}{2}$, which has the two projections, $m_j = \pm\frac{1}{2}$. For a p-electron, j can have the values, $j = 1 \pm \frac{1}{2} = \frac{3}{2}$ or $\frac{1}{2}$; for a d-state, $j = 2 \pm \frac{1}{2} = \frac{5}{2}$ or $\frac{3}{2}$.

The new angular momentum eigenfunctions to replace those of Equation 8.1 are not obtained so easily, since they must be the appropriate linear combinations of the old functions which diagonalize the matrices $\mathbf{J}^2$ and $\mathbf{J}_z$. We know, in principle, how to determine these functions but it will suffice for the time being merely to denote the eigenfunctions by their eigenvalues in ket notation as follows:

Orbital state	j value	Eigenkets $\lvert j, m_j \rangle$	
s	$\frac{1}{2}$	$\lvert \frac{1}{2}, \frac{1}{2} \rangle, \lvert \frac{1}{2}, -\frac{1}{2} \rangle$	
p	$\frac{3}{2}$	$\lvert \frac{3}{2}, \frac{3}{2} \rangle, \lvert \frac{3}{2}, \frac{1}{2} \rangle, \lvert \frac{3}{2}, -\frac{1}{2} \rangle, \lvert \frac{3}{2}, -\frac{3}{2} \rangle$	
	$\frac{1}{2}$	$\lvert \frac{1}{2}, \frac{1}{2} \rangle, \lvert \frac{1}{2}, -\frac{1}{2} \rangle$	(8.10)
d	$\frac{5}{2}$	$\lvert \frac{5}{2}, \frac{5}{2} \rangle, \lvert \frac{5}{2}, \frac{3}{2} \rangle, \lvert \frac{5}{2}, \frac{1}{2} \rangle, \lvert \frac{5}{2}, -\frac{1}{2} \rangle,$	
		$\lvert \frac{5}{2}, -\frac{3}{2} \rangle, \lvert \frac{5}{2}, -\frac{5}{2} \rangle$	
	$\frac{3}{2}$	$\lvert \frac{3}{2}, \frac{3}{2} \rangle, \lvert \frac{3}{2}, \frac{1}{2} \rangle, \lvert \frac{3}{2}, -\frac{1}{2} \rangle, \lvert \frac{3}{2}, -\frac{3}{2} \rangle$	

PROBLEM 8-8

Assume that the spin-orbit interaction can be "turned off" for the orbital states enumerated in Equation 8.10. Since m_ℓ and m_s would then be good quantum numbers, write the eigenkets for the above states in either the form $\lvert \ell, m_\ell \rangle \lvert s, m_s \rangle$ or $\lvert m_\ell, m_s \rangle$.

We will now summarize the properties of the angular momentum operators, using the symbol $\mathbf{J}$ which can have either integral or half-integral values. Thus, all of the orbital and spin cases are incorporated in one formalism.

$$
\left.
\begin{aligned}
& \mathbf{J}_\pm = \mathbf{J}_x \pm i\mathbf{J}_y \\
& \mathbf{J}^2 = \mathbf{J}_x^2 + \mathbf{J}_y^2 + \mathbf{J}_z^2 = \tfrac{1}{2}(\mathbf{J}_+\mathbf{J}_- + \mathbf{J}_-\mathbf{J}_+) + \mathbf{J}_z^2 \\
& \vec{\mathbf{J}} \times \vec{\mathbf{J}} = i\hbar\vec{\mathbf{J}} \\
& [\mathbf{J}^2, \mathbf{J}_x] = [\mathbf{J}^2, \mathbf{J}_y] = [\mathbf{J}^2, \mathbf{J}_z] = 0 \\
& [\mathbf{J}_+, \mathbf{J}_-] = 2\hbar\mathbf{J}_z \\
& [\mathbf{J}_z, \mathbf{J}_+] = \hbar\mathbf{J}_+ \\
& [\mathbf{J}_z, \mathbf{J}_-] = -\hbar\mathbf{J}_- \\
& [\mathbf{J}^2, \mathbf{J}_+] = [\mathbf{J}^2, \mathbf{J}_-] = 0 \\
& \mathbf{J}_\pm \lvert j, m_j \rangle = \hbar\sqrt{(j \mp m_j)(j \pm m_j + 1)} \, \lvert j, m_j \pm 1 \rangle
\end{aligned}
\right\}
\quad (8.11)
$$

Hereafter, we will drop the subscript j on m, since it is clear that when j is a good quantum number, m refers to one of the allowed projections of j.

The matrix elements of these operators are:

$$\langle j', m' | \mathbf{J}^2 | j, m \rangle = \hbar^2 j(j + 1)\delta_{j',j}\delta_{m',m}$$

$$\langle j', m' | \mathbf{J}_z | j, m \rangle = m\hbar\delta_{j',j}\delta_{m',m}$$

$$\langle j', m' | \mathbf{J}_\pm | j, m \rangle = \hbar\sqrt{(j \mp m)(j \pm m + 1)}\,\delta_{j',j}\delta_{m',m\pm1}.$$

Then the matrices are as follows:

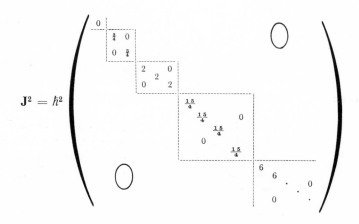

$$\mathbf{J}_+ = \hbar$$

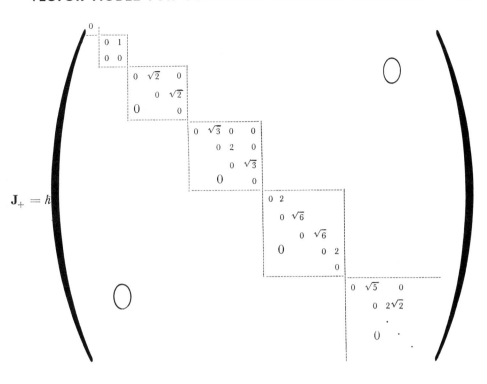

$$\mathbf{J}_- = \hbar$$

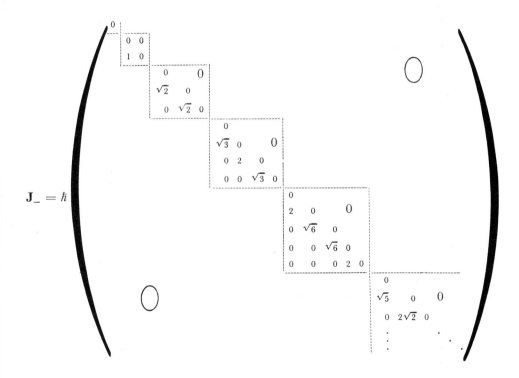

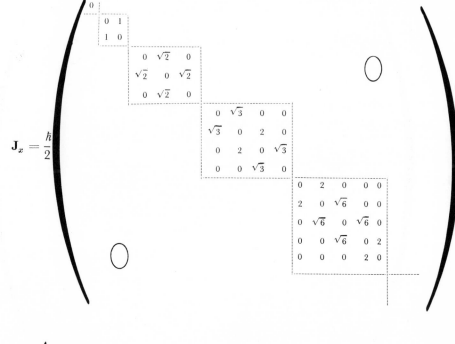

$$\mathbf{J}_x = \frac{\hbar}{2}$$

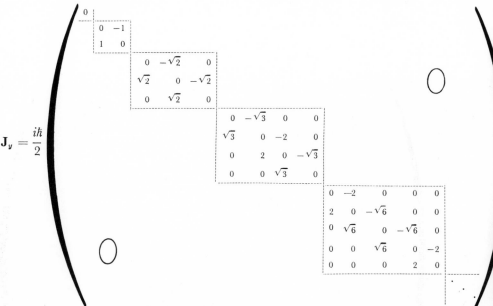

$$\mathbf{J}_y = \frac{i\hbar}{2}$$

4. ADDITIONAL INTERACTIONS IN THE ONE-ELECTRON PROBLEM

If one wishes to account for the observed fine structure of the level splittings in the spectra of one-electron atoms, there are two other effects which should be considered. The relativistic correction is one of these, since its contribution is the same order of magnitude as the spin-orbit splitting. Although the Dirac relativistic theory is the proper theory to use, it is possible to obtain the relativistic corrections from the classical Hamiltonian by writing the kinetic energy in the relativistic form,

$$T = E - m_0 c^2 = m_0 c^2 \left(1 + \frac{p^2}{m_0^2 c^2}\right)^{\frac{1}{2}} - m_0 c^2,$$

where Equations 1.16 and 1.18 have been used. For $p \ll m_0 c$, the radical may be expanded to give

$$T = m_0 c^2 \left(1 + \frac{p^2}{2m_0^2 c^2} - \frac{p^4}{8m_0^4 c^4} + \cdots \right) - m_0 c^2$$

$$\approx \frac{p^2}{2m_0} - \frac{p^4}{8m_0^3 c^2}.$$

Then the approximate Hamiltonian becomes,

$$\mathscr{H} \approx \left(\frac{p^2}{2m_0} + V\right) - \frac{p^4}{8m_0^3 c^2}.$$

The relativistic correction can be obtained by treating the last term as a perturbation (see Problem 9-15). The result is

$$\Delta E_{\text{rel}} = \frac{Z^2 \alpha^2 |E_n|}{4n^2} \left(3 - \frac{4n}{\ell + \frac{1}{2}}\right). \tag{8.12}$$

Note that this would be in exact agreement with Equation 4.13 for $k = \ell + \frac{1}{2}$.

The second effect, the nuclear hyperfine interaction, is several orders of magnitude smaller than the spin-orbit and relativistic corrections. Although it is often difficult to detect in atomic spectroscopy, the hyperfine interaction is of considerable importance in electron and nuclear magnetic resonance experiments. It was pointed out in section 9 of Chapter 3 that nucleons possess an intrinsic spin of $\frac{1}{2}\hbar$. Consequently, nuclei have a net spin except for the special cases in which the nucleon spins pair off to give a total spin of zero. The interaction of the nuclear spin with the spin of the electron is known as the hyperfine interaction. As in the spin-orbit case, the energy associated with the hyperfine interaction may be expressed by the vector coupling model as

$$\Delta E_{\text{hyperfine}} = A\vec{I} \cdot \vec{S}, \tag{8.13}$$

where $\vec{I}$ is the nuclear spin and A is the hyperfine constant. Although the nuclear spin is frequently larger than the spin and the orbital angular momentum of the electron, it should be recalled that the nuclear magneton is smaller than the Bohr magneton by a factor of about 2000. Hence, the hyperfine energy is smaller than the spin-orbit energy by roughly a factor of 10^{-3}.

The nuclear spin variables, operators, and eigenvalues may be incorporated into the Schrödinger formulation in the same way as the spin for the electron. Thus, we may define a nuclear spin state by the ket $|I, m_I\rangle$, where I is the nuclear spin quantum number, and m_I is the spin projection. Degeneracies of high order can occur here, since the nuclear spin can assume large values. For example, a nuclear spin of $\frac{9}{2}$ has 10 possible projections, which are all degenerate in the absence of an interaction that would remove the degeneracy. The total wave function, Equation 8.1, including nuclear spin is modified as follows:

$$\Psi_{\text{total}} = |R_{n\ell} \cdot e^{-iE_n t/\hbar}\rangle \, |\ell, m_\ell\rangle \, |s, m_s\rangle \, |I, m_I\rangle. \tag{8.14}$$

However, if the hyperfine interaction is included in the Hamiltonian, Equation 8.14 is no longer an eigenfunction, since $\vec{I} \cdot \vec{S}$ does not commute with either $\vec{I}$ or $\vec{S}$. This is again analogous to the spin-orbit case discussed in section 2. We must now enlarge our definition of the total angular momentum of the system, since it is that quantity which is conserved, even in the presence of internal interactions. Therefore, we use the vector coupling model to define the total angular momentum,

$$\vec{F} = \vec{J} + \vec{I} = \vec{L} + \vec{S} + \vec{I},$$

where F and M_F are the good quantum numbers and where M_F can take on the set of values $|j + F|, |j + F - 1|, \ldots, |j - F|$. The form of the new eigenfunctions which replace Equation 8.14 is then,

$$\Psi_{\text{total}} = |R_{n\ell} \, e^{-iE_n t/\hbar}\rangle \, |F, M_F\rangle. \tag{8.15}$$

As in the spin-orbit case, the kets $|F, M_F\rangle$ may be expressed by linear combinations of the kets $|\ell, m_\ell\rangle \, |s, m_s\rangle \, |I, m_I\rangle$.

5. IDENTICAL PARTICLES, EXCHANGE, AND SYMMETRY

In quantum mechanics we regard identical particles as indistinguishable. This is an important departure from classical mechanics, where it is possible not only to specify the coordinates of each particle at a given instant but also to follow the subsequent motion of each. Thus, even identical particles can be "tagged" in classical mechanics. Identical particles are far more elusive in quantum mechanics, as the following example will show. Suppose that we observe two electrons in the same orbital state and we ignore all interactions between them. Then we can express the total spin state as the product of the two one-particle states. In particular, if we know that their total spin is zero, we could write the two-particle spin function as either $\chi_+(1) \cdot \chi_-(2)$ or as $\chi_-(1) \cdot \chi_+(2)$. Since we do not know which electron has its spin up and which has spin down, either function is as good as the other. Furthermore, for all we know, the electrons might even be exchanging positions and spin states while we make our measurements. Therefore, the most general spin function is a linear combination of these two states, namely,

$$\psi_{\text{spin}}^{\pm} = \frac{1}{\sqrt{2}} \left[\chi_+(1) \cdot \chi_-(2) \pm \chi_-(1) \cdot \chi_+(2) \right]. \tag{8.16}$$

The reader should note that ψ_{spin}^{+} does not change sign under exchange of the two electrons, while ψ_{spin}^{-} does change sign under particle exchange. ψ_{spin}^{+} is said to be symmetric and ψ_{spin}^{-} antisymmetric under particle exchange.

The foregoing ideas may be extended to more general functions by defining the two-electron wave functions,

$$\psi_{ab}(1, 2) = \psi_a(1) \cdot \psi_b(2) \quad \text{and} \quad \psi_{ba}(1, 2) = \psi_b(1) \cdot \psi_a(2). \tag{8.17}$$

Here the letter a or b represents the set of quantum numbers (n, ℓ, m, m_s) which defines a specific one-particle state described by both spatial and spin coordinates. The integer 1 or 2 refers to the set of spatial and spin coordinates associated with a given electron. Then, the symmetric and antisymmetric linear combinations of these functions are,

and

$$\left. \begin{array}{c} \psi_S = \dfrac{1}{\sqrt{2}} \left(\psi_{ab} + \psi_{ba} \right) \\[2em] \psi_A = \dfrac{1}{\sqrt{2}} \left(\psi_{ab} - \psi_{ba} \right). \end{array} \right\} \tag{8.18}$$

We now define a *particle exchange operator*, P_{12}, which exchanges the coordinates of the two electrons. Thus:

$$P_{12}\psi_{ab}(1, 2) = \psi_a(2) \cdot \psi_b(1) = \psi_{ba}(1, 2),$$

and

$$P_{12}\psi_{ba}(1, 2) = \psi_b(2) \cdot \psi_a(1) = \psi_{ab}(1, 2).$$

Performing the same operations on the functions given in Equation 8.18, we find that:

$$\begin{array}{c} P_{12}\psi_S = +\psi_S \\[0.5em] P_{12}\psi_A = -\psi_A. \end{array} \tag{8.19}$$

From the foregoing we see that although the functions ψ_{ab} and ψ_{ba} are *not* eigenfunctions of the particle exchange operator, P_{12}, the linear combinations defined by ψ_S and ψ_A *are* eigenfunctions. A symmetric eigenfunction of P_{12} has the eigenvalue $+1$, while an antisymmetric eigenfunction has the eigenvalue -1. Furthermore, if the operator P_{12} commutes with the two-particle Hamiltonian, then the exchange symmetry of each eigenfunction is constant for all time.

Operations with the particle exchange operator should not be confused with those of the parity operator Π which was discussed in section 10 of

Chapter 7. Although the eigenvalues of both operators are ± 1, Π is concerned with symmetry under inversion of the coordinates (reflection through the origin), whereas P is concerned with symmetry under particle exchange.

It has been found experimentally that *the wave function for a system of electrons must be antisymmetric in the exchange of the coordinates of any two electrons.* This is a general statement of the *Pauli principle*, which was first enunciated in 1925 in a form something like the following: *no two electrons can occupy simultaneously the same quantum state on the same atom.* In order to reconcile these two statements let us address the specific problem of two electrons on the same atom. From Equations 8.17 and 8.18 note that if the two electrons were in the same spatial and spin quantum states, that is, if $\psi_a = \psi_b$ in violation of the Pauli principle, then the antisymmetric wave function would be zero, whereas the symmetric wave function would not. Thus:

$$\psi_A = \frac{1}{\sqrt{2}}\,(\psi_{aa} - \psi_{aa}) = 0,$$

$$\psi_S = \frac{1}{\sqrt{2}}\,(\psi_{aa} + \psi_{aa}) = \sqrt{2}\psi_{aa}.$$

It follows that the Pauli exclusion principle will be automatically satisfied if we require that only an antisymmetric total wave function be used to describe a system of electrons. The antisymmetric wave function for two electrons (Equation 8.18) can be written conveniently as the 2×2 determinant:

$$\psi_A = \frac{1}{\sqrt{2}} \begin{vmatrix} \psi_a(1) & \psi_a(2) \\ \psi_b(1) & \psi_b(2) \end{vmatrix} = \frac{1}{\sqrt{2}} [\psi_a(1) \cdot \psi_b(2) - \psi_b(1) \cdot \psi_a(2)].$$

The generalization of this form, which is called the *Slater determinant*,[8] can be used to construct the antisymmetric total wave function for any system of N electrons. Thus:

$$\psi_A = \frac{1}{\sqrt{N!}} \begin{vmatrix} \psi_a(1) & \psi_a(2) & \cdots & \psi_a(N) \\ \psi_b(1) & \psi_b(2) & \cdots & \psi_b(N) \\ \cdot & & & \\ \cdot & & & \\ \cdot & & & \\ \psi_n(1) & \psi_n(2) & \cdots & \psi_n(N \end{vmatrix} \qquad (8.20)$$

Note that the rule for expanding the determinant insures that the wave function is antisymmetric; the fact that the determinant is zero if any two rows or

---

[8] See, for example, J. C. Slater, *Quantum Theory of Matter*, 2nd ed. McGraw-Hill Book Co., New York, 1968, Chapter 11.

columns are equal guarantees that the Pauli principle is obeyed. An alternative definition which is equivalent to Equation 8.20 is,

$$\psi_A = \frac{1}{\sqrt{N!}} \sum_{n=1}^{n!} \epsilon P_{1,2,\dots,N}[\psi_a(1)\psi_b(2) \cdots \psi_n(N)], \qquad (8.21)$$

where $P_{1,2,\dots,N}$ permutes the coordinates labels $1, 2, \dots, N$. The factor ϵ is $+1$ when the permuted arrangement of these integers is equivalent to an *even* number of two-particle permutations; $\epsilon = -1$ when the arrangement is equivalent to an *odd* number of two-particle permutations.

Electrons are not the only particles that require antisymmetric total wave functions. It is an experimentally observed fact that all particles with half-integral spin are in this category, which includes positrons, protons, neutrons, mesons, neutrinos, and nuclei with odd mass numbers. As a class these particles are called *fermions*, since they are found to obey Fermi-Dirac statistics.* On the other hand, all particles with zero or integral spin—which includes photons, phonons, pions, alpha particles, and nuclei with even mass numbers—must have symmetric total wave functions. Particles of this class are called *bosons* since they obey Bose-Einstein statistics.*

As in the case of a one-particle wave function, the wave function for a system of particles can be written as a product of a spatial part and a spin part whenever the spatial and spin coordinates are independent. Then, for the total wave function we have the following possibilities:

Fermion states:

$$\psi_A = \begin{cases} \text{(symmetric } N\text{-particle spatial function} \times \text{antisymmetric} \\ N\text{-particle spin function)} \\ \text{(antisymmetric } N\text{-particle spatial function} \times \text{symmetric} \\ N\text{-particle spin function)} \end{cases}$$

Boson states:

$$\psi_S = \begin{cases} \text{(symmetric } N\text{-particle spatial function} \times \text{symmetric} \\ N\text{-particle spin function)} \\ \text{(antisymmetric } N\text{-particle spatial function} \times \text{antisymmetric} \\ N\text{-particle spin function)} \end{cases}$$

To illustrate the above with a specific example, let us separate the spatial and spin coordinates of the two one-electron wave functions, ψ_a and ψ_b, as follows:

$$\psi_a = \psi_n \chi_\pm \quad \text{and} \quad \psi_b = \psi_{n'} \chi_\pm,$$

where n and n' represent different sets of the quantum numbers n, ℓ, m. Then

---

* See Appendix A of this chapter.

we can obtain the two spatial functions,

$$\psi_{\text{spatial}} = \frac{1}{\sqrt{2}} [\psi_n(1)\psi_{n'}(2) \pm \psi_{n'}(1)\psi_n(2)],$$

where the function having the plus sign is symmetric and the one having the minus sign is antisymmetric in the exchange of the electrons. Likewise, we can write the spin functions:

$$\chi_+(1) \cdot \chi_+(2)$$

$$\chi_-(1) \cdot \chi_-(2)$$

$$\frac{1}{\sqrt{2}} [\chi_+(1) \cdot \chi_-(2) \pm \chi_-(1) \cdot \chi_+(2)],$$

where the last function containing the minus sign is antisymmetric and all others are symmetric. Therefore, four antisymmetric product functions may be written as follows:

$$\frac{1}{\sqrt{2}} [\psi_n(1) \cdot \psi_{n'}(2) - \psi_{n'}(1)\psi_n(2)] \cdot \begin{cases} \chi_+(1) \cdot \chi_+(2) \\ \frac{1}{\sqrt{2}} [\chi_+(1) \cdot \chi_-(2) + \chi_-(1) \cdot \chi_+(2)] \\ \chi_-(1) \cdot \chi_-(2) \end{cases}$$

$$\frac{1}{\sqrt{2}} [\psi_n(1) \cdot \psi_{n'}(2) + \psi_{n'}(1)\psi_n(2)] \cdot \frac{1}{\sqrt{2}} [\chi_+(1) \cdot \chi_-(2) - \chi_-(1) \cdot \chi_+(2)].$$

$$(8.22)$$

PROBLEM 8-9

(a) Write an expression similar to Equation 8.21 for the most general *symmetric* wave function for N bosons.
(b) Show that Equations 8.20 and 8.21 are equivalent for the case of three electrons.

PROBLEM 8-10

(a) Apply the particle exchange operator P_{12} to each of the probability densities $\psi_{ab}^*\psi_{ab}$ and $\psi_{ba}^*\psi_{ba}$. Do they remain unchanged?
(b) Operate on $\psi_S^*\psi_S$ and $\psi_A^*\psi_A$ with P_{12}. Do they remain unchanged?
(c) Do the above results suggest a physical reason for choosing ψ_S and ψ_A as wave functions rather than ψ_{ab} and ψ_{ba}?

PROBLEM 8-11

Show that P_{12} commutes with the Hamiltonian,

$$\mathscr{H} = \frac{p_1^2}{2m_1} + \frac{p_2^2}{2m_2} - \frac{Ze^2}{r_1} - \frac{Ze^2}{r_2} + \frac{e^2}{r_{12}}.$$

6. EXCHANGE DEGENERACY AND EXCHANGE ENERGY

If the Hamiltonian for the two-electron system commutes with the particle exchange operator, then all four of the functions of Equation 8.22 are energy eigenfunctions. This will occur when the Hamiltonian itself is symmetric in the coordinates of the two electrons. By way of illustration, consider the following Hamiltonian for two *non-interacting* electrons:*

$$\mathscr{H}(1, 2) = \left[-\frac{\hbar^2}{2m} \nabla_1^2 + V_1 \right] + \left[-\frac{\hbar^2}{2m} \nabla_2^2 + V_2 \right] = \mathscr{H}(1) + \mathscr{H}(2).$$
(8.23)

$$[P_{12}, \mathscr{H}(1, 2)] = 0.$$
(8.24)

As a result of Equation 8.24 we know that the symmetry of each eigenfunction of $\mathscr{H}(1, 2)$ will be constant for all time (see Equation 6.18). Furthermore, the functions given in Equation 8.22 can be shown to be *degenerate* eigenfunctions of $\mathscr{H}(1, 2)$. Assume that the one-electron functions satisfy the following equations:

$$\mathscr{H}(1)\psi_n(1) = E_n\psi_n(1)$$
$$\mathscr{H}(2)\psi_n(2) = E_n\psi_n(2)$$
$$\mathscr{H}(1)\psi_{n'}(1) = E_{n'}\psi_{n'}(1)$$
$$\mathscr{H}(2)\psi_{n'}(2) = E_{n'}\psi_{n'}(2).$$

Then:

$$\mathscr{H}(1, 2)\psi_A = (E_n + E_{n'})\psi_A,$$

where ψ_A is any one of the four functions in Equation 8.22. Although these four functions are degenerate when there are no spin-dependent forces or interactions between the electrons, we will see that the motions of the two electrons are correlated with their spin states. Thus, for the $S = 0$ state (singlet state), when the spins are antiparallel, the spatial eigenfunction is symmetric and the electrons can come very close together. In fact, the probability density for this state has a maximum value when the electrons are so close together that their coordinates are essentially the same. However, for the $S = 1$ state (triplet

---

* Although this is physically unrealizable, we will even neglect, for the time being, the Coulomb interactions between the two electrons.

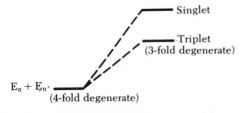

$E_n + E_{n'}$ ———

(4-fold degenerate)

Singlet

Triplet
(3-fold degenerate)

Figure 8–3 The effect of the Coulomb repulsion on the exchange degeneracy for two electrons.

state), when the spins are parallel, the antisymmetric spatial function goes to zero as the electrons approach each other. Thus, in the triplet state the motions of the electrons are correlated so that they avoid each other as much as possible, while in the singlet state they attract each other. That is, they move as if they were under the influence of a force that is repulsive for parallel spins and attractive for antiparallel spins. This force is called the *exchange force* since it has its origin in the symmetry requirements on the wave function under the exchange of indistinguishable particles. The exchange force is strictly a quantum mechanical phenomenon with no classical analog.

In spite of the introduction of this new force, the exchange force, the functions in Equation 8.22 are still degenerate when the Hamiltonian is written as in Equation 8.23. Suppose we now add one term to the Hamiltonian, namely, the Coulomb repulsion between the two electrons, $+e^2/r_{12}$. Although this term does not depend upon the spin coordinates, it is evident that it will partially remove the exchange degeneracy of Equation 8.22. Since the electrons are closer together, on the average, in the singlet state than in the triplet state, the Coulomb repulsion increases the energy of the singlet state more than it increases the energy of the triplet state (see Figure 8-3). The energy difference between the singlet and triplet states is called the *exchange energy*; it is really an electrostatic energy, but its very existence depends upon the fact that the exchange force causes the aligned spin state to have lower energy. Accordingly, it is energetically more favorable for the spins of electrons in the same orbital state to assume a parallel alignment. It is in this sense that we say the exchange force aligns the spins. This topic will be discussed further in Chapter 10 in connection with the helium atom and ferromagnetism.

PROBLEM 8-12

Compute the ground state energy for 3 identical fermions confined to a cubical box of edge a. How does this energy compare with the ground state energy for 3 identical bosons having the same mass?

PROBLEM 8-13

Consider a system consisting of a large number of identical particles, of the order of Avogadro's number, distributed over a great number of closely spaced energy levels in a big box. Would you expect the heat capacity of this system to depend upon whether these particles are bosons or fermions? Explain.

What is the minimum total energy of 5 neutrons and 3 protons confined to a cubical box of edge a? Neglect all interactions between particles.

7. THE PERIODIC CHART AND MANY-ELECTRON ATOMS

In Chapter 7 we obtained the spatial states for a single electron in the presence of an atomic nucleus, and we saw how these states can be specified by the three quantum numbers n, ℓ, and m_ℓ. In the present chapter we have seen how the spin of the electron is included by adding a fourth quantum number, m_s, which gives the projection of the spin along the axis of quantization. In ket form, the state of a single atomic electron may be represented by $|n, \ell, m_\ell, m_s\rangle$, or by $|n, j, m_j, m_s\rangle$, where the latter takes into account the effect of the spin-orbit interaction. When a second electron is added to the atom we must consider not only the strong Coulomb repulsion between the two electrons, but also the symmetry properties of indistinguishable particles of spin $\frac{1}{2}$.

In a neutral, unexcited atom the electons occupy the lowest possible energy states consistent with the Pauli principle. Thus, for hydrogen the electron occupies a ψ_{1s} state (see section 8 of Chapter 7) with its spin in either the $+m_s$ or the $-m_s$ state. In helium the second electron also occupies the ψ_{1s} orbital state but its spin state must be opposite to that of the first electron. Since these two electrons exhaust the possible electronic states for $n = 1$, we say that the first shell is full or "closed." Electrons associated with the same n value are said to be in the same *shell*, and those having the same value of ℓ (as well as n) are said to be in the same *subshell*. For $n = 1$ there is only the s subshell ($\ell = 0$); for $n = 2$ there are two subshells, the s and p, corresponding to $\ell = 0$ and $\ell = 1$, respectively. In the case of lithium, with $Z = 3$, the third electron must go into the $2s$ subshell. Normally, the s states have slightly lower energies than the states with non-zero angular momentum associated with the same principal quantum number. For beryllium, with $Z = 4$, the fourth electron closes the $2s$ subshell. The fifth electron of boron goes into a $2p$ state. Since there are three p orbitals, there are six possible p states, including spin, which are successively filled by the elements from $Z = 5$ to $Z = 10$. Neon, for which $Z = 10$, has a completely closed-shell configuration since the $1s$, $2s$, and $2p$ subshells are all full. Note that a closed shell or subshell has a net angular momentum of zero and a net spin of zero, since all of the orbital momenta and spins cancel in pairs.

In addition to the s and p subshells, the $n = 3$ shell contains a d subshell which can hold ten electrons, making a possible total of 18 electrons in that shell. In general, the total number of electrons that the nth shell can hold is $2n^2$. The periodic chart is built up in this fashion, although there are important exceptions to the orderly filling of the shells and subshells illustrated above. For example, in the transition metal series ($Z = 21$ to 28) the $4s$ subshell fills

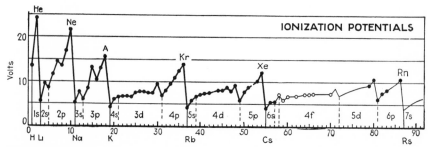

Figure 8-4 Ionization potential for the outermost electron of each element. (From *Introduction to Atomic Spectra* by Harvey E. White. Copyright 1934 by McGraw-Hill Book Company. Used with permission.)

before the $3d$ subshell, and in the rare earths ($Z = 58$ to 71) the $5s$ and $5p$ subshells fill before the $4f$ subshell. These unfilled, inner subshells account for the unusual magnetic properties of these metals (see section 3 of Chapter 10) as well as many other aspects of their behavior in the solid state.

The specification of the quantum states for all of the electrons of a given atom (or ion) is known as the *electronic configuration* for that atom. Thus, for copper the electronic configuration is $1s^2\ 2s^2\ 2p^6\ 3s^2\ 3p^6\ 3d^{10}\ 4s^1$, where the superscript gives the number of electrons of each type. The configurations for all of the elements are given in tabular form in Appendix B of this chapter.

The outermost electrons of an atom, called the *valence* electrons, are the electrons that take part in the formation of chemical bonds. In a solid, these are the electrons that form the conduction bands of metals and the valence and conduction bands of semiconductors. Many aspects of the physical and chemical behavior of substances depend upon how firmly these valence electrons are bound to their atom. Figure 8-4 shows a plot of ionization energy for the last electron of the configuration versus atomic number. Note the great stability of the electronic configurations for the noble gases. Since they have no valence electrons, they are almost chemically inert. However, the element just above and just below each noble gas is quite active chemically. The additional electron of the element just above the noble gas is easily removed in chemical reactions. Likewise, the hole in the closed-shell configuration of each element just below a noble gas is easily filled. The energy given up by the atom when an electron fills the hole is called the *electron affinity*. For a detailed discussion of the relationship of the properties of the elements to their electronic configurations, the reader is referred to a text on general chemistry.

We will now look briefly at the many-electron problem. In order to obtain accurate values for the ground state energy of the atom, one would have to include in the Hamiltonian all of the following contributions to the energy: (1) the single particle kinetic energies, (2) the single electron interactions with the nucleus, (3) the electrostatic interactions (including exchange correlation) of the electrons, (4) the spin-orbit interactions, (5) the spin-spin interactions between electrons, (6) the dipole-dipole interactions between the orbital magnetic moments of the electrons, (7) the hyperfine interactions between the nuclear spin and the electron spin, and (8) other effects such as relativistic corrections and quadrupole interactions. We know how to handle (1) and (2) for any number of independent particles by merely writing the Hamiltonian as the sum of all of the single particle Hamiltonians; the total wave function is then expressed as the properly symmetrized combination of products of the one-particle wave functions. (3) is often neglected in the first approximation, although it can be a large effect. It can be treated by perturbation theory for a small number of electrons. In the case of a large number of electrons, one generally regards each electron as moving in an effective potential which describes its average interaction with all of the other electrons. This topic will be discussed further in Chapter 10. (7) and (8) were discussed in section 4. Since they represent small corrections to the energies they are usually neglected unless a refined calculation is required. The remaining interactions, (4), (5), and (6), will be treated by means of one of the vector coupling models of the next section.

8. ANGULAR MOMENTUM VECTOR COUPLING SCHEMES

Thus far we have considered only the coupling of the spin and orbital momentum of a single electron by means of the spin-orbit interaction. A question naturally arises as to how one should proceed in the case of two electrons where there are four constituent momenta to be considered. One model, known as the *j-j coupling model*, assumes that the spin-orbit interaction dominates the electrostatic interactions between the particles. Thus, we write $\vec{J}_1 = \vec{L}_1 + \vec{S}_1$ and $\vec{J}_2 = \vec{L}_2 + \vec{S}_2$ for each of the two particles, where the quantum numbers j_i are obtained as before. Then the total angular momentum is obtained by combining $\vec{J}_1$ and $\vec{J}_2$. That is,

$$\vec{J} = \vec{J}_1 + \vec{J}_2$$

and $j = |j_1 + j_2|, |j_1 + j_2 - 1|, \ldots, |j_1 - j|_2$.

We will illustrate *j-j* coupling by applying it to two inequivalent* p electrons. For each electron we found above that $j_1 = j_2 = \frac{1}{2}$ or $\frac{3}{2}$. Then the possible ways of combining these are shown in Table 8-1. In a weak magnetic

TABLE 8-1 *j-j* Coupling of Two Inequivalent p Electrons

j_1	j_2	j	Spectral Terms	Number of States in a Magnetic Field (Number of m_j Values)
$\frac{3}{2}$	$\frac{3}{2}$	3, 2, 1, 0	$\left(\frac{3}{2}, \frac{3}{2}\right)_{3,2,1,0}$	16
$\frac{3}{2}$	$\frac{1}{2}$	2, 1	$\left(\frac{3}{2}, \frac{1}{2}\right)_{2,1}$	8
$\frac{1}{2}$	$\frac{1}{2}$	1, 0	$\left(\frac{1}{2}, \frac{1}{2}\right)_{1,0}$	4
$\frac{1}{2}$	$\frac{3}{2}$	2, 1	$\left(\frac{1}{2}, \frac{3}{2}\right)_{2,1}$	8
		10 states		36 states

*Inequivalent electrons occupy different atoms or different orbits of the same atom. Hence, they automatically satisfy the Pauli principle.

field, each state of a given j will split into $2j + 1$ states corresponding to the allowed values of m_j.

Although j-j coupling is used extensively for the description of the nuclear states observed in nuclear spectroscopy, it is not appropriate for many atomic systems because of the strong electrostatic (and other) interactions between the two electrons. The model that has been most successful in accounting for atomic spectra of all but the heavier atoms is known as the Russell-Saunders coupling scheme.[9] In essence, this model simply assumes that the electrostatic interaction (including exchange forces) between two electrons dominates the spin-orbit interaction. In this case, the orbital momenta and the spins of the two electrons couple separately to form $\vec{L} = \vec{L}_1 + \vec{L}_2$ and $\vec{S} = \vec{S}_1 + \vec{S}_2$. Then the total angular momentum is given by $\vec{J} = \vec{L} + \vec{S}$, as before. For two inequivalent p electrons we have: $\ell = 2$, 1, or 0 and $s = 1$ or 0. For each ℓ and s the j values are $|\ell + s|$, $|\ell + s - 1|$, $\ldots$, $|\ell - s|$ and for each j value there are $(2j + 1)$ values of m_j. The combinations are shown in Table 8-2. Once again there are 36 states in a weak magnetic field, although of course, their energies are not the same as those in the j-j coupling scheme.

Suppose we now subject the two electrons to an intense magnetic field such that $\mu_B \mathscr{B}$ is much greater than the spin-orbit interaction as well as all interactions between the electrons. This will uncouple the momenta, regardless of whether Russell-Saunders or j-j coupling holds in weak fields, so that $\vec{L}_1$, $\vec{L}_2$, $\vec{S}_1$, and $\vec{S}_2$ will precess independently about $\mathscr{B}$. The good quantum numbers will no longer be j and m but will be m_{ℓ_1}, m_{ℓ_2}, m_{s_1}, and m_{s_2}. The total number of states—that is, the possible combinations of these quantum numbers—will be $(2\ell_1 + 1)(2\ell_2 + 1)(2s_1 + 1)(2s_2 + 1) = 36$, as before.

We have thus far discussed only inequivalent electrons, whereas *equivalent* electrons are of great physical importance in atomic spectroscopy. Since equivalent electrons occupy the same orbital state on the same atom, the Pauli

TABLE 8-2 Russell-Saunders Coupling of Two Inequivalent p Electrons

ℓ	s	j	Spectral Terms	Number of States in a Magnetic Field (Number of m_j Values)
2	1	3, 2, 1	$^3D_{1,2,3}$	15
2	0	2	$^1D_2$	5
1	1	2, 1, 0	$^3P_{0,1,2}$	9
1	0	1	$^1P_1$	3
0	1	1	$^3S_1$	3
0	0	0	$^1S_0$	1
		10 states		36 states

[9] H. N. Russell and F. A. Saunders, *Astrophys. J.* **61**, 38 (1925). For a more detailed discussion, see J. H. VanVleck, *The Theory of Electric and Magnetic Susceptibilities*, Oxford University Press, Oxford, 1932.

TABLE 8-3 The Fifteen States for Two Equivalent p Electrons

m_{l_1}	m_{l_2}	m_{s_1}	m_{s_2}	m_j	m_l	m_s	Spectral Term
1	1	$\frac{1}{2}$	$-\frac{1}{2}$	2	2	0	$^1D_2$
1	0	$\frac{1}{2}$	$-\frac{1}{2}$	1	1	0	$^1D_2$
1	-1	$\frac{1}{2}$	$-\frac{1}{2}$	0	0	0	$^1D_2$
0	-1	$\frac{1}{2}$	$-\frac{1}{2}$	-1	-1	0	$^1D_2$
-1	-1	$\frac{1}{2}$	$-\frac{1}{2}$	-2	-2	0	$^1D_2$
1	0	$\frac{1}{2}$	$\frac{1}{2}$	2	1	1	$^3P_2$
1	-1	$\frac{1}{2}$	$\frac{1}{2}$	1	0	1	$^3P_2$
1	0	$-\frac{1}{2}$	$-\frac{1}{2}$	0	1	-1	$^3P_2$
1	-1	$-\frac{1}{2}$	$-\frac{1}{2}$	-1	0	-1	$^3P_2$
0	-1	$-\frac{1}{2}$	$-\frac{1}{2}$	-2	-1	-1	$^3P_2$
1	0	$-\frac{1}{2}$	$\frac{1}{2}$	1	1	0	$^3P_1$
0	-1	$\frac{1}{2}$	$\frac{1}{2}$	0	-1	1	$^3P_1$
0	-1	$\frac{1}{2}$	$-\frac{1}{2}$	-1	-1	0	$^3P_1$
1	-1	$\frac{1}{2}$	$-\frac{1}{2}$	0	0	0	$^3P_0$
0	0	$\frac{1}{2}$	$-\frac{1}{2}$	0	0	0	$^1S_0$

exclusion principle puts restrictions on the allowed values of m_l and m_s. Let us now re-examine the two coupling schemes just discussed and apply the exclusion principle.

In the case of the Russell-Saunders coupling we find that for two equivalent p electrons we must exclude the following terms: $^3D_{1,2,3}$, $^1P_1$, and $^3S_1$. The exclusion of the ^3D terms is easily understood, since a D term requires that $l = 2$, which means that m_{l_1} and m_{l_2} must be the same; then the Pauli principle permits only a singlet spin state (opposite spins), or a ^1D term. Similarly, the ^3S term requires that $m_{l_1} = m_{l_2} = 0$ and that the spins be parallel, which violates the Pauli principle. The exclusion of the ^1P term is not quite so obvious. However, if a table is constructed such as Table 8-3, which shows all of the allowed states, every state is accounted for by the $^1D_2$, $^3P_{2,1,0}$, and $^1S_0$ terms. There are no additional distinguishable states which can be attributed to a ^1P term. In a weak magnetic field these five states split into 15 states. It is easy to see why only 15 of the original 36 states for two inequivalent electrons remain. Six of the 36 states violate the Pauli principle, and of the remaining 30 states only half are distinguishable.

When applying j-j coupling to two equivalent p electrons, note first that for $j_1 = j_2 = \frac{3}{2}$ we must exclude the following cases:

$$j_1 = \tfrac{3}{2}, \quad j_2 = \tfrac{3}{2}, \quad j = 3$$
$$j_1 = \tfrac{3}{2}, \quad j_2 = \tfrac{3}{2}, \quad j = 1$$
$$j_1 = \tfrac{1}{2}, \quad j_2 = \tfrac{1}{2}, \quad j = 1.$$

The first case is obvious since it would require parallel orbital moments and parallel spins, which would violate the Pauli principle. The third case would require $l_1 = l_2 = 0$, as well as parallel spins. The middle case is not so

TABLE 8-4 j-j Coupling of Two Equivalent p Electrons

j_1	j_2	j	Spectral Terms	Number of States in a Magnetic Field
$\frac{3}{2}$	$\frac{3}{2}$	2, 0	$(\frac{3}{2}, \frac{3}{2})_{2,0}$	6
$\frac{3}{2}$	$\frac{1}{2}$	2, 1	$(\frac{3}{2}, \frac{1}{2})_{2,1}$	8
$\frac{1}{2}$	$\frac{1}{2}$	0	$(\frac{1}{2}, \frac{1}{2})_0$	1
		5 states		15 states

obvious, but as in the case of the ^1P term in the example from Russell-Saunders coupling, it can be shown that no additional distinguishable states are available to provide for the $j = 1$ level when $j_1 = j_2 = \frac{3}{2}$. Also, the states given by $j_1 = \frac{3}{2}, j_2 = \frac{1}{2}$ are indistinguishable from those given by $j_1 = \frac{1}{2}, j_2 = \frac{3}{2}$. Then, for equivalent electrons, the distinguishable allowed states are shown in Table 8-4.

The above coupling schemes, augmented with the empirical rules of Hund,[10] enable one to account for many features of the atomic spectra of many-electron atoms without actually solving the many-body problem. A quantum mechanical calculation is necessary, however, to obtain the energies of the states and to verify Hund's rules.[11] Hund's rules for the ground state of a collection of equivalent electrons may be summarized as follows:

(1) Electron spins align so as to produce the largest value of S consistent with the Pauli principle.

(2) The states with largest ℓ values are filled first.

(3) $j = \ell - s$ for less than half-filled shells, but $j = \ell + s$ for more than half-filled shells.

To see how these rules are applied, consider the ground state of samarium, which has 6 electrons in the $4f$ shell and no other partially filled shell. Since the $4f$ shell can hold $2(2 \cdot 3 + 1) = 14$ electrons, it is less than half full in Sm, and all six spins are parallel. Thus, $s = 3$. A possible set of different m_{ℓ_i} values is 3, 2, 1, 0, -1, -2, which combine to give $m_\ell = 3$, corresponding to $\ell = 3$; that is, an F term. By the third rule above, $j = \ell - s = 0$, and the term symbol is $^7F_0$. The superscript is the multiplicity, $2s + 1$, while the subscript is the value of j.

As a second example, consider the ground state of iron, which has 6 electrons in the $3d$ shell, the only partially filled shell. Since this shell can hold $2(2 \cdot 2 + 1) = 10$ electrons, it is more than half full. Then 5 electrons have parallel spins and the sixth must be antiparallel, producing a total spin of 2. The m_ℓ values of the 5 parallel electrons are 2, 1, 0, -1, -2 which combine to

[10] See, for example, G. Herzberg, *Atomic Spectra and Atomic Structure*, Dover Publications, New York, 1944, or H. E. White, *Introduction to Atomic Spectra*, McGraw-Hill Book Co., New York, 1934.

[11] E. U. Condon and G. H. Shortley, *The Theory of Atomic Spectra*, Cambridge University Press, Cambridge, 1953.

zero; the sixth electron has a momentum of 2. Hence, $\ell = 2$, which corresponds to a D term. From the third rule above, $j = \ell + s = 4$, and the ground state term for Fe is $^5D_4$.

We will close this section by merely stating the selection rules for electric dipole transitions which are appropriate for the two vector coupling models. More will be said about selection rules in section 6 of Chapter 9. It is assumed that only one electron at a time makes a transition.

Russell-Saunders coupling:

For the electron making the transition, $\Delta\ell = \pm 1$, (that is, the parity must change).

For the atom as a whole,

$$\Delta s = 0$$

$$\Delta \ell = 0, \pm 1$$

$$\Delta j = 0, \pm 1 \text{ (but a transition from } j = 0 \text{ to } j = 0 \text{ is forbidden)}$$

$$\Delta m_j = 0, \pm 1 \text{ (but if } \Delta j = 0, \text{ a transition from } m_j = 0 \text{ to } m_j = 0 \text{ is forbidden)}$$

j-j coupling:

For the electron making the transition, $\Delta\ell = \pm 1$ (parity must change), and $\Delta j = 0, \pm 1$.

For the atom as a whole,

$$\Delta j = 0, \pm 1 \text{ (a transition from } j = 0 \text{ to } j = 0 \text{ is forbidden).}$$

$$\Delta m_j = 0, \pm 1 \text{ (if } \Delta j = 0, \text{ a transition from } m_j = 0 \text{ to } m_j = 0 \text{ is forbidden).}$$

9. THE LANDÉ g FACTOR AND THE ZEEMAN EFFECT

We have previously obtained the result that the orbital and spin contributions to the magnetic moment are given by (section 9 of Chapter 3),

$$\vec{\mu}_\ell = -\frac{g_\ell e}{2m_0 c}\vec{L} = -g_\ell \mu_B \sqrt{\ell(\ell+1)}\,\hat{\ell}$$

and

$$\vec{\mu}_s = -\frac{g_s e}{2m_0 c}\vec{S} = -g_s \mu_B \sqrt{s(s+1)}\,\hat{s},$$

where $g_\ell = 1$ and $g_s = 2.0024 \sim 2$. Now, when $\vec{L}$ and $\vec{S}$ are coupled we have

$$\vec{J} = \vec{L} + \vec{S}$$

and

$$\vec{\mu} = \vec{\mu}_\ell + \vec{\mu}_s = -\frac{\mu_B}{\hbar}(\vec{L} + 2\vec{S}). \qquad (8.25)$$

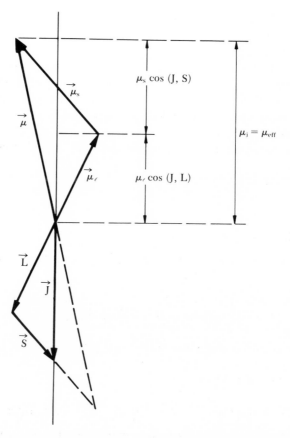

Figure 8-5 The *effective* magnetic moment of an atom is the component of $\vec{\mu}$ which lies along the direction of the total angular momentum.

It is evident from the last two expressions that *the total magnetic moment is not, in general, colinear with the total angular momentum.* This is illustrated in Figure 8-5. Since $\vec{L}$ and $\vec{S}$ precess about $\vec{J}$, it is apparent from the figure that $\vec{\mu}$ also precesses about $\vec{J}$. However, the *effective* magnetic moment, that is, the component of $\vec{\mu}$ along $\vec{J}$, maintains the constant value,

$$\mu_j = \frac{\vec{\mu} \cdot \vec{J}}{|\vec{J}|} = -\frac{\mu_B}{\hbar} \frac{\vec{L} \cdot \vec{J} + 2\vec{S} \cdot \vec{J}}{|\vec{J}|}$$

$$= -\frac{\mu_B}{\hbar} \frac{\vec{L} \cdot (\vec{L} + \vec{S}) + 2\vec{S} \cdot (\vec{L} + \vec{S})}{|\vec{J}|}$$

$$= -\frac{\mu_B}{\hbar} \frac{\vec{L}^2 + 2\vec{S}^2 + 3\vec{L} \cdot \vec{S}}{|\vec{J}|}$$

$$= -\frac{\mu_B}{\hbar} \frac{\vec{L}^2 + 2\vec{S}^2 + \frac{3}{2}(\vec{J}^2 - \vec{L}^2 - \vec{S}^2)}{|\vec{J}|}$$

where we have used a relationship in the last step that will be verified in Problem 8-16.

Then,

$$\mu_j = -\frac{\mu_B |\vec{J}|}{\hbar}\left(\frac{3\vec{J}^2 + \vec{S}^2 - \vec{L}^2}{2\vec{J}^2}\right)$$

$$= -\frac{\mu_B}{\hbar}|\vec{J}| \cdot \left(1 + \frac{\vec{J}^2 + \vec{S}^2 - \vec{L}^2}{2\vec{J}^2}\right) = -\mu_B\sqrt{j(j+1)}$$

$$\times\left[1 + \frac{j(j+1) + s(s+1) - \ell(\ell+1)}{2j(j+1)}\right].$$

Defining the Landé g factor as

$$g = 1 + \frac{j(j+1) + s(s+1) - \ell(\ell+1)}{2j(j+1)}, \tag{8.26}$$

the *effective* magnetic moment becomes,

$$|\mu_j| = g\mu_B\sqrt{j(j+1)}. \tag{8.27}$$

Note that for zero spin, Equation 8.26 reduces to the classical case of $g = 1$; in a similar fashion $g = 2$ if $\ell = 0$.

Now it is possible to account for the so-called anomalous Zeeman effect mentioned in Chapter 3. In a weak magnetic field, the angular momentum $\vec{J}$ will precess about $\mathscr{B}$ such that the projection of $\vec{J}$ along the field direction will be one of the allowed values, $m_j\hbar$. The corresponding magnetic moment along the field direction (taken to be the z-direction) will then be,

$$\mu_z = -g\mu_B m_j,$$

having a magnetic dipolar energy of

$$E = gm_j\mu_B\mathscr{B}. \tag{8.28}$$

The feature that distinguishes Equation 8.28 from the case of the normal Zeeman effect is that here we see that g is no longer unity, but depends upon the quantum numbers ℓ, s, and j. By way of illustration let us calculate g for an electron in a p state and an s state:

orbital state	ℓ	j	g
p	1	$\frac{3}{2}$	$\frac{4}{3}$
p	1	$\frac{1}{2}$	$\frac{2}{3}$
s	0	$\frac{1}{2}$	2

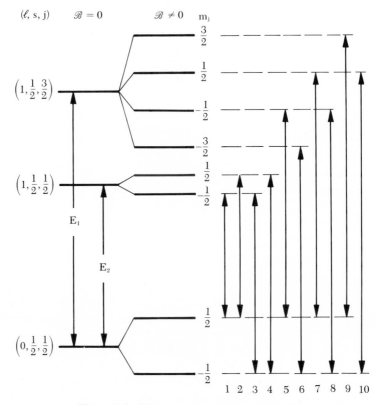

Figure 8-6   Zeeman splittings for p and s states.

In a magnetic field $\mathscr{B}$, such that $\mu_B\mathscr{B}$ is less than the spin-orbit energy, j and m_j are good quantum numbers and the energies of the states split, as shown in Table 8-5. These splittings are shown schematically in Figure 8-6, and the allowed spectral lines due to electric dipole transitions are shown in Figure 8-7. The lines corresponding to transitions in which m changes by ± 1 are labeled σ, and those corresponding to $\Delta m = 0$ are labeled π. The π lines are plane polarized with the direction of polarization parallel to the field. The σ lines

TABLE 8-5   Energy Splitting in a Magnetic Field

Orbital State	j	m_j	g	$\Delta E = gm_j$ (in units of $\mu_B\mathscr{B}$)
p	$\frac{3}{2}$	$\frac{3}{2}$	$\frac{4}{3}$	2
p	$\frac{3}{2}$	$\frac{1}{2}$	$\frac{4}{3}$	$\frac{2}{3}$
p	$\frac{3}{2}$	$-\frac{1}{2}$	$\frac{4}{3}$	$-\frac{2}{3}$
p	$\frac{3}{2}$	$-\frac{3}{2}$	$\frac{4}{3}$	-2
p	$\frac{1}{2}$	$\frac{1}{2}$	$\frac{2}{3}$	$\frac{1}{3}$
p	$\frac{1}{2}$	$-\frac{1}{2}$	$\frac{2}{3}$	$-\frac{1}{3}$
s	$\frac{1}{2}$	$\frac{1}{2}$	2	1
s	$\frac{1}{2}$	$-\frac{1}{2}$	2	-1

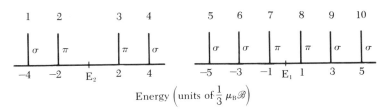

Energy $\left(\text{units of }\frac{1}{3}\,\mu_B\mathscr{B}\right)$

Figure 8-7 The spectral lines resulting from the electronic dipole transitions shown in Figure 8-6. If the emitted light is observed perpendicular to the magnetic field, all of the above lines are seen (transverse Zeeman effect). However, if the emitted light is observed parallel to the field only the σ lines are seen (longitudinal Zeeman effect).

are circularly polarized when observed parallel to the field, and linearly polarized (perpendicular to the field) when observed at right angles to the field. The selection rules for these transitions will be derived in the next chapter.

It is evident from the foregoing that the so-called "anomalous" Zeeman effect is, in reality, what would normally be expected for an electron having half-integral spin in a weak magnetic field. The "normal" or classical Zeeman effect discussed in section 10 of Chapter 3 cannot occur for a single electron in a weak field because of the spin term in Equation 8.26. However, in atoms in which the spins are paired so that the total spin is zero, the g value for all spectroscopic states is the classical value and only three spectral lines are observed. The pair of lines whose energies are shifted by $\pm\mu_B\mathscr{B}$ are circularly polarized, and the unshifted line is plane polarized.

In a sufficiently strong magnetic field such that the spin–orbit splitting is less than the magnetic splitting, the anomalous Zeeman effect converts to the normal Zeeman effect. This phenomenon, known as the Paschen-Bach effect, can be explained in the following manner. When the spin and the orbital magnetic moments interact with the applied field more strongly than they

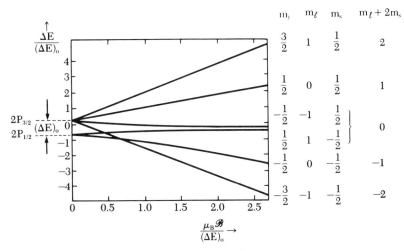

Figure 8-8 Energy level splittings in a magnetic field for the $2p$ states associated with $j = 3/2$ and $j = 1/2$. The zero-field fine structure splitting $(\Delta E)_0$ is largely due to the spin-orbit interaction. The weak-field Zeeman effect occurs for values of $\mathscr{B}$ below the point where the levels would first cross. The Paschen-Bach effect occurs for very large fields where the splitting produces five levels that are nearly equally spaced. In the latter case, three spectral lines are seen as in the normal Zeeman effect.

interact with each other, $\vec{L}$ and $\vec{S}$ are effectively uncoupled and they precess independently about $\vec{\mathscr{B}}$. The good quantum numbers are then m_ℓ and m_s, instead of j and m_j which hold in weak fields. Hence, in strong fields the projections of the magnetic moment on the field direction are proportional to $m_\ell + 2m_s$ rather than m_j. This is illustrated graphically for p electrons in Figure 8-8, where the energy level splittings are shown as a function of applied magnetic field strength. Using the selection rules, $\Delta m_\ell = 0, \pm 1$ and $\Delta m_s = 0$, only three spectral lines are observed, as in the normal Zeeman effect, for transitions from the $2p$ states to the $2s$ states in high magnetic fields.

PROBLEM 8-15

(a)   Using Russell-Saunders coupling, obtain the spectral terms for the ground states of Nd^{3+}, Gd^{3+}, and Ho^{3+}, which have, respectively, 3, 7, and 10 electrons in the unfilled $4f$ shell.[12]

(b)   Calculate the g values and effective magnetic moments for these ions.

(Ans.: $^4I_{\frac{9}{2}}$, $^8S_{\frac{7}{2}}$, $^5I_8$;   0.738,   2.000,   1.250;   3.68,   7.94, 10.6μ_B.)

PROBLEM 8-16

(a)   Show that in the vector coupling model one may define an operator,

$$\vec{L} \cdot \vec{S} = \tfrac{1}{2}(\mathbf{J}^2 - \mathbf{L}^2 - \mathbf{S}^2).$$

This operator will be used later to define the *spin–orbit operator* (see Equation 8.8).

(b)   Does this operator commute with $\mathbf{J}^2$, $\mathbf{L}^2$, and $\mathbf{S}^2$?

PROBLEM 8-17

(a)   For a single d electron, calculate the g values and the Zeeman splittings in units of $\mu_B \mathscr{B}$.

(b)   Do the same for a single f electron.

PROBLEM 8-18

(a)   Consider a system for which $\mathscr{H}_0$, $\mathbf{L}^2$, $\mathbf{L}_z$, $\mathbf{S}^2$, and $\mathbf{S}_z$ form a set of mutually commuting operators. What are the good quantum numbers for the system? Write a possible eigenfunction in symbolic ket form. With these kets as a basis, which operators have diagonal matrix representations?

(b)   Now assume that a spin-orbit interaction is turned on so that the Hamiltonian of the system becomes $\mathscr{H} = \mathscr{H}_0 + \alpha \vec{L} \cdot \vec{S}$. Evaluate the commutators, $[\mathscr{H}, \mathbf{L}_z]$

---

[12] See *American Institute of Physics Handbook*, 2nd Ed. McGraw-Hill Book Co., New York, 1963, page 5–222, Table 5g–29.

and $[\mathscr{H}, \mathbf{S}_z]$. Do $\mathscr{H}$, $\mathbf{L}_z$, and $\mathbf{S}_z$ commute? Are m_ℓ and m_s still good quantum numbers?

(c) Defining $\mathbf{J}_z = \mathbf{L}_z + \mathbf{S}_z$, show that $[\mathscr{H}, \mathbf{J}_z] = 0$, where $\mathscr{H} = \mathscr{H}_0 + \alpha \vec{L} \cdot \vec{S}$ as in part (b). Can we now claim that $\mathscr{H}$, $\mathbf{L}^2$, $\mathbf{S}^2$, $\mathbf{J}^2$, and $\mathbf{J}_z$ form a set of mutually commuting operators? What are the good quantum numbers for this more complex system? Write a possible eigenfunction in ket form. With these kets as a basis, which operators have diagonal matrix representations?

10. EIGENFUNCTIONS OF COUPLED ANGULAR MOMENTA

In section 8 we saw how several angular momenta can be combined using one of the vector coupling models. This approach, augmented by Hund's rules, enables one to account for many details of the spectra of multi-electron atoms without knowing the exact eigenfunctions of the total angular momentum. However, it is frequently necessary to know these eigenfunctions in order to use a representation in which the square of the total angular momentum is diagonal.

Consider two angular momenta $\vec{J}_1$ and $\vec{J}_2$, each of which may be either a spin or an orbital momentum, or a combination of these. $\vec{J}_1$ is defined in a space of dimension $(2j_1 + 1)$, where j_1 is the maximum projection of $\vec{J}_1$, that is, the maximum value of m_1. The eigenkets $|j_1, m_1\rangle$ constitute the basis on which $\mathbf{J}_1^2$ and $\mathbf{J}_{1z}$ are diagonal matrices. Likewise, $\vec{J}_2$ resides in a space of dimension $(2j_2 + 1)$ and the eigenkets $|j_2, m_2\rangle$ permit a diagonal representation for $\mathbf{J}_2^2$ and $\mathbf{J}_{2z}$. We now must find out what happens when we define a new space by combining $\vec{J}_1$ and $\vec{J}_2$ to write $\vec{J} = \vec{J}_1 + \vec{J}_2$. The new space is called a *product space* and it has dimension $(2j_1 + 1) \cdot (2j_2 + 1)$. This product space contains subspaces of dimension $N_1, N_2, \ldots$, where $N_1 = 2j + 1 = 2(j_1 + j_2) + 1$, $N_2 = 2(j - 1) + 1 = 2(j_1 + j_2) - 1, \ldots, N_{2j_2+1} = 2(j_1 - j_2) + 1$, and we have assumed that $j_1 > j_2$. For example, if $j_1 = 3$ and $j_2 = 1$, $N = 21$; there are three subspaces which have dimensions of 9, 7, and 5, respectively.

A simpler example, but one of considerable importance, is the case of coupling two unit angular momenta, that is, $j_1 = j_2 = 1$. The product space in this case has dimension $3 \times 3 = 9$, consisting of the subspaces of dimension 5, 3, and 1. The matrix representations of $\mathbf{J}^2$ and $\mathbf{J}_z$ can be portrayed schematically as follows:

$$
\begin{pmatrix}
5 \times 5 & & \\
 & 3 \times 3 & \\
 & & 1 \times 1
\end{pmatrix}
$$

9×9 matrix

In order to diagonalize these matrices completely we must obtain the simultaneous eigenfunctions of $\mathbf{J}^2 = (\mathbf{J}_1 + \mathbf{J}_2)^2$ and $\mathbf{J}_z = \mathbf{J}_{1z} + \mathbf{J}_{2z}$. Our task, then, will be to express the new eigenfunctions in terms of functions which we already know, namely, $|j_1 m_1\rangle$, the eigenkets of $\mathbf{J}_1^2$ and $\mathbf{J}_{1z}$, and $|j_2 m_2\rangle$, the eigenkets of $\mathbf{J}_2^2$ and $\mathbf{J}_{2z}$. Since the coordinates of $\mathbf{J}_1$ and $\mathbf{J}_2$ are independent in the old representation, the eigenkets may be written as the product functions,

$$|j_1 j_2 m_1 m_2\rangle \equiv |j_1 m_1\rangle\,|j_2 m_2\rangle.$$

These kets form a complete, orthonormal basis so that we may define a projection operator,

$$\sum_{m_1, m_2} |j_1 j_2 m_1 m_2\rangle\langle j_1 j_2 m_1 m_2| = \left(\sum_{m_1} |j_1 m_1\rangle\langle j_1 m_1|\right)\left(\sum_{m_2} |j_2 m_2\rangle\langle j_2 m_2|\right) = 1. \quad (8.29)$$

Let us designate the *new* eigenkets by $|j_1 j_2 jm\rangle$, where

$$\mathbf{J}^2\,|j_1 j_2 jm\rangle = j(j+1)\hbar^2\,|j_1 j_2 jm\rangle$$

and

$$\mathbf{J}_z\,|j_1 j_2 jm\rangle = m\hbar\,|j_1 j_2 jm\rangle.$$

The new eigenkets may be expressed in terms of the original eigenkets by means of the projection operator, Equation 8.29. Thus,

$$|j_1 j_2 jm\rangle = \sum_{m_1, m_2} |j_1 j_2 m_1 m_2\rangle\langle j_1 j_2 m_1 m_2\,|\,j_1 j_2 jm\rangle. \quad (8.30)$$

Since j_1 and j_2 are fixed numbers for a specific calculation, the notation may be simplified by omitting them from the kets. It should be remembered, however, that the maximum values of m_1 and m_2 are j_1 and j_2, respectively, and that the maximum value of m is $j_1 + j_2$. Let us also include the requirement that m always be equal to the sum $m_1 + m_2$. Then we have

$$|jm\rangle = \sum_{m_1 + m_2 = m} |m_1 m_2\rangle\langle m_1 m_2\,|\,jm\rangle. \quad (8.31)$$

The scalars, $\langle m_1 m_2\,|\,jm\rangle$, are merely coefficients that tell how much each old eigenket $|m_1 m_2\rangle$ contributes to the new eigenket $|jm\rangle$. These coefficients are called the *angular momentum coupling coefficients* or the *Clebsch-Gordon coefficients*. They are not difficult to evaluate for small values of j_1 and j_2, but the general expression for them is quite complicated. Many of the Clebsch-Gordon coefficients are available in tabular form.[13] We will illustrate how they may be obtained by means of the angular momentum raising and lowering operators for a few special cases.

---

[13] See E. U. Condon and G. H. Shortley, *op, cit.*; G. J. Nijgh, A. H. Wapstra and R. van Lieshout, *Nuclear Spectroscopy Tables*, North-Holland Publishing Co., Amsterdam, 1969; M. Rotenberg, R. Bivins, N. Metropolis, and J. K. Wooten, Jr., *The 3-j and 6-j Symbols*, Technology Press, Cambridge, Mass., 1959.

By way of illustration, let us consider the case of $j_1 = j_2 = 1$; that is, we wish to combine two unit angular momenta. Since the space has dimensionality of $(2 + 1) \cdot (2 + 1) = 9$, there are nine eigenkets in both the old and the new bases. The old kets, $|j_1 j_2, m_1 m_2\rangle \equiv |m_1 m_2\rangle$ are:

$$|1, 1\rangle, |1, 0\rangle, |1, -1\rangle, |0, 1\rangle, |0, 0\rangle, |0, -1\rangle, |-1, 1\rangle, |-1, 0\rangle, |-1, -1\rangle. \tag{8.32}$$

The new kets, $|jm\rangle$, which are primed for convenience when j and m are replaced by numbers, are:

For

$$j = 2: |2, 2\rangle', |2, 1\rangle', |2, 0\rangle', |2, -1\rangle', |2, -2\rangle'$$

$$j = 1: |1, 1\rangle', |1, 0\rangle', |1, -1\rangle'$$

$$j = 0: |0, 0\rangle'.$$

Note that state $|2, 2\rangle' \equiv |1, 1\rangle$, since they both correspond to the case where $m_1 = m_2 = 1$ and m_1 and m_2 are parallel. Likewise, $|2, -2\rangle' \equiv |-1, -1\rangle$. Thus, we can find all of the states for $j = 2$ by multiple applications of the lowering operator on $|2, 2\rangle'$ or by using the raising operator on $|2, -2\rangle'$. Choosing the former approach, we utilize Equation 8.11 and proceed as follows:

$$J_- |2, 2\rangle' = (J_{1-} + J_{2-}) |1, 1\rangle.$$

$$\sqrt{(2 + 2)(2 - 2 + 1)} |2, 1\rangle' = \sqrt{(1 + 1)(1 - 1 + 1)} |0, 1\rangle$$
$$+ \sqrt{(1 + 1)(1 - 1 + 1)} |1, 0\rangle$$

$$\therefore \quad |2, 1\rangle' = \frac{1}{\sqrt{2}} (|0, 1\rangle + |1, 0\rangle).$$

$$J_- |2, 1\rangle' = \frac{1}{\sqrt{2}} (J_{1-} + J_{2-})(|0, 1\rangle + |1, 0\rangle)$$

$$\sqrt{(2 + 1)(2 - 1 + 1)} |2, 0\rangle' = \frac{1}{\sqrt{2}} [\sqrt{(1 + 0)(1 - 0 + 1)} |-1, 1\rangle$$
$$+ \sqrt{(1 + 1)(1 - 1 + 1)} |0, 0\rangle$$
$$+ \sqrt{(1 + 1)(1 - 1 + 1)} |0, 0\rangle$$
$$+ \sqrt{(1 + 0)(1 - 0 + 1)} |1, -1\rangle]$$

$$\therefore \quad |2, 0\rangle' = \frac{1}{\sqrt{6}} (|-1, 1\rangle + 2 |0, 0\rangle + |1, -1\rangle),$$

and in like manner,

$$|2, -1\rangle' = \frac{1}{\sqrt{2}} (|0, -1\rangle + |-1, 0\rangle),$$

$$|2, -2\rangle' = |-1, -1\rangle.$$

Notice that all of the nine original kets have been used in forming the first five primed kets. It turns out that all new kets having the same m value will be linear combinations of the same group of old kets, each of which satisfies the condition $m_1 + m_2 = m$. We thus choose $|1, 1\rangle'$ to be the linear combination of $|0, 1\rangle$ and $|1, 0\rangle$ which is orthogonal to $|2, 1\rangle'$, namely,

$$|1, 1\rangle' = \frac{1}{\sqrt{2}} (|0, 1\rangle - |1, 0\rangle).$$

Using the lowering operator we readily obtain:

$$|1, 0\rangle' = \frac{1}{\sqrt{2}} (|1, -1\rangle - |-1, 1\rangle)$$

and

$$|1, -1\rangle' = \frac{1}{\sqrt{2}} (|0, -1\rangle - |-1, 0\rangle).$$

The last function, $|0, 0\rangle'$, must be a linear combination of $|1, -1\rangle$, $|-1, 1\rangle$, and $|0, 0\rangle$, and it must be orthogonal to both $|2, 0\rangle'$ and $|1, 0\rangle'$. If we denote the coefficients of $|0, 0\rangle'$ as (a, b, c), then its scalar products with $|2, 0\rangle'$ and $|1, 0\rangle'$

Table 8-6 Clebsch-Gordon Coefficients for $j_1 = j_2 = 1$

New Ket $\lvert jm\rangle$	Old Ket $\lvert m_1 m_2\rangle$	Clebsch-Gordon Coefficient $\langle m_1 m_2 \mid jm\rangle$
$\lvert 2, 2\rangle'$	$\lvert 1, 1\rangle$	1
$\lvert 2, 1\rangle'$	$\lvert 0, 1\rangle$	$1/\sqrt{2}$
	$\lvert 1, 0\rangle$	$1/\sqrt{2}$
$\lvert 2, 0\rangle'$	$\lvert 1, -1\rangle$	$1/\sqrt{6}$
	$\lvert 0, 0\rangle$	$2/\sqrt{6}$
	$\lvert -1, 1\rangle$	$1/\sqrt{6}$
$\lvert 2, -1\rangle'$	$\lvert 0, -1\rangle$	$1/\sqrt{2}$
	$\lvert -1, 0\rangle$	$1/\sqrt{2}$
$\lvert 2, -2\rangle'$	$\lvert -1, -1\rangle$	1
$\lvert 1, 1\rangle'$	$\lvert 0, 1\rangle$	$1/\sqrt{2}$
	$\lvert 1, 0\rangle$	$-1/\sqrt{2}$
$\lvert 1, 0\rangle'$	$\lvert 1, -1\rangle$	$1/\sqrt{2}$
	$\lvert 0, 0\rangle$	0
	$\lvert -1, 1\rangle$	$-1/\sqrt{2}$
$\lvert 1, -1\rangle'$	$\lvert 0, -1\rangle$	$1/\sqrt{2}$
	$\lvert -1, 0\rangle$	$-1/\sqrt{2}$
$\lvert 0, 0\rangle'$	$\lvert 1, -1\rangle$	$1/\sqrt{3}$
	$\lvert 0, 0\rangle$	$-1/\sqrt{3}$
	$\lvert -1, 1\rangle$	$1/\sqrt{3}$

yield the following relations:

$$a + 2b + c = 0$$
$$a - c = 0.$$

Since these must be simultaneously satisfied we obtain,

$$a = -b = c.$$

Therefore,

$$|0, 0\rangle' = \frac{1}{\sqrt{3}} (|1, -1\rangle - |0, 0\rangle + |-1, 1\rangle).$$

We may summarize the results of the above example by tabulating the Clebsch-Gordon coefficients that we have obtained (refer to Equation 8.31).

PROBLEM 8-19

Show that the nine kets obtained above are eigenkets of both $J^2 = (J_1 + J_2)^2$ and $J_z = J_{1z} + J_{2z}$.

PROBLEM 8-20

Write the six state functions $|jm\rangle$ for a single p electron in terms of the kets $|m_\ell, m_s\rangle \equiv |\ell, m_\ell\rangle |s, m_s\rangle$. (Answers in Appendix C of this Chapter.)

PROBLEM 8-21

Show that the spin-orbit coupling factor $\vec{L} \cdot \vec{S}$ may be represented by the following operators:

$$\vec{L} \cdot \vec{S} = L_z S_z + \tfrac{1}{2}(L_+ S_- + L_- S_+).$$

PROBLEM 8-22

Find the matrix representation of the spin-orbit operator of the previous problem using:

(a) the six states $|m_\ell, m_s\rangle$ for a p electron as the basis functions,

(b) the six states $|j, m\rangle$ for a p electron as the basis functions. (Answers in Appendix C of this chapter.)

(a)   Write the fifteen kets $|jm\rangle$ for two equivalent p electrons in terms of the one-electron kets,

$$|m_1 m_2\rangle \equiv |j_1 m_1\rangle \, |j_2 m_2\rangle.$$

II. APPLICATION TO TWO-PARTICLE SPIN FUNCTIONS

The method of the previous section is readily applied to the task of obtaining the spin wave functions for two or more particles. Let us first consider the case of two identical particles of spin $\frac{1}{2}$, where we will now label the spinors of Equation 8.2 with a 1 or a 2 in order to distinguish the particles. Thus, $\chi_+(1)$ means that particle 1 has its spin up and $\chi_-(1)$ means that particle 1 has its spin down. In ket form these spinors are written as $|m_{s_1}, m_{s_2}\rangle \equiv |s_1 m_{s_1}\rangle \, |s_2 m_{s_2}\rangle$, so that we have for two identical spin $\frac{1}{2}$ particles:

$$
\begin{aligned}
|\tfrac{1}{2}, \tfrac{1}{2}\rangle &= \chi_+(1) \cdot \chi_+(2) \\
|\tfrac{1}{2}, -\tfrac{1}{2}\rangle &= \chi_+(1) \cdot \chi_-(2) \\
|-\tfrac{1}{2}, \tfrac{1}{2}\rangle &= \chi_-(1) \cdot \chi_+(2) \\
|-\tfrac{1}{2}, -\tfrac{1}{2}\rangle &= \chi_-(1) \cdot \chi_-(2).
\end{aligned}
\tag{8.33}
$$

If we consider the *total* spin operators $\mathbf{S}^2 = (\mathbf{S}_1 + \mathbf{S}_2)^2$ and $\mathbf{S}_z = \mathbf{S}_{1z} + \mathbf{S}_{2z}$, it is easy to show that the kets of Equation 8.33 are *not* eigenkets of $\mathbf{S}^2$, although they are still eigenkets of $\mathbf{S}_z$.

Show that the operator S^2 for two spin $\frac{1}{2}$ particles may be expressed as:

$$\mathbf{S}^2 = \tfrac{3}{2}\hbar^2 + 2\mathbf{S}_{1z}\mathbf{S}_{2z} + (\mathbf{S}_{1+}\mathbf{S}_{2-} + \mathbf{S}_{1-}\mathbf{S}_{2+}). \tag{8.34}$$

Given that the kets of Equation 8.33 are eigenkets of $\mathbf{S}_1^2$, $\mathbf{S}_2^2$, $\mathbf{S}_{1z}$ and $\mathbf{S}_{2z}$, show that they are *not* all eigenkets of $\mathbf{S}^2 = (\mathbf{S}_1 + \mathbf{S}_2)^2$.

Now let us find the spinors $|s, m_s\rangle$ defined in the product spin space. This is a 4-dimensional space containing a 3-dimensional subspace corresponding to $s = 1$, and a 1-dimensional subspace, for $s = 0$. The new kets are $|1, 1\rangle'$, $|1, 0\rangle'$, $|1, -1\rangle'$, and $|0, 0\rangle'$. (Primes are included here although these kets can be distinguished from those of Equation 8.33.) It should be apparent that $|1, 1\rangle' = \chi_+(1) \cdot \chi_+(2)$ and $|1, -1\rangle' = \chi_-(1) \cdot \chi_-(2)$. To find $|1, 0\rangle'$ we operate as follows:

$$\mathbf{S}_- |1, 1\rangle' = (\mathbf{S}_{1-} + \mathbf{S}_{2-})\chi_+(1) \cdot \chi_+(2) = \chi_-(1)\chi_+(2) + \chi_+(1)\chi_-(2)$$

$$\therefore \quad |1, 0\rangle' = \frac{1}{\sqrt{2}} (|-\tfrac{1}{2}, \tfrac{1}{2}\rangle + |\tfrac{1}{2}, -\tfrac{1}{2}\rangle).$$

The last ket, $|0, 0\rangle'$, must be orthogonal to $|1, 0\rangle'$. Therefore, we have:

$$|1, 1\rangle' = |\tfrac{1}{2}, \tfrac{1}{2}\rangle$$

$$|1, 0\rangle' = \frac{1}{\sqrt{2}} (|-\tfrac{1}{2}, \tfrac{1}{2}\rangle + |\tfrac{1}{2}, -\tfrac{1}{2}\rangle) \qquad (8.35)$$

$$|1, -1\rangle' = |-\tfrac{1}{2}, -\tfrac{1}{2}\rangle$$

$$|0, 0\rangle' = \frac{1}{\sqrt{2}} (|-\tfrac{1}{2}, \tfrac{1}{2}\rangle - |\tfrac{1}{2}, -\tfrac{1}{2}\rangle).$$

The first three eigenkets in Equation 8.35 correspond to the triplet state ($s = 1$), while the last is a singlet state ($s = 0$).

PROBLEM 8-26

Show that the kets in Equation 8.35 satisfy the eigen-equations,

$$\mathbf{S}^2 |s, m_s\rangle = s(s + 1)\hbar^2 |s, m_s\rangle$$

$$\mathbf{S}_z |s, m_s\rangle = m_s \hbar |s, m_s\rangle.$$

The generalization to particles having higher spin values is straightforward. Thus, for a spin-one particle there are three spin states which may be symbolized as follows:

$$\chi_+ = \chi(\text{up}) \qquad = \begin{pmatrix} 1 \\ 0 \\ 0 \end{pmatrix} = |1, 1\rangle$$

$$\chi_0 = \chi(\text{horizontal}) = \begin{pmatrix} 0 \\ 1 \\ 0 \end{pmatrix} = |1, 0\rangle \qquad (8.36)$$

$$\chi_- = \chi(\text{down}) \qquad = \begin{pmatrix} 0 \\ 0 \\ 1 \end{pmatrix} = |1, -1\rangle.$$

For two spin-one particles there are nine product functions that can be formed from the functions given in Equation 8.36. These are: $\chi_+(1) \cdot \chi_+(2)$, $\chi_+(1) \cdot \chi_0(2)$, $\chi_+(1) \cdot \chi_-(2)$, $\chi_0(1) \cdot \chi_+(2)$, and so on. The appropriate linear combinations of these product functions which form simultaneous eigenfunctions of $\mathbf{J}^2$ and $\mathbf{J}_z$ have already been obtained in the previous section for the case of $j_1 = j_2 = 1$.

PROBLEM 8-27

(a) Write the expressions for the nine basis functions for two spin-one particles which provide a diagonal representation for $\mathbf{J}^2$ and $\mathbf{J}_z$.

(b) Verify that these functions are the correct eigenfunctions of $\mathbf{J}^2$ and $\mathbf{J}_z$.

12. APPLICATIONS TO SYSTEMS OF IDENTICAL PARTICLES

Exchange degeneracy and the characteristic symmetry of indistinguishable particles are responsible for many interesting physical phenomena as diverse as ferromagnetism, the superfluidity of liquid helium, and the saturation of chemical bonds. As an example of the role played by the statistics of indistinguishable particles, let us consider the states of the hydrogen molecule. Since the total wave function may be written as a product of the wave functions for each type of particle, the molecular wave function may be expressed as,

$$\psi_{\text{molecules}} = \psi_{\text{protons}} \cdot \psi_{\text{electrons}}$$

$$= [\psi(\text{spatial}) \cdot \psi(\text{spin})]_{\text{protons}} \cdot [\psi(\text{spatial}) \cdot \psi(\text{spin})]_{\text{electrons}} \quad (8.37)$$

We already have seen that the two-electron wave function must be antisymmetric in the exchange of the two electrons. Likewise, the two-proton wave function must be antisymmetric in the exchange of the protons, since protons are also Fermi particles. Then the possible proton functions have the same form as the four electron functions given in Equation 8.22. Hence, ψ_{molecule} has 16 different forms, each of which is antisymmetric in the exchange of two identical particles.

Let us now suppose that we are performing an experiment that would detect a change in the spatial state of the protons. For example, we might be looking at the infrared spectrum of the molecule. Since the interesting electronic transitions are in the ultraviolet or visible spectral regions, we can essentially ignore the electronic part of ψ_{molecule} in the analysis of this experiment. The principal contributions to ψ_{proton} (spatial) are from the translational, rotational and vibrational states of the dumbbell-shaped molecule. We will ignore the translational states since we are interested in changes of the internal energy of the molecules. The vibrational states will be ignored for two reasons. First, they are symmetric in the exchange of the protons; second, they require rather large excitation energies.* Our molecular wave function now reduces to

$$\psi_{\text{molecule}} = \psi_{\text{protons}}(\text{rotation}) \cdot \psi_{\text{protons}}(\text{spin}),$$

which must be antisymmetric. The rotational states are the spherical harmonics which are characterized by the angular momentum quantum number j, where

$$\psi_{\text{protons}}(\text{rotation}) = \psi_j(1) \cdot \psi_j(2),$$

and

$$E_j = \frac{j(j+1)\hbar^2}{2A},$$

A being the moment of inertia of the molecule. Since for a dumbbell molecule the exchange of the two protons is equivalent to an inversion through the origin, the symmetry of each state is given by the parity, $(-1)^j$. This leads to the important result (for hydrogen) that *only odd values of j are allowed when the proton spins are parallel and only even values of j are allowed when the proton spins are antiparallel.*

Hydrogen molecules which are symmetric in the nuclear spin are called *ortho hydrogen*, while those that are antisymmetric are called *para hydrogen*. Assuming the same statistical weight for all states and a uniform distribution over the rotation states, there should be three times as many ortho molecules as para molecules. This follows from the fact that there are three symmetric spin functions which can be used with each odd rotation function, while only one antisymmetric spin function can be used with each even rotation state. This surprising prediction is confirmed experimentally for hydrogen at ordinary temperatures. The infrared spectra consist of bands of lines corresponding to odd-odd or even-even transitions, in keeping with the fact that transitions are forbidden between states of different exchange symmetry (see Equation 8.24), and each line associated with an odd-odd transition is roughly three times as intense as each line in the even-even band.

At lower temperatures, however, when kT becomes comparable with the energy intervals between rotation levels, the relative populations of the rotation levels are greatly altered. In fact, the system tends to go completely into the para state—corresponding to zero rotational energy—at about 20°K. This ortho-para conversion‡ is an amazing verification of the role of symmetry and statistics in quantum processes. Since the excited states of para hydrogen lie lower than the corresponding ortho states, the heat capacity at low temperatures is much greater for para hydrogen than for ortho hydrogen or for any mixture of the two forms. So-called "ordinary" hydrogen is a mixture whose concentration of para molecules can be as low as 25 per cent at room temperature and as high as 100 per cent at low temperatures.

---

* See section 8 of Chapter 5, particularly Problem 5-18.

‡ The conversion process is slow except when catalyzed by adsorption on carbon. The carbon dissociates the metastable ortho molecule; when the atoms recombine at low temperatures they form the stable para molecule.

Ortho-para processes are not limited to the hydrogen molecule. In general, if I is the nuclear spin of a homonuclear molecule, then each nuclear spin has $2I + 1$ possible orientations. Of these, there are $2I + 1$ states that are symmetric where both nuclei have the *same* spin projections. In addition, there are $I(2I + 1)$ possible combinations of two different spin projections, for each of which there is a symmetric and an antisymmetric function. Then the ratio of symmetric to antisymmetric states is:

$$\frac{\text{Number ortho states}}{\text{Number para states}} = \frac{(2I + 1) + I(2I + 1)}{I(2I + 1)} = \frac{I + 1}{I}. \quad (8.38)$$

For the hydrogen molecule, $I = \frac{1}{2}$ and the ratio of ortho to para states is $3:1$ as we obtained above.

A further example is provided by the deuterium molecule. Although the deuteron is composed of two fermions—a neutron and a proton—it is a spin 1 boson. The deuteron does not exist in the spin-zero state. From Equation 8.38, there should be twice as many ortho states as para states. The spin functions for this case were obtained in section 10 (see Table 8-6). It is evident that there are six symmetric spin functions corresponding to $S = 2$ and $S = 0$; the three functions corresponding to $S = 1$ are antisymmetric.

Since deuterons are bosons, the total wave function for the deuterium molecule must be symmetric. Using arguments similar to those for hydrogen, we discover that the ortho states are associated with the even rotation states and para states are associated with odd rotation states. Accordingly, spectral lines corresponding to transitions between even rotation states have twice the intensity of those corresponding to odd rotation states. At very low temperatures, deuterium converts to the ortho form for which the zero rotational state is allowed.

Although a rigorous discussion of the superfluid state of He^4 is beyond the scope of this book, it should be noted that the differences in behavior of liquid He^3 and He^4 are due to the fact that the former consists of Fermi particles and the latter of Bose particles. The repulsive exchange forces prevent any further condensation of liquid He^3, and accordingly it behaves like an "ordinary" liquid. However, in liquid He^4 the exchange forces are attractive, and at $2.18°K$ a transition occurs to a lower state which no longer contributes to the entropy or the specific heat. This is the state that is known as the superfluid state.

PROBLEM 8-28

(a) Write the relative energies of the eight lowest rotation states for a diatomic molecule.

(b) If the allowed transitions are given by $\Delta j = \pm 2$, what are the relative energies of the lines of the emission spectrum?

(c) How would you expect the heat capacities of ortho, para, and "normal" hydrogen to compare?

PROBLEM 8-29

Apply the particle exchange operator to the nine kets for two spin-one particles (see Table 8-6 in section 10) and show that the ratio of the number of ortho to para states is 2:1.

PROBLEM 8-30

Describe the characteristics of the rotation spectrum of oxygen molecules composed only of O^{16} atoms.

PROBLEM 8-31

Explain how one can easily determine the spin of a nuclear species provided that it forms diatomic molecules.

SUMMARY

In this chapter we have seen how spin can be incorporated into the Schrödinger formalism by postulating spin operators which act upon spin functions in a manner analogous to the angular momentum operators. When the spin and spatial coordinates are independent, the total wave function is a product of the spin function and the solution to the spin-independent Schrödinger equation. If any spin-dependent forces are to be included in the Hamiltonian, they must be expressed in terms of the spin operators. In particular, the spin-orbit interaction is discussed and it is shown that it can be expressed in terms of the operator scalar product $\vec{L} \cdot \vec{S}$. The vector coupling models are then introduced and applied to simple systems. The product space formed by coupling angular momenta is described and a procedure given for obtaining the eigenfunctions in the new space.

The concept of the exchange degeneracy of indistinguishable particles was introduced. It was shown that a satisfactory wave function for two or more identical particles must be either a symmetric or an antisymmetric linear combination of the possible product states. This derives from the requirement that the wave function be an eigenfunction of the particle exchange operator and that its probability density remain unchanged under an exchange of particles. The experimental connection between particle spin and the exchange symmetry of wave functions is summarized by the statements: a collection of bosons (identical particles having zero or integral spin) is described by a wave function that is symmetric under the exchange of any two particles; a collection of fermions (identical particles having half-integral spin) is described by a wave function that is antisymmetric under the exchange of any two particles. Fermions tend to *avoid* the same quantum state, while bosons tend to condense into the *same* quantum state. Furthermore, the energy density of a collection of fermions is greater than that for a collection of the same number of bosons. It is in this sense that bosons attract one another and fermions repel each other via an *exchange force* which has no classical analog. These concepts are shown to account for the infrared spectra of diatomic homonuclear molecules and for the electronic exchange energy which separates singlet and triplet states in atomic spectra.

CHAPTER 8   APPENDICES A to C

APPENDIX A.   THE DISTRIBUTION FUNCTIONS RESULTING
FROM QUANTUM STATISTICS

The quantum statistics associated with the two species of identical particles provide us with two new distribution functions to replace the classical Maxwell-Boltzmann distribution function discussed in Chapter 2, Appendix B. Recall that in that derivation the quantity g_i was called the *intrinsic probability* of the ith state. We will now interpret the ith cell as the energy level ϵ_i and will regard g_i as the *degeneracy* of that level. If we also invoke the Pauli exclusion principle, we constrain the number of Fermi particles in an energy level to a value that is either less than or equal to the degeneracy. That is, for fermions,

$$n_i \leq g_i.$$

Then the total number of ways of filling g_i states of energy ϵ_i with n_i fermions is simply the number of combinations of g_i identical states taken n_i at a time,

$$\frac{g_i!}{n_i!(g_i - n_i)!},$$

and

$$W = \prod_i \frac{g_i!}{n_i!(g_i - n_i)!}.$$

This is the distribution given in Problem 8-32.

For bosons the same combinatorial argument cannot be used, since a state is no longer simply occupied or unoccupied but can contain any number of particles. Using a somewhat different approach, the distribution of Problem 8-33 is obtained.[14]

A more sophisticated method of deriving the distribution functions is to use the grand partition function[15]

$$Z = \sum_i e^{-n_i \alpha}, \tag{A1}$$

where $\alpha = (\epsilon_i + \lambda_1)/kT$. Then, for fermions, n_i can be only 0 or 1, so

$$Z = 1 + e^{-\alpha},$$

[14] R. B. Leighton, *Principles of Modern Physics*, McGraw-Hill Book Co., New York, 1959, Chapter 10.

[15] T. L. Hill, *An Introduction to Statistical Thermodynamics*, Addison-Wesley Publishing Co., Reading, Mass., 1960, section 22-1.

and the occupation number of fermions per state is

$$\bar{n}_i = -kT \frac{\partial}{\partial \lambda_1} \ln Z = -\frac{\partial}{\partial \alpha} \ln Z = \frac{e^{-\alpha}}{1 + e^{-\alpha}} = \frac{1}{e^{\alpha} + 1}.$$ (A2)

Defining the Fermi energy by $\epsilon_F = -\lambda_1$, this becomes

$$\bar{n}_i = \frac{1}{\exp\left(\frac{\epsilon_i - \epsilon_F}{kT}\right) + 1}.$$ (A3)

For bosons the grand partition function becomes

$$Z = 1 + e^{-\alpha} + e^{-2\alpha} + e^{-3\alpha} + \cdots = \frac{1}{1 - e^{-\alpha}}.$$

Then,

$$\bar{n}_i = -\frac{\partial}{\partial \alpha} \ln Z = \frac{e^{-\alpha}}{1 - e^{-\alpha}} = \frac{1}{e^{\alpha} - 1} = \frac{1}{\exp\left(\frac{\epsilon_i + \lambda_1}{kT}\right) - 1}.$$ (A4)

PROBLEM 8-32:

Using the method of Lagrange's multipliers (See Appendix B of Chapter 2), obtain the Fermi-Dirac distribution function from the probability function

$$W = \prod_i \frac{g_i!}{n_i!(g_i - n_i)!}.$$

PROBLEM 8-33:

In like manner, obtain the Bose-Einstein distribution function from the probability function

$$W = \prod_i \frac{(n_i + g_i - 1)!}{n_i!(g_i - 1)!}.$$

APPENDIX B. GROUND STATE ELECTRON CONFIGURATIONS FOR THE ELEMENTS[16]

Z	Element	K (1) s	L (2) s	L (2) p	M (3) s	M (3) p	M (3) d	N (4) s	N (4) p	N (4) d	N (4) f	O (5) s	O (5) p	O (5) d	O (5) f	P (6) s	P (6) p	P (6) d	Q (7) s	Ground State Term
1	H hydrogen	1																		$^2S_{1/2}$
2	He helium	2																		$^1S_0$
3	Li lithium	2	1																	$^2S_{1/2}$
4	Be beryllium	2	2																	$^1S_0$
5	B boron	2	2	1																$^2P_{1/2}$
6	C carbon	2	2	2																$^3P_0$
7	N nitrogen	2	2	3																$^4S_{3/2}$
8	O oxygen	2	2	4																$^3P_2$
9	F fluorine	2	2	5																$^2P_{3/2}$
10	Ne neon	2	2	6																$^1S_0$
11	Na sodium	2	2	6	1															$^2S_{1/2}$
12	Mg magnesium	2	2	6	2															$^1S_0$
13	Al aluminum	2	2	6	2	1														$^2P_{1/2}$
14	Si silicon	2	2	6	2	2														$^3P_0$
15	P phosphorus	2	2	6	2	3														$^4S_{3/2}$
16	S sulfur	2	2	6	2	4														$^3P_2$
17	Cl chlorine	2	2	6	2	5														$^2P_{3/2}$
18	Ar argon	2	2	6	2	6														$^1S_0$
19	K potassium	2	2	6	2	6	.	1												$^2S_{1/2}$
20	Ca calcium	2	2	6	2	6	.	2												$^1S_0$
21	Sc scandium	2	2	6	2	6	1	2												$^2D_{3/2}$
22	Ti titanium	2	2	6	2	6	2	2												$^3F_2$
23	V vanadium	2	2	6	2	6	3	2												$^4F_{3/2}$
24	Cr chromium	2	2	6	2	6	5	1												$^7S_3$
25	Mn manganese	2	2	6	2	6	5	2												$^6S_{5/2}$
26	Fe iron	2	2	6	2	6	6	2												$^5D_4$
27	Co cobalt	2	2	6	2	6	7	2												$^4F_{9/2}$
28	Ni nickel	2	2	6	2	6	8	2												$^3F_4$
29	Cu copper	2	2	6	2	6	10	1												$^2S_{1/2}$
30	Zn zinc	2	2	6	2	6	10	2												$^1S_0$
31	Ga gallium	2	2	6	2	6	10	2	1											$^2P_{1/2}$
32	Ge germanium	2	2	6	2	6	10	2	2											$^3P_0$
33	As arsenic	2	2	6	2	6	10	2	3											$^4S_{3/2}$
34	Se selenium	2	2	6	2	6	10	2	4											$^3P_2$
35	Br bromine	2	2	6	2	6	10	2	5											$^2P_{3/2}$
36	Kr krypton	2	2	6	2	6	10	2	6											$^1S_0$
37	Rb rubidium	2	2	6	2	6	10	2	6	.	.	1								$^2S_{1/2}$
38	Sr strontium	2	2	6	2	6	10	2	6	.	.	2								$^1S_0$
39	Y yttrium	2	2	6	2	6	10	2	6	1	.	2								$^2D_{3/2}$
40	Zr zirconium	2	2	6	2	6	10	2	6	2	.	2								$^3F_2$
41	Nb niobium	2	2	6	2	6	10	2	6	4	.	1								$^6D_{1/2}$
42	Mo molybdenum	2	2	6	2	6	10	2	6	5	.	1								$^7S_3$
43	Tc technetium	2	2	6	2	6	10	2	6	6	.	1								$^6S_{5/2}$
44	Ru ruthenium	2	2	6	2	6	10	2	6	7	.	1								$^5F_5$
45	Rh rhodium	2	2	6	2	6	10	2	6	8	.	1								$^4F_{9/2}$
46	Pd palladium	2	2	6	2	6	10	2	6	10	.	.								$^1S_0$
47	Ag silver	2	2	6	2	6	10	2	6	10	.	1								$^2S_{1/2}$
48	Cd cadmium	2	2	6	2	6	10	2	6	10	.	2								$^1S_0$
49	In indium	2	2	6	2	6	10	2	6	10	.	2	1							$^2P_{1/2}$
50	Sn tin	2	2	6	2	6	10	2	6	10	.	2	2							$^3P_0$
51	Sb antimony	2	2	6	2	6	10	2	6	10	.	2	3							$^4S_{3/2}$
52	Te tellurium	2	2	6	2	6	10	2	6	10	.	2	4							$^3P_2$

[16] *Handbook of Chemistry and Physics*, 51st. ed., The Chemical Rubber Co., Cleveland, 1970; also, *American Institute of Physics Handbook*, 2nd ed., McGraw-Hill Book Co., New York, 1963, Table 7c–1, p. 7–14.

APPENDIX B. (*continued*)

Z	Element		K (1) s	L (2) s p	M (3) s p d	N (4) s p d f	O (5) s p d f	P (6) s p d	Q (7) s	Ground State Term
53	I	iodine	2	2 6	2 6 10	2 6 10 .	2 5			$^2P_{\frac{3}{2}}$
54	Xe	xenon	2	2 6	2 6 10	2 6 10 .	2 6			$^1S_0$
55	Cs	cesium	2	2 6	2 6 10	2 6 10 .	2 6 . .	1		$^2S_{\frac{1}{2}}$
56	Ba	barium	2	2 6	2 6 10	2 6 10 .	2 6 . .	2		$^1S_0$
57	La	lanthanum	2	2 6	2 6 10	2 6 10 .	2 6 1 .	2		$^2D_{\frac{3}{2}}$
58	Ce	cerium	2	2 6	2 6 10	2 6 10 1	2 6 1 .	2		$^3H_5$
59	Pr	praseodymium	2	2 6	2 6 10	2 6 10 3	2 6 . .	2		$^4I_{\frac{9}{2}}$
60	Nd	neodymium	2	2 6	2 6 10	2 6 10 4	2 6 . .	2		$^5I_4$
61	Pm	promethium	2	2 6	2 6 10	2 6 10 5	2 6 . .	2		$^6H_{\frac{5}{2}}$
62	Sm	samarium	2	2 6	2 6 10	2 6 10 6	2 6 . .	2		$^7S_3$
63	Eu	europium	2	2 6	2 6 10	2 6 10 7	2 6 . .	2		$^8S_{\frac{7}{2}}$
64	Gd	gadolinium	2	2 6	2 6 10	2 6 10 7	2 6 1 .	2		$^9D_2$
65	Tb	terbium	2	2 6	2 6 10	2 6 10 9	2 6 . .	2		
66	Dy	dysprosium	2	2 6	2 6 10	2 6 10 10	2 6 . .	2		
67	Ho	holmium	2	2 6	2 6 10	2 6 10 11	2 6 . .	2		
68	Er	erbium	2	2 6	2 6 10	2 6 10 12	2 6 . .	2		
69	Tm	thulium	2	2 6	2 6 10	2 6 10 13	2 6 . .	2		$^2F_{\frac{7}{2}}$
70	Yb	ytterbium	2	2 6	2 6 10	2 6 10 14	2 6 . .	2		$^1S_0$
71	Lu	lutetium	2	2 6	2 6 10	2 6 10 14	2 6 1 .	2		$^2D_{\frac{3}{2}}$
72	Hf	hafnium	2	2 6	2 6 10	2 6 10 14	2 6 2 .	2		$^3F_2$
73	Ta	tantalum	2	2 6	2 6 10	2 6 10 14	2 6 3 .	2		$^4F_{\frac{3}{2}}$
74	W	wolfram (tungsten)	2	2 6	2 6 10	2 6 10 14	2 6 4 .	2		$^5D_0$
75	Re	rhenium	2	2 6	2 6 10	2 6 10 14	2 6 5 .	2		$^6S_{\frac{5}{2}}$
76	Os	osmium	2	2 6	2 6 10	2 6 10 14	2 6 6 .	2		$^5D_4$
77	Ir	iridium	2	2 6	2 6 10	2 6 10 14	2 6 7 .	2		$^4F_{\frac{9}{2}}$
78	Pt	platinum	2	2 6	2 6 10	2 6 10 14	2 6 9 .	1		$^3D_3$
79	Au	gold	2	2 6	2 6 10	2 6 10 14	2 6 10 .	1		$^2S_{\frac{1}{2}}$
80	Hg	mercury	2	2 6	2 6 10	2 6 10 14	2 6 10 .	2		$^1S_0$
81	Tl	thallium	2	2 6	2 6 10	2 6 10 14	2 6 10 .	2 1		$^2P_{\frac{1}{2}}$
82	Pb	lead	2	2 6	2 6 10	2 6 10 14	2 6 10 .	2 2		$^3P_0$
83	Bi	bismuth	2	2 6	2 6 10	2 6 10 14	2 6 10 .	2 3		$^4S_{\frac{3}{2}}$
84	Po	polonium	2	2 6	2 6 10	2 6 10 14	2 6 10 .	2 4		$^3P_2$
85	At	astatine	2	2 6	2 6 10	2 6 10 14	2 6 10 .	2 5		
86	Rn	radon	2	2 6	2 6 10	2 6 10 14	2 6 10 .	2 6		$^1S_0$
87	Fr	francium	2	2 6	2 6 10	2 6 10 14	2 6 10 .	2 6 .	1	$^2S_{\frac{1}{2}}$
88	Ra	radium	2	2 6	2 6 10	2 6 10 14	2 6 10 .	2 6 .	2	$^1S_0$
89	Ac	actinium	2	2 6	2 6 10	2 6 10 14	2 .6 10 .	2 6 1	2	$^2D_{\frac{3}{2}}$
90	Th	thorium	2	2 6	2 6 10	2 6 10 14	2 6 10 .	2 6 2	2	$^3F_2$
91	Pa	protactinium	2	2 6	2 6 10	2 6 10 14	2 6 10 1	2 6 2	2	
92	U	uranium	2	2 6	2 6 10	2 6 10 14	2 6 10 3	2 6 1	2	$^5L_6$
93	Np	neptunium	2	2 6	2 6 10	2 6 10 14	2 6 10 4	2 6 1	2	
94	Pu	plutonium	2	2 6	2 6 10	2 6 10 14	2 6 10 6	2 6 .	2	
95	Am	americium	2	2 6	2 6 10	2 6 10 14	2 6 10 7	2 6 .	2	
96	Cm	curium	2	2 6	2 6 10	2 6 10 14	2 6 10 7	2 6 1	2	
97	Bk	berkelium	2	2 6	2 6 10	2 6 10 14	2 6 10 8	2 6 1	2	
98	Cf	californium	2	2 6	2 6 10	2 6 10 14	2 6 10 10	2 6 .	2	
99	Es	einsteinium	2	2 6	2 6 10	2 6 10 14	2 6 10 11	2 6 .	2	
100	Rm	fermium	2	2 6	2 6 10	2 6 10 14	2 6 10 12	2 6 .	2	
101	Md	mendelevium	2	2 6	2 6 10	2 6 10 14	2 6 10 13	2 6 .	2	
102	No	nobelium	2	2 6	2 6 10	2 6 10 14	2 6 10 14	2 6 .	2	
103	Lw	lawrencium	2	2 6	2 6 10	2 6 10 14	2 6 10 14	2 6 1	2	

APPENDIX C. ANSWERS TO SELECTED PROBLEMS

The results of these problems will be needed for problems in later chapters.

PROBLEM 8-20

The new kets $|j, m\rangle$ are primed for convenience; the unprimed kets are $|m_l, m_s\rangle$.

$$|\tfrac{3}{2}, \tfrac{3}{2}\rangle' = |1, \tfrac{1}{2}\rangle$$

$$|\tfrac{3}{2}, \tfrac{1}{2}\rangle' = \frac{1}{\sqrt{3}} \left(\sqrt{2}\, |0, \tfrac{1}{2}\rangle + |1, -\tfrac{1}{2}\rangle \right)$$

$$|\tfrac{3}{2}, -\tfrac{1}{2}\rangle' = \frac{1}{\sqrt{3}} \left(|-1, \tfrac{1}{2}\rangle + \sqrt{2}\, |0, -\tfrac{1}{2}\rangle \right)$$

$$|\tfrac{3}{2}, -\tfrac{3}{2}\rangle' = |-1, -\tfrac{1}{2}\rangle$$

$$|\tfrac{1}{2}, \tfrac{1}{2}\rangle' = \frac{1}{\sqrt{3}} \left(|0, \tfrac{1}{2}\rangle - \sqrt{2}\, |1, -\tfrac{1}{2}\rangle \right)$$

$$|\tfrac{1}{2}, -\tfrac{1}{2}\rangle' = \frac{1}{\sqrt{3}} \left(\sqrt{2}\, |-1, \tfrac{1}{2}\rangle - |0, -\tfrac{1}{2}\rangle \right).$$

PROBLEM 8-22

(a) On the basis $|1, \tfrac{1}{2}\rangle$, $|1, -\tfrac{1}{2}\rangle$, $|0, \tfrac{1}{2}\rangle$, $|0, -\tfrac{1}{2}\rangle$, $|-1, \tfrac{1}{2}\rangle$, $|-1, -\tfrac{1}{2}\rangle$,

$$(\vec{L} \cdot \vec{S}) = \frac{\hbar^2}{2} \begin{pmatrix} 1 & 0 & 0 & 0 & 0 & 0 \\ 0 & -1 & \sqrt{2} & 0 & 0 & 0 \\ 0 & \sqrt{2} & 0 & 0 & 0 & 0 \\ 0 & 0 & 0 & 0 & \sqrt{2} & 0 \\ 0 & 0 & 0 & \sqrt{2} & -1 & 0 \\ 0 & 0 & 0 & 0 & 0 & 1 \end{pmatrix}$$

(b) On the basis ordered as in Problem 8-20 above,

$$(\vec{L} \cdot \vec{S}) = \frac{\hbar^2}{2} \begin{pmatrix} 1 & & & & & \\ & 1 & & & & \\ & & 1 & & & \\ & & & 1 & & \\ & & & & -2 & \\ & & & & & -2 \end{pmatrix}$$

PROBLEM 8-27

See Table 8-6 in section 10.

SUGGESTED
REFERENCES

David Bohm, *Quantum Theory*. Prentice-Hall, Inc., New York 1951.

E. U. Condon and G. H. Shortley, *The Theory of Atomic Spectra*. Cambridge University Press, Cambridge, 1953.

R. H. Dicke and J. P. Wittke, *Introduction to Quantum Mechanics*. Addison-Wesley Publishing Co., Inc., 1960.

P. A. M. Dirac, *The Principles of Quantum Mechanics, 3rd ed.* Oxford University Press, London, 1947.

A. R. Edmonds, *Angular Momentum in Quantum Mechanics*. Princeton University Press, Princeton, 1957.

R. M. Eisberg, *Fundamentals of Modern Physics*. John Wiley and Sons, Inc., New York, 1961.

W. R. Hindmarsh, *Atomic Spectra*. Pergamon Press, New York, 1967.

R. B. Leighton, *Principles of Modern Physics*. McGraw-Hill Book Co., Inc., New York, 1959.

E. Merzbacher, *Quantum Mechanics*. John Wiley and Sons, Inc., New York, 1961.

Albert Messiah, *Quantum Mechanics*. North-Holland Publishing Co., Amsterdam, 1958.

D. Park, *Introduction to the Quantum Theory*. McGraw-Hill Book Co., Inc., New York, 1964.

J. L. Powell and B. Crasemann, *Quantum Mechanics*. Addison-Wesley Publishing Co., Inc., Reading, Mass., 1961.

M. E. Rose, *Elementary Theory of Angular Momentum*. John Wiley and Sons, Inc., New York, 1957.

D. S. Saxon, *Elementary Quantum Mechanics*. Holden-Day, Inc., San Francisco, 1968.

Leonard I. Schiff, *Quantum Mechanics, 3rd ed.* McGraw-Hill Book Co., New York, 1969.

John C. Slater, *Quantum Theory of Matter*. McGraw-Hill Book Co., New York, 1968.

H. E. White, *Introduction to Atomic Spectra*. McGraw-Hill Book Co., New York, 1934.

Robert L. White, *Basic Quantum Mechanics*. McGraw-Hill Book Co., New York, 1966.

CHAPTER 9

APPROXIMATION METHODS AND APPLICATIONS

It may come as a surprise to the reader to learn that we have already treated most of the physical systems that have exact solutions. Our repertoire consists of the solutions for the harmonic oscillator, the rigid rotator, the hydrogen atom, and the bound states of a particle in a rectangular well or box. Even here we have begun to detect discrepancies due to additional interactions such as the spin-orbit and spin-spin interactions. Complications may also arise when an electromagnetic field is present. A further difficulty appears when additional particles are added to the system. In such cases we have obtained an *exact* solution only for the unrealistic case of completely non-interacting particles. This might be a good approximation if the only interactions were the weak intermolecular forces; but if the particles are charged, the addition of the Coulomb forces could have a tremendous effect upon the nature of the solutions.

If we are to make further headway in the application of quantum mechanics to real systems we will need to develop techniques for arriving at approximate solutions. The most widely used of these approximate methods for bound systems is perturbation theory.[1]

I. PERTURBATION THEORY FOR STATIONARY STATES

The easiest and most direct approach for solving a real physical system is to regard it as a modification of one of the model systems whose solutions we already know. In particular, we wish to express the Hamiltonian for the real system, $\mathcal{H}$, as the sum of $\mathcal{H}_0$, the idealized Hamiltonian whose solutions we know, and an additional part $\mathcal{H}^{(1)}$ which contains the new interactions. If $\mathcal{H}^{(1)}$ is small compared to $\mathcal{H}_0$, the corrections to both the eigenfunctions and

---

[1] E. Schrödinger, Ann. Physik **80,** 437 (1926).

the eigenvalues resulting from $\mathscr{H}^{(1)}$ will also be small. In such cases it is possible to use the methods of perturbation theory.

We begin by writing the Hamiltonian for the real system as

$$\mathscr{H} = \mathscr{H}_0 + \lambda \mathscr{H}^{(1)}, \tag{9.1}$$

where $\mathscr{H}_0$ is the part that can be solved exactly and $\mathscr{H}^{(1)}$ is regarded as the perturbation. The parameter λ is not fundamental but is added for convenience in "turning on" and "turning off" the perturbation. It is restricted to values between 0 and 1, the former corresponding to the ideal Hamiltonian, and the latter corresponding to the physical problem to be solved.

The sets of known functions and eigenvalues are given by

$$\mathscr{H}_0 \psi_i^{(0)} = E_i^{(0)} \psi_i^{(0)}, \tag{9.2}$$

where the $\psi_i^{(0)}$ form an orthonormal set such that $\langle i \mid j \rangle = \delta_{ij}$. The equation we wish to solve is

$$(\mathscr{H}_0 + \lambda \mathscr{H}^{(1)}) \psi_n = E_n \psi_n. \tag{9.3}$$

It is assumed in this section that both the $\psi_i^{(0)}$ and the ψ_n are discrete, non-degenerate sets of time-independent functions. We further assume that there is a one-to-one correspondence between the members of each set. Thus, ψ_1 goes into $\psi_1^{(0)}$ as λ varies from 1 to 0, $\psi_2 \to \psi_2^{(0)}$, etc. Stated mathematically,

$$\lim_{\lambda \to 0} \psi_n = \psi_n^{(0)} \quad \text{and} \quad \lim_{\lambda \to 0} E_n = E_n^{(0)}. \tag{9.4}$$

In keeping with the spirit of these limits, let us represent the new eigenfunctions and eigenvalues by the following power series in λ:

$$\begin{aligned} \psi_n &= \psi_n^{(0)} + \lambda \psi_n^{(1)} + \lambda^2 \psi_n^{(2)} + \cdots \\ E_n &= E_n^{(0)} + \lambda E_n^{(1)} + \lambda^2 E_n^{(2)} + \cdots \end{aligned} \tag{9.5}$$

It is assumed that successive terms of these series get smaller so that the series converge. Our task will be to find these corrections to the eigenfunctions and the eigenenergies to any desired order. Substituting Equation 9.5 into Equation 9.3, we obtain:

$$(\mathscr{H}_0 + \lambda \mathscr{H}^{(1)})(\psi_n^{(0)} + \lambda \psi_n^{(1)} + \lambda^2 \psi_n^{(2)} + \cdots)$$
$$= (E_n^{(0)} + \lambda E_n^{(1)} + \lambda^2 E_n^{(2)} + \cdots)(\psi_n^{(0)} + \lambda \psi_n^{(1)} + \lambda^2 \psi_n^{(2)} + \cdots).$$

$$\mathscr{H}_0 \psi_n^{(0)} + \lambda(\mathscr{H}^{(1)} \psi_n^{(0)} + \mathscr{H}_0 \psi_n^{(1)}) + \lambda^2(\mathscr{H}_0 \psi_n^{(2)} + \mathscr{H}^{(1)} \psi_n^{(1)}) + \cdots$$
$$= E_n^{(0)} \psi_n^{(0)} + \lambda(E_n^{(1)} \psi_n^{(0)} + E_n^{(0)} \psi_n^{(1)})$$
$$+ \lambda^2(E_n^{(0)} \psi_n^{(2)} + E_n^{(1)} \psi_n^{(1)} + E_n^{(2)} \psi_n^{(0)}) + \cdots$$

Or,

$$(\mathscr{H}_0 \psi_n^{(0)} - E_n^{(0)} \psi_n^{(0)}) + \lambda(\mathscr{H}^{(1)} \psi_n^{(0)} + \mathscr{H}_0 \psi_n^{(1)} - E_n^{(1)} \psi_n^{(0)} - E_n^{(0)} \psi_n^{(1)})$$
$$+ \lambda^2(\mathscr{H}_0 \psi_n^{(2)} + \mathscr{H}^{(1)} \psi_n^{(1)} - E_n^{(0)} \psi_n^{(2)} - E_n^{(1)} \psi_n^{(1)} - E_n^{(2)} \psi_n^{(0)}) + \cdots = 0.$$

Since the parameter λ is arbitrary, the coefficient of each power of λ can be set equal to zero. Therefore, we have the equations:

$$\mathcal{H}_0 \psi_n^{(0)} = E_n^{(0)} \psi_n^{(0)}$$

$$\mathcal{H}^{(1)} \psi_n^{(0)} + \mathcal{H}_0 \psi_n^{(1)} = E_n^{(1)} \psi_n^{(0)} + E_n^{(0)} \psi_n^{(1)} \tag{9.6}$$

$$\mathcal{H}_0 \psi_n^{(2)} + \mathcal{H}^{(1)} \psi_n^{(1)} = E_n^{(0)} \psi_n^{(2)} + E_n^{(1)} \psi_n^{(1)} + E_n^{(2)} \psi_n^{(0)}$$

$$\cdot$$
$$\cdot$$
$$\cdot$$

The first of these is simply Equation 9.2, our starting point. The second may be solved by expanding the first-order correction to the wave function in terms of the unperturbed eigenfunctions. That is,

$$\psi_n^{(1)} = \sum_i a_{ni} \psi_i^{(0)}.$$

Then the first order equation becomes:

$$\mathcal{H}^{(1)} \psi_n^{(0)} + \mathcal{H}_0 \sum_i a_{ni} \psi_i^{(0)} = E_n^{(1)} \psi_n^{(0)} + E_n^{(0)} \sum_i a_{ni} \psi_i^{(0)}.$$

Multiplying by $\psi_k^{(0)*}$ and integrating,

$$\langle k| \, \mathcal{H}^{(1)} \, |n\rangle + \sum_i a_{ni} \langle k| \, \mathcal{H}_0 \, |i\rangle = E_n^{(1)} \langle k \mid n \rangle + \sum_i a_{ni} E_n^{(0)} \langle k \mid i \rangle.$$

$$\langle k| \, \mathcal{H}^{(1)} \, |n\rangle + \sum_i a_{ni} E_i^{(0)} \delta_{ki} = E_n^{(1)} \delta_{kn} + \sum_i a_{ni} E_n^{(0)} \delta_{ki},$$

$$\langle k| \, \mathcal{H}^{(1)} \, |n\rangle + a_{nk} E_k^{(0)} = E_n^{(1)} \delta_{kn} + a_{nk} E_n^{(0)}.$$

The quantity $\langle k| \, \mathcal{H}^{(1)} \, |n\rangle$ should be recognized as the matrix element, $\mathcal{H}_{kn}^{(1)}$, of the operator $\mathcal{H}^{(1)}$ with the unperturbed eigenfunctions as the basis. Then we may write,

$$\mathcal{H}_{kn}^{(1)} + a_{nk}(E_k^{(0)} - E_n^{(0)}) = E_n^{(1)} \delta_{kn}. \tag{9.7}$$

For $k = n$, Equation 9.7 gives us the important result,

$$E_n^{(1)} = \mathcal{H}_{nn}^{(1)}. \tag{9.8}$$

In words, Equation 9.8 says: *the first order correction to the energy of the nth eigenstate is the diagonal matrix element corresponding to the nth row and the nth column of the matrix of the perturbation.* Note that the *unperturbed* wave functions are used for this calculation. For $k \neq n$, Equation 9.7 becomes

$$\mathcal{H}_{kn}^{(1)} + a_{nk}(E_k^{(0)} - E_n^{(0)}) = 0,$$

or,

$$a_{nk} = \frac{\mathcal{H}_{kn}^{(1)}}{E_n^{(0)} - E_k^{(0)}}. \tag{9.9}$$

Equation 9.9 tells us how to find the first order corrections to the wave function for the nth state. Thus,

$$\psi_n = \psi_n^{(0)} + \sum_{k \neq n} a_{nk} \psi_k^{(0)}$$

$$\psi_n = \psi_n^{(0)} + \frac{\mathscr{H}_{1n}^{(1)}}{E_n^{(0)} - E_1^{(0)}} \, \psi_1^{(0)} + \frac{\mathscr{H}_{2n}^{(1)}}{E_n^{(0)} - E_2^{(0)}} \, \psi_2^{(0)} + \cdots . \qquad (9.10)$$

Each basis function that has a non-zero matrix element with $\psi_n^{(0)}$ contributes to the new wave function for the nth state. Hence, when the matrix of the perturbation has non-vanishing, off-diagonal elements, this means that the perturbation has produced some interference or mixing of the original wave functions so that they are no longer orthogonal. Note that the amount of this mixing depends not only on the magnitude of the matrix element in the numerator, but also on the "energy denominator." Thus, states that are widely separated in energy are not expected to interfere to any large degree. States that are close together will be expected to have a large amount of mixing. The denominator of Equation 9.9 can never be zero, since we have excluded the possibility of any degeneracy in this derivation.

To obtain the second-order corrections we will use the last expression given in Equation 9.6,

$$\mathscr{H}_0 \psi_n^{(2)} + \mathscr{H}^{(1)} \psi_n^{(1)} = E_n^{(0)} \psi_n^{(2)} + E_n^{(1)} \psi_n^{(1)} + E_n^{(2)} \psi_n^{(0)}.$$

Once again we utilize the expansion postulate and write

$$\psi_n^{(2)} = \sum_j b_{nj} \psi_j^{(0)},$$

as well as our previous assumption that

$$\psi_n^{(1)} = \sum_i a_{ni} \psi_i^{(0)}.$$

Making these substitutions,

$$\mathscr{H}_0 \sum_j b_{nj} \psi_j^{(0)} + \mathscr{H}^{(1)} \sum_i a_{ni} \psi_i^{(0)} = E_n^{(0)} \sum_j b_{nj} \psi_j^{(0)} + E_n^{(1)} \sum_i a_{ni} \psi_i^{(0)} + E_n^{(2)} \psi_n^{(0)}.$$

Multiplying by $\psi_k^{(0)*}$ and integrating:

$$\sum_j b_{nj} \langle k| \, \mathscr{H}_0 \, |j\rangle + \sum_i a_{ni} \langle k| \, \mathscr{H}^{(1)} \, |i\rangle$$
$$= \sum_j b_{nj} E_n^{(0)} \langle k \, | \, j\rangle + \sum_i a_{ni} E_n^{(1)} \langle k \, | \, i\rangle + E_n^{(2)} \langle k \, | \, n\rangle.$$

$$\sum_j b_{nj} E_j^{(0)} \delta_{kj} + \sum_i a_{ni} \mathscr{H}_{ki}^{(1)} = \sum_j b_{nj} E_n^{(0)} \delta_{kj} + \sum_i a_{ni} E_n^{(1)} \delta_{ki} + E_n^{(2)} \delta_{kn},$$

$$b_{nk} E_k^{(0)} + \sum_i a_{ni} \mathscr{H}_{ki}^{(1)} = b_{nk} E_n^{(0)} + a_{nk} E_n^{(1)} + E_n^{(2)} \delta_{kn},$$

or,

$$b_{nk}(E_k^{(0)} - E_n^{(0)}) + \sum_i a_{ni} \mathscr{H}_{ki}^{(1)} - a_{nk} E_n^{(1)} = E_n^{(2)} \delta_{kn}. \qquad (9.11)$$

For $k = n$, Equation 9.11 gives the second-order energy correction as

$$E_n^{(2)} = \sum_i a_{ni} \mathscr{H}_{ni}^{(1)} - a_{nn} E_n^{(1)}$$

$$= \sum_i a_{ni} \mathscr{H}_{ni}^{(1)} - a_{nn} \mathscr{H}_{nn}^{(1)}$$

$$= \sum_{i \neq n} a_{ni} \mathscr{H}_{ni}^{(1)}.$$

Then,

$$E_n^{(2)} = \sum_{i \neq n} \frac{\mathscr{H}_{in}^{(1)} \mathscr{H}_{ni}^{(1)}}{E_n^{(0)} - E_i^{(0)}} = \sum_{i \neq n} \frac{|\mathscr{H}_{ni}^{(1)}|^2}{E_n^{(0)} - E_i^{(0)}}, \tag{9.12}$$

where the results of the first-order theory expressed by Equations 9.8 and 9.9 have been used. The second-order correction to the energy has been obtained using only the first-order wave function as expressed by the expansion coefficients a_{ni}.

To obtain the wave function to second order, we return to Equation 9.11 and seek solutions for $k \neq n$. Then,

$$b_{nk}(E_k^{(0)} - E_n^{(0)}) + \sum_i a_{ni} \mathscr{H}_{ki}^{(1)} - a_{nk} \mathscr{H}_{nn}^{(1)} = 0$$

$$b_{nk}(E_n^{(0)} - E_k^{(0)}) = \sum_i a_{ni} \mathscr{H}_{ki}^{(1)} - a_{nk} \mathscr{H}_{nn}^{(1)}$$

and,

$$b_{nk} = \sum_{i \neq n} \frac{\mathscr{H}_{in}^{(1)} \mathscr{H}_{ki}^{(1)}}{(E_n^{(0)} - E_k^{(0)})(E_n^{(0)} - E_i^{(0)})} - \frac{\mathscr{H}_{kn}^{(1)} \mathscr{H}_{nn}^{(0)}}{(E_n^{(0)} - E_k^{(0)})^2}, \tag{9.13}$$

where $n \neq k$.

It is evident that the corrected wave functions rapidly get quite complicated. To second order, the nth eigenfunction is

$$\psi_n = \psi_n^{(0)} + \sum_{k \neq n} a_{nk} \psi_k^{(0)} + \sum_{k \neq n} b_{nk} \psi_k^{(0)}, \tag{9.14}$$

where the expansion coefficients are given by Equations 9.9 and 9.13. For this reason, perturbation theory is rarely used for wave function corrections beyond the first order and for energy corrections beyond the second order. In principle, however, higher order corrections can be obtained by proceeding to solve, in a similar manner, the equations corresponding to higher powers of λ as in Equation 9.6.

2. APPLICATIONS OF THE PERTURBATION METHOD TO NON-DEGENERATE STATES

As our first example, let us determine the effect of an anharmonic term on the ground state energy of a harmonic oscillator. The unperturbed Hamiltonian is

$$\mathscr{H}_0 = \frac{p_x^2}{2m} + \tfrac{1}{2} kx^2, \tag{9.15}$$

with energy eigenvalues given by $E_n^{(0)} = \omega\hbar(n + \frac{1}{2})$. The unperturbed energy eigenfunctions of this Hamiltonian are tabulated in section 7 of Chapter 5. Now if we add an anharmonic term of the form bx^3 to the potential, the Hamiltonian becomes

$$\mathcal{H} = \frac{p_x^2}{2m} + \frac{1}{2}kx^2 + bx^3 = \mathcal{H}_0 + \mathcal{H}^{(1)},$$

where $\mathcal{H}^{(1)} = bx^3$ may be regarded as a perturbation if b is small. Then the first order correction to the ground state energy is, from Equation 9.8,

$$E_0^{(1)} = \mathcal{H}_{00}^{(1)} = \langle\psi_0^{(0)}| bx^3 |\psi_0^{(0)}\rangle = 0.$$

This result follows immediately from parity arguments without performing the integration. Since each oscillator state has definite parity, the square of any state has even parity. For an odd perturbation, then, the integrand is odd and the integral over symmetric limits will vanish.

When the first-order perturbation vanishes, it is necessary to proceed to second order in order to find the energy correction. A glance at Equation 9.12 tells us that we are going to need the off-diagonal matrix elements

$$\mathcal{H}_{ni}^{(1)} = \langle n| bx^3 |i\rangle.$$

It will be shown below that when $\langle n|$ is the unperturbed ground state wave function there are only two non-vanishing matrix elements, namely,

$$\mathcal{H}_{01}^{(1)} = \frac{3b}{\sqrt{8\alpha^3}}$$

and

$$\mathcal{H}_{03}^{(1)} = \frac{b}{2}\sqrt{\frac{3}{\alpha^3}}.$$

Then,

$$E_0^{(2)} = -\frac{11\,b^2}{8\alpha^3\omega\hbar} = -\frac{11\,b^2}{8\alpha^2k},$$

and

$$E_0 = \frac{1}{2}\omega\hbar - \frac{11\,b^2}{8\alpha^2k}.$$

PROBLEM 9-1

Verify the above values for $\mathcal{H}_{01}^{(1)}$ and $\mathcal{H}_{03}^{(1)}$ by direct integration.

We will now use the raising and lowering operators developed in Chapter 5 to obtain the matrix of the perturbation. This will enable us to simply write down the energy correction for any level without performing the integrations. Recall that the unperturbed Hamiltonian for the harmonic oscillator can be written in the forms (see Equations 5.43 and 5.45),

$$\mathcal{H}_0 = p^2 + q^2 = aa^\dagger + a^\dagger a, \tag{9.16}$$

where $q = \sqrt{\alpha}\, x$, $p = -i(\partial/\partial q)$, and $\alpha = \omega m/\hbar$. From Equation 5.58 we have

$$q = \frac{1}{\sqrt{2}} (a + a^\dagger),$$

from which we obtain

$$q^3 = \frac{1}{2\sqrt{2}} [a^3 + (a^\dagger)^3 + a^2 a^\dagger + a^\dagger a^2 + a(a^\dagger)^2 + (a^\dagger)^2 a + aa^\dagger a + a^\dagger aa^\dagger]. \tag{9.17}$$

It is readily shown that:

$$\begin{aligned}
\sqrt{8}\langle n' \mid q^3 n\rangle = &\sqrt{(n+1)(n+2)(n+3)}\ \delta_{n',n+3} \\
&+ \sqrt{n(n-1)(n-2)}\ \delta_{n',n-3} \\
&+ 3(n+1)\sqrt{n+1}\ \delta_{n',n+1} + 3n\sqrt{n}\ \delta_{n',n-1}.
\end{aligned} \tag{9.18}$$

PROBLEM 9-2

> Verify Equation 9.18 with the aid of the operator equations,
>
> $$a^\dagger \mid n\rangle = \sqrt{n+1} \mid n+1\rangle,$$
>
> and
>
> $$a \mid n\rangle = \sqrt{n} \mid n-1\rangle.$$

We can readily construct the matrix $\mathcal{H}_{n'n}^{(1)}$ from Equation 9.18, since

$$\mathcal{H}_{n'n}^{(1)} = \langle n' \mid bx^3 \mid n\rangle = \frac{b}{\sqrt{8\alpha^3}} \langle n' \mid q^3 \mid n\rangle.$$

Thus:

$$(\mathcal{H}_{n'n}^{(1)}) = \frac{b}{\sqrt{8\alpha^3}} \begin{pmatrix}
0 & 3 & 0 & \sqrt{6} & 0 & 0 & \cdots \\
3 & 0 & 6\sqrt{2} & 0 & 2\sqrt{6} & 0 \\
0 & 6\sqrt{2} & 0 & 9\sqrt{3} & 0 & 2\sqrt{15} \\
\sqrt{6} & 0 & 9\sqrt{3} & 0 & 24 & 0 \\
0 & 2\sqrt{6} & 0 & 24 & 0 & 15\sqrt{5} \\
0 & 0 & 2\sqrt{15} & 0 & 15\sqrt{5} & 0 \\
\cdot \\
\cdot \\
\cdot
\end{pmatrix} \tag{9.19}$$

PROBLEM 9-3

Using the matrix elements given in Equation 9.19, calculate the second-order energy shifts in the first two excited states of a harmonic oscillator due to the perturbation bx^3. Would you expect perturbation theory to be valid for states corresponding to large n values?

$$\left(\text{Ans.: } E_1^{(2)} = -\frac{71\ b^2}{8\alpha^2 k}\ ;\ E_2^{(2)} = -\frac{191\ b^2}{8\alpha^2 k}\ .\right)$$

PROBLEM 9-4

Write the wave functions corrected to first order in the perturbation bx^3 for the ground state and the first excited state of a harmonic oscillator.

PROBLEM 9-5

Find the first-order corrections to the energies of the ground state and the first two excited states of a harmonic oscillator, arising from the perturbation $\mathcal{H}^{(1)} = cx^4$.

$$\left(\text{Ans.: } E_0^{(1)} = \frac{3c}{4\alpha^2}\ ;\ E_1^{(1)} = \frac{15c}{4\alpha^2}\ ;\ E_2^{(1)} = \frac{39c}{4\alpha^2}\ .\right)$$

PROBLEM 9-6

Using the raising and lowering operators for the harmonic oscillator obtain an expression analogous to Equation 9.18 for the matrix elements, $\langle n'|\ q^4\ |n\rangle$.

PROBLEM 9-7

From the result of the previous problem, construct the matrix $\mathcal{H}_{n'n}^{(1)}$.

PROBLEM 9-8

Use the matrix of the previous problem to obtain the energies of the ground state and first excited state of the harmonic oscillator corrected to second order in the perturbation

$\mathcal{H}^{(1)} = cx^4$. Would you expect perturbation theory to be valid for states corresponding to large n values?

Ans.: $E_0 = \frac{1}{2}\omega\hbar + \dfrac{3c}{4\alpha^2}\left(1 - \dfrac{7c}{2\alpha^2\omega\hbar}\right)$

$E_1 = \frac{3}{2}\omega\hbar + \dfrac{15c}{4\alpha^2}\left(1 - \dfrac{11c}{2\alpha^2\omega\hbar}\right).$

PROBLEM 9-9

Consider the infinite well having a V bottom as shown in the figure. If ϵ is small, the quantity $V^{(1)} = (\epsilon/a)|x|$ may be regarded as a perturbation on the square well potential in

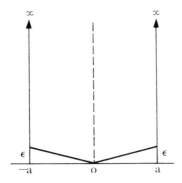

the region $-a \leq x \leq a$. Calculate the ground state energy and wave function correct to first order in this perturbation. (Three or four terms are sufficient for the wave function.)

PROBLEM 9-10

Calculate the first-order correction to the ground state energy of a harmonic oscillator due to a perturbation of the form $\mathcal{H}^{(1)} = ve^{-\beta x^2}$.

$\left(\text{Ans.: } E_0^{(1)} = v\left(\dfrac{\alpha}{\alpha + \beta}\right)^{\frac{1}{2}}.\right)$

PROBLEM 9-11

An electron is performing simple harmonic motion in one dimension. If a weak electric field $\mathscr{E}$ is applied in the direction of its motion, find the shift in the energies of the ground state and first excited state in the presence of the electric field (the Stark shift).

$$\left(\text{Ans.:} \quad E_0^{(1)} = E_1^{(1)} = 0; \quad E_0^{(2)} = E_1^{(2)} = -\frac{e^2\mathscr{E}^2}{2k}.\right)$$

PROBLEM 9-12

Use Equations 5.41 or the method of Problem 5-25 to find the wave function correct to first order for the nth state of the charged harmonic oscillator in an electric field discussed in the previous problem.

$$\left(\text{Ans.:} \quad \psi_n = |n\rangle + \frac{e\mathscr{E}}{\omega\hbar\sqrt{2\alpha}}\left[\sqrt{n+1}\,|n+1\rangle - \sqrt{n}\,|n-1\rangle\right].\right)$$

PROBLEM 9-13

Show that the potential, $V = \frac{1}{2}kx^2 - e\mathscr{E}x$, of the two previous problems can be solved exactly. (Hint: complete the square.) Find the energies of the system and interpret your results.

PROBLEM 9-14

Derive the energy shift of a harmonic oscillator to second order in the perturbation bx^3 for the nth level,

$$E_n^{(2)} = -\frac{b^2}{\alpha^2 k}(30n^2 + 30n + 11).$$

PROBLEM 9-15

(a)   Calculate the relativistic correction for the ground state of hydrogen by means of the perturbation term leading to Equation 8.12,

$$\mathscr{H}^{(1)} = -\frac{\hbar^4}{8m_0^3c^2}\nabla^4.$$

(b)   Do the same for the $2s$ state of hydrogen. (Hint: make use of the formulas given in Appendix A of Chapter 7.)

3. INTERACTIONS OF A CHARGED PARTICLE WITH STATIC ELECTRIC AND MAGNETIC FIELDS

The classical Hamiltonian for a particle of mass m and charge q in the presence of static electric and magnetic fields is given by

$$\mathcal{H} = \frac{1}{2m}\left(\vec{p} - \frac{q}{c}\vec{A}\right)^2 + V + q\Phi, \tag{9.20}$$

where $\vec{p}$ is the ordinary momentum, $\vec{A}$ is the vector potential, V is the mechanical potential energy, and Φ is the electrostatic potential. The potentials and fields are related by the equations,

$$\vec{\mathcal{B}} = \vec{\nabla} \times \vec{A} \quad \text{and} \quad \vec{\mathcal{E}} = -\nabla\Phi - \frac{1}{c}\frac{\partial\vec{A}}{\partial t}. \tag{9.21}$$

Since the potentials Φ and $\vec{A}$ do not uniquely specify the fields $\vec{\mathcal{E}}$ and $\vec{\mathcal{B}}$, we are at liberty to make an additional assumption relating them. For convenience, we assume that

$$\vec{\nabla} \cdot \vec{A} + \frac{1}{c}\frac{\partial\Phi}{\partial t} = 0, \tag{9.22}$$

which is known as the Lorentz condition. This choice arises quite naturally during the simplification of Maxwell's equations.

PROBLEM 9-16

Substitute Equation 9.21 into Maxwell's equations and show where the Lorentz condition arises. How does it simplify the field equations?

The Hamiltonian (Equation 9.20) may be readily derived from the Lorentz force,

$$\vec{F} = q\left[\vec{\mathcal{E}} + \frac{1}{c}(\vec{v} \times \vec{\mathcal{B}})\right], \tag{9.23}$$

where $\vec{v}$ is the velocity of the particle. Substituting Equation 9.21 into 9.23,

$$\vec{F} = q\left\{-\vec{\nabla}\Phi - \frac{1}{c}\left[\frac{\partial\vec{A}}{\partial t} - \vec{v} \times (\vec{\nabla} \times \vec{A})\right]\right\}. \tag{9.24}$$

Using a standard vector identity and regarding the components of $\vec{v}$ (as well as those of $\vec{r}$) as independent variables, we have

$$\vec{\nabla}(\vec{v} \cdot \vec{A}) = (\vec{v} \cdot \vec{\nabla})\vec{A} + \vec{v} \times (\vec{\nabla} \times \vec{A}). \tag{9.25}$$

It can also be easily shown that

$$(\vec{v} \cdot \vec{\nabla})\vec{A} = \frac{d\vec{A}}{dt} - \frac{\partial \vec{A}}{\partial t}.$$ (9.26)

Making use of these relations the Lorentz force becomes,

$$\vec{F} = -\vec{\nabla}\left(q\Phi - \frac{q}{c}\vec{v} \cdot \vec{A}\right) - \frac{q}{c}\frac{d\vec{A}}{dt},$$

where we interpret the quantity in parentheses as a generalized potential function U. Then the Lagrangian for the system is[2]

$$L = T - U = \tfrac{1}{2}m\vec{v}^2 - q\Phi + \frac{q}{c}\vec{v} \cdot \vec{A}$$ (9.27)

$$= \frac{1}{2m}\left(m\vec{v} + \frac{q}{c}\vec{A}\right)^2 - \frac{q^2}{2mc^2}\vec{A}^2 - q\Phi.$$

From Equation 9.27 it follows that the generalized momenta are:

$$p_x = \frac{\partial L}{\partial \dot{x}} = m\dot{x} + \frac{q}{c}A_x$$

$$p_y = \frac{\partial L}{\partial \dot{y}} = m\dot{y} + \frac{q}{c}A_y$$

$$p_z = \frac{\partial L}{\partial \dot{z}} = m\dot{z} + \frac{q}{c}A_z,$$

where it is assumed that the components of $\vec{A}$ are not velocity dependent. Then we can write

$$\vec{p} = m\vec{v} + \frac{q}{c}\vec{A}.$$ (9.28)

Thus, the canonical momentum is now composed of the ordinary linear momentum plus a contribution from the magnetic field.

PROBLEM 9-17

Verify Equations 9.25 and 9.26.

The Hamiltonian function is defined as

$$\mathscr{H} = \sum_i p_i \dot{q}_i - L = \vec{p} \cdot \vec{v} - \frac{1}{2m}\left(m\vec{v} + \frac{q}{c}\vec{A}\right)^2 + \frac{q^2}{2mc^2}\vec{A}^2 + q\Phi,$$

---

[2] H. Goldstein, *Classical Mechanics*, Addison-Wesley Publishing Co., Reading, Mass., 1950.

wherein we substitute the expression for $\vec{v}$ from Equation 9.28, namely,

$$\vec{v} = \frac{1}{m}\left(\vec{p} - \frac{q}{c}\vec{A}\right).$$

Thereupon, the Hamiltonian simplifies to

$$\mathcal{H} = \frac{1}{2m}\left(\vec{p} - \frac{q}{c}\vec{A}\right)^2 + q\Phi, \tag{9.29}$$

plus any mechanical potentials which may be present. Note that the kinetic energy term for a particle in a magnetic field can be obtained by merely substituting the quantity $(\vec{p} - q/c\,\vec{A})$ for p in the expression $p^2/2m$.

In order to utilize the above Hamiltonian in quantum mechanics, we will regard $\vec{p}$ and $\vec{A}$ as operators and will replace $\vec{p}$ with $-i\hbar\vec{\nabla}$. It is therefore useful to know under what conditions $\vec{p}$ and $\vec{A}$ commute. Since each component behaves as follows,

$$[A_x, p_x] = -i\hbar\frac{\partial}{\partial x}A_x,$$

then

$$[\vec{A}\cdot\vec{p}, \vec{p}\cdot\vec{A}] = i\hbar\vec{\nabla}\cdot\vec{A}. \tag{9.30}$$

Referring to the Lorentz condition, Equation 9.22, we note that $\vec{\nabla}\cdot\vec{A} = 0$ *whenever the electric field is constant in time.* Under this circumstance $\vec{p}$ and $\vec{A}$ commute and the Hamiltonian may be written,

$$\mathcal{H} = \frac{p^2}{2m} + q\Phi - \frac{q}{mc}\vec{p}\cdot\vec{A} + \frac{q^2}{2mc^2}\vec{A}^2. \tag{9.31}$$

It is common practice to represent a uniform magnetic field by the vector potential,

$$\vec{A} = -\tfrac{1}{2}\vec{r}\times\vec{\mathscr{B}}. \tag{9.32}$$

To show that this satisfies our original definition of the vector potential, Equation 9.21, we operate as follows:

$$\vec{\nabla}\times\vec{A} = -\tfrac{1}{2}\vec{\nabla}\times(\vec{r}\times\vec{\mathscr{B}})$$

$$= -\tfrac{1}{2}[\vec{r}(\vec{\nabla}\cdot\vec{\mathscr{B}}) + (\vec{\mathscr{B}}\cdot\vec{\nabla})\vec{r} - \vec{\mathscr{B}}(\vec{\nabla}\cdot\vec{r}) - (\vec{r}\cdot\vec{\nabla})\vec{\mathscr{B}}].$$

Since the divergence of $\vec{\mathscr{B}}$ is always zero, the first term vanishes. The second term is $(\vec{\mathscr{B}}\cdot\vec{\nabla})\vec{r} = \vec{\mathscr{B}}$. Since the divergence of $\vec{r}$ is 3, the third term becomes $3\vec{\mathscr{B}}$. The last term is zero because $\vec{\mathscr{B}}$ is a constant vector. Thus,

$$\vec{\nabla}\times\vec{A} = -\tfrac{1}{2}(\vec{\mathscr{B}} - 3\vec{\mathscr{B}}) = \vec{\mathscr{B}},$$

as required. Equation 9.32 must also satisfy the Lorentz condition:

$$\vec{\nabla} \cdot \vec{A} = -\tfrac{1}{2}[\vec{\mathscr{B}} \cdot \vec{\nabla} \times \vec{r} - \vec{r} \times \vec{\nabla} \cdot \vec{\mathscr{B}}] = 0,$$

since the curl of $\vec{r}$ is zero and the divergence of $\vec{\mathscr{B}}$ is zero. Substituting this vector potential into the Hamiltonian (Equation 9.31),

$$\mathscr{H} = \frac{p^2}{2m} + q\Phi - \frac{q}{2mc}\,\vec{\mathscr{B}} \cdot \vec{L} + \frac{q^2}{8mc^2}\,(\vec{r} \times \vec{\mathscr{B}})^2. \tag{9.33}$$

Letting q be the electronic charge e, we see that the linear term in the magnetic field is simply the Zeeman term. If the z-direction is taken along the field direction, the magnetic dipole or Zeeman energy is just

$$\Delta E = m_\ell \mu_B \mathscr{B},$$

as discussed in section 10 of Chapter 3. On the other hand, we may incorporate spin in Equation 9.33 by rewriting the second term as

$$\frac{e}{2mc}\,\vec{\mathscr{B}} \cdot (\vec{L} + 2\vec{S}),$$

and by adding the spin-orbit term, Equation 8.8. The Zeeman energy then becomes

$$\Delta E = (m_\ell + 2m_s)\mu_B \mathscr{B} = g m_j \mu_B \mathscr{B},$$

in agreement with section 9 of Chapter 8. The quadratic term in $\mathscr{B}$ is the *diamagnetic* term which is important for the calculation of the diamagnetic susceptibility. Since it is considerably smaller than the Zeeman term, it can usually be neglected, unless either the Zeeman term vanishes or extremely high magnetic fields are used.

As our next example let us treat the weak-field Zeeman effect in hydrogen by perturbation theory. The Zeeman energy (see section 9 of Chapter 8) is given by

$$\Delta E = \frac{\mu_B \mathscr{B}}{\hbar}\,(\mathbf{L}_z + 2\mathbf{S}_z) = \frac{g\mu_B \mathscr{B}}{\hbar}\,\mathbf{J}_z, \tag{9.34}$$

where the field $\mathscr{B}$ is taken to be in the z-direction. Since we must regard the spin and orbital momenta as strongly coupled in weak magnetic fields, the appropriate representation is $|j, m_j\rangle$. Then, by first order perturbation theory, the Zeeman energy corrections are

$$\mathscr{H}^{(1)}_{j,m_j;j,m_j} = \frac{g\mu_B \mathscr{B}}{\hbar}\,\langle j, m_j | J_z | j, m_j \rangle = g\mu_B \mathscr{B} m_j. \tag{9.35}$$

These energies have already been calculated for p and s states of hydrogen, and are given in Table 8-5 in section 9 of Chapter 8.

The reader might well question the utility of the middle expression in Equation 9.34 above. Its usefulness arises when the states are expressed in the $|m_\ell, m_s\rangle \equiv |\ell, m_\ell\rangle |s, m_s\rangle$ representation. Thus, it would be used for the strong-field Zeeman effect (Paschen-Bach effect) when the dipole energy is comparable to or greater than the spin-orbit energy. It can also be used for the weak-field Zeeman effect if the kets $|j, m_j\rangle$ are written in terms of the kets $|m_\ell, m_s\rangle$. It will no doubt be instructive to show this in detail for a few cases. Using the kets for a p-electron given in Problem 8-20 we have:

$$\mathcal{H}^{(1)}_{\frac{3}{2},\frac{3}{2};\frac{3}{2},\frac{3}{2}} = \frac{\mu_B \mathcal{B}}{\hbar} \langle 1, \tfrac{1}{2}| \mathbf{L}_z + 2\mathbf{S}_z |1, \tfrac{1}{2}\rangle = 2\mu_B \mathcal{B},$$

$$\mathcal{H}^{(1)}_{\frac{3}{2},\frac{1}{2};\frac{3}{2},\frac{3}{2}} = \frac{\mu_B \mathcal{B}}{\hbar} \cdot [\tfrac{2}{3}\langle 0, \tfrac{1}{2}| \mathbf{L}_z + 2\mathbf{S}_z |0, \tfrac{1}{2}\rangle + \tfrac{1}{3}\langle 1, -\tfrac{1}{2}| \mathbf{L}_z + 2\mathbf{S}_z |1, -\tfrac{1}{2}\rangle]$$

$$= \tfrac{2}{3}\mu_B \mathcal{B},$$

and so forth, in agreement with the values calculated from Equation 9.35.

The spin-orbit energy correction for hydrogen can also be calculated readily by means of perturbation theory. Using Equation 8.8 and the operator given in Problem 8-16, we can write

$$\mathcal{H}^{(1)} = -\tfrac{1}{4}\left(\frac{e}{m_0 c}\right)^2 \frac{1}{r_3} (\mathbf{J}^2 - \mathbf{L}^2 - \mathbf{S}^2), \tag{9.36}$$

where $\partial V/\partial r$ has been replaced by e^2/r^2. This form of the operator $\vec{L} \cdot \vec{S}$ was chosen since it has a diagonal representation when the kets $|j, m_j\rangle$ are used as basis vectors. Then, for the p states of hydrogen,

$$\langle j, m_j| \mathcal{H}^{(1)} |j, m_j\rangle = -\left(\frac{e\hbar}{2m_0 c}\right)^2 [j(j+1) - \ell(\ell+1) - s(s+1)]$$

$$\langle j, m_j| \frac{1}{r^3} |j, m_j\rangle$$

$$= -\left(\frac{e\hbar}{2m_0 c}\right)^2 \cdot \left[j(j+1) - \frac{11}{4}\right] \cdot \frac{1}{3n^3 a_0^3} = \frac{E_n}{6n}\left(\frac{\lambda_c}{a_0}\right)^2 \left[j(j+1) - \frac{11}{4}\right],$$

where Appendix A of Chapter 7 was used to evaluate $\langle 1/r^3\rangle$. Then, for $j = \tfrac{3}{2}$ and $j = \tfrac{1}{2}$:

$$\langle \tfrac{3}{2}, m_j| \mathcal{H}^{(1)} |\tfrac{3}{2}, m_j\rangle = \frac{E_n}{6n}\left(\frac{\lambda_c}{a_0}\right)^2 \sim + 2 \times 10^{-4}E_n,$$

$$\langle \tfrac{1}{2}, m_j| \mathcal{H}^{(1)} |\tfrac{1}{2}, m_j\rangle = -\frac{E_n}{3}\left(\frac{\lambda_c}{a_0}\right)^2 \sim - 4 \times 10^{-4}E_n,$$

in good agreement with our previous calculation. The spin-orbit correction cannot be obtained so easily for s-states, but requires the relativistic theory.

4. PERTURBATION THEORY FOR DEGENERATE STATES

In the event that degeneracies exist among the unperturbed states, it is still possible to calculate the first-order perturbation corrections by means of Equation 9.8. The effect of the first-order correction is frequently such as to remove some or all of the original degeneracy. This was illustrated by the examples treating the Zeeman effect and the spin-orbit interaction in the previous section. On the other hand, it might turn out that the first-order correction is zero for all states—as in the Stark effect for states of definite parity—so that perturbation theory fails unless a correction is obtained in the second-order calculation. It might also be desirable to proceed to second order in a case where the first-order correction is not zero but where the degeneracy remains. However, any attempt to apply the second-order theory of the previous section to a system where degeneracy persists in the first-order correction is doomed to fail. This is most readily seen by examining Equation 9.7. If, for instance, the kth and n^{th} states are degenerate, then for $k \neq n$, both of the quantities δ_{kn} and $(E_k^{(0)} - E_n^{(0)})$ are zero simultaneously. This gives the trivial result that $\mathscr{H}_{kn}^{(1)} = 0$, which may or may not be true. The difficulty lies in the fact that we are unable to find out what the actual value of $\mathscr{H}_{kn}$ is by means of the non-degenerate theory when the states n and k are degenerate.

In order to develop a procedure for treating degeneracies, let us alter our previous approach somewhat, and assume that we can find solutions to the eigenequation

$$(\mathscr{H}_0 + \mathscr{H}^{(1)})\psi_n = E_n\psi_n, \tag{9.37}$$

where $\psi_n = \sum_i c_{ni} |i\rangle$ and $\mathscr{H}_0 |i\rangle = E_i^{(0)} |i\rangle$. The vectors $|i\rangle = \psi_i^{(0)}$ are the eigenfunctions of the unperturbed Hamiltonian, as before. Multiplying by another vector of the set, $\langle k|$, we obtain:

$$c_{nk}E_k^{(0)} + \sum_i c_{ni}\mathscr{H}_{ki}^{(1)} = c_{nk}E_n,$$

or

$$c_{nk}(E_n - E_k^{(0)}) = \sum_i c_{ni}\mathscr{H}_{ki}^{(1)}. \tag{9.38}$$

For convenience let us assume that we have only two degenerate states, the nth and ℓ^{th}. Then the contributions to the energies of these states for $i \neq n, \ell$ are negligible, and we may write

$$\psi_n = c_{nn} |n\rangle + c_{n\ell} |\ell\rangle$$

and

$$\psi_\ell = c_{\ell n} |n\rangle + c_{\ell\ell} |\ell\rangle.$$

First setting $k = n$ and then $k = \ell$ in Equation 9.38, we obtain the two equations:

$$c_{nn}(\mathscr{H}_{nn}^{(1)} + E_n^{(0)} - E_n) + c_{n\ell}\mathscr{H}_{n\ell}^{(1)} = 0$$
$$c_{nn}\mathscr{H}_{\ell n}^{(1)} + c_{n\ell}(\mathscr{H}_{\ell\ell}^{(1)} + E_\ell^{(0)} - E_n) = 0. \tag{9.39}$$

The necessary condition for these equations to have a solution is that the determinant of their coefficients be zero. That is,

$$\begin{vmatrix} \mathscr{H}_{nn}^{(1)} + E_n^{(0)} - E_n & \mathscr{H}_{n\ell}^{(1)} \\ \mathscr{H}_{\ell n}^{(1)} & \mathscr{H}_{\ell\ell}^{(1)} + E_\ell^{(0)} - E_n \end{vmatrix} = 0, \qquad (9.40)$$

which has the solutions

$$E_n = \tfrac{1}{2}[(\mathscr{H}_{nn}^{(1)} + E_n^{(0)}) + (\mathscr{H}_{\ell\ell}^{(1)} + E_\ell^{(0)})] \pm \tfrac{1}{2}\{[(\mathscr{H}_{nn}^{(1)} + E_n^{(0)})$$
$$- (\mathscr{H}_{\ell\ell}^{(1)} + E_\ell^{(0)})]^2 + 4\,|\mathscr{H}_{n\ell}^{(1)}|^2\}^{\frac{1}{2}}. \qquad (9.41)$$

If, as we assumed above, $E_n^{(0)} = E_\ell^{(0)}$, then Equation 9.41 reduces to

$$E_n = E_n^{(0)} + \tfrac{1}{2}(\mathscr{H}_{nn}^{(1)} + \mathscr{H}_{\ell\ell}^{(1)}) \pm \tfrac{1}{2}[(\mathscr{H}_{nn}^{(1)} - \mathscr{H}_{\ell\ell}^{(1)})^2 + 4\,|\mathscr{H}_{n\ell}^{(1)}|^2]^{\frac{1}{2}}.$$

The meaning of what has been done in solving the degenerate case will become apparent if one but compares Equation 9.40 with the diagonalization procedure of Chapter 6. Thus, a matrix which has degenerate diagonal elements and non-zero off-diagonal elements in the representation of the unperturbed eigenfunctions has a diagonal representation when expressed in the new basis of the eigenfunctions of the perturbed Hamiltonian.

As an illustrative example, let us calculate the Stark splitting for the $n = 2$ level of the hydrogen atom by an electric field $\mathscr{E}$ in the z-direction. The perturbation term of the Hamiltonian is $\mathscr{H}^{(1)} = e\mathscr{E}z = e\mathscr{E}r\cos\theta$. There are four degenerate wave functions corresponding to $n = 2$, ψ_{200}, ψ_{210}, ψ_{211}, and $\psi_{21\bar{1}}$. Since each of these functions has definite parity and the perturbation is odd, all diagonal matrix elements are zero; that is, there is no first-order correction to the energy and the degeneracy persists. Now let us obtain the full matrix of the perturbation, whose elements are given by

$$\langle \psi_{2\ell'm'}|\,e\mathscr{E}r\cos\theta\,|\psi_{2\ell m}\rangle.$$

Using parity arguments, one of the quantities ℓ and ℓ' must be odd and the other must be even in order to have a non-zero matrix element. Furthermore, we learned from the properties of the spherical harmonics that m' must equal m for a non-zero integral. Thus the only non-zero matrix elements are:

$$\mathscr{H}_{200,210}^{(1)} = \mathscr{H}_{210,200}^{(1)} = e\mathscr{E}\,\langle \psi_{200}|\,r\cos\theta\,|\psi_{210}\rangle$$
$$= \frac{e\mathscr{E}}{16a_0^4}\int_0^\infty r^4\left(2 - \frac{r}{a_0}\right)e^{-r/a_0}\,dr\int_0^\pi \sin\theta\cos^2\theta\,d\theta$$
$$= 3a_0 e\mathscr{E}.$$

The perturbation matrix is:

$$
(\mathcal{H}^{(1)}) = \left(\begin{array}{cc|cc}
0 & 3a_0 e\mathcal{E} & 0 & 0 \\
3a_0 e\mathcal{E} & 0 & 0 & 0 \\
\hline
0 & 0 & 0 & 0 \\
0 & 0 & 0 & 0
\end{array}\right).
$$

Since the degeneracy persists, we must use degenerate perturbation theory for the second-order calculation. This simply amounts to diagonalizing the 2×2 submatrix. Then,

$$
\begin{vmatrix}
-E_2^{(2)} & 3a_0 e\mathcal{E} \\
3a_0 e\mathcal{E} & -E_2^{(2)}
\end{vmatrix} = 0,
$$

and

$$
E_2^{(2)} = \pm 3a_0 e\mathcal{E}.
$$

The diagonalized matrix of the Hamiltonian is:

$$
(\mathcal{H}) = (\mathcal{H}_0) + (\mathcal{H}^{(1)}) = \begin{pmatrix}
E_2^{(0)} + 3a_0 e\mathcal{E} & 0 & 0 & 0 \\
0 & E_2^{(0)} - 3a_0 e\mathcal{E} & 0 & 0 \\
0 & 0 & E_2^{(0)} & 0 \\
0 & 0 & 0 & E_2^{(0)}
\end{pmatrix}.
$$

$$(9.42)$$

Equation 9.42 shows that the four-fold degeneracy is partially lifted by the Stark effect, in that a single level is split into three levels by the electric field. There is still a two-fold degeneracy associated with the unperturbed energy $E_2^{(0)}$. The shifted levels are associated with states that are described by linear combinations of the original wave functions. Thus, the state with energy $E_2^{(0)} + 3a_0 e\mathcal{E}$ has the wave function $(1/\sqrt{2})(\psi_{200} + \psi_{210})$, while the state $(1/\sqrt{2})(\psi_{200} - \psi_{210})$ has energy $E_2^{(0)} - 3a_0 e\mathcal{E}$. This behavior can be described by attributing a permanent electric dipole moment of magnitude $3a_0 e$ to the $n = 2$ state of hydrogen. It then follows that the three energy states correspond to parallel, antiparallel, and transverse orientations of this dipole with respect to the electric field.

PROBLEM 9-18

Calculate the Stark effect for the $n = 3$ state of hydrogen. [Ans.: $E_3^{(0)}$, $E_3^{(0)} \pm \frac{9}{2}a_0 e\mathcal{E}$, $E_3^{(0)} \pm 9a_0 e\mathcal{E}$.]

PROBLEM 9-19

Find the new basis functions for the diagonal representation of the complete Hamiltonian in the preceding problem.

PROBLEM 9-20

A two-dimensional, isotropic oscillator has the Hamiltonian

$$\mathcal{H} = -\frac{\hbar^2}{2m}\left(\frac{\partial^2}{\partial x^2} + \frac{\partial^2}{\partial y^2}\right) + \frac{k}{2}(1 + bxy)(x^2 + y^2).$$

(a) If $b = 0$, what are the wave functions and energies of the three lowest states?

(b) If b is a small positive number such that $b \ll 1$, find the perturbation theory corrections to the energies of the three lowest states.

PROBLEM 9-21

A certain electric potential is described by the spin Hamiltonian,

$$\mathcal{H}^{(1)} = DS_z^2 + E(S_x^2 - S_y^2).$$

Treating this as a perturbation on an ion with three degenerate spin-one states, find the energy splitting that results.

5. TIME-DEPENDENT PERTURBATION THEORY

We will now consider the case in which the perturbing part of the Hamiltonian is time-dependent; that is, when

$$\mathcal{H}(\vec{r}, t) = \mathcal{H}_0 + \mathcal{H}^{(1)}(\vec{r}, t).$$

As before, $\mathcal{H}^{(1)}$ is regarded as small with respect to $\mathcal{H}_0$. We require solutions of the Schrödinger equation,

$$\mathcal{H}\Psi = i\hbar\frac{\partial\Psi}{\partial t},$$

where

$$\mathcal{H}_0\Psi_i = E_i^{(0)}\Psi_i \quad \text{and} \quad \Psi_i = \psi_i e^{-(i/\hbar)E_i^{(0)}t}.$$

The ψ_i satisfy the requirement $\langle i\,|\,j\rangle = \delta_{ij}$ for a complete, orthonormal basis for the expansion of an arbitrary wave function. In contrast with previous practice, however, we will now regard the expansion coefficients themselves as time-dependent. Thus,

$$\Psi = \sum_n a_n(t)\Psi_n \tag{9.43}$$

and

$$\frac{\partial\Psi}{\partial t} = \sum_n \left(\dot{a}_n - \frac{i}{\hbar}E_n^{(0)}a_n\right)\Psi_n. \tag{9.44}$$

Substituting Equations 9.43 and 9.44 into the Schrödinger equation, we obtain

$$(\mathcal{H}_0 + \mathcal{H}^{(1)}) \sum_n a_n \Psi_n = \sum_n (i\hbar \dot{a}_n + E_n^{(0)} a_n) \Psi_n,$$

or

$$\sum_n a_n (\mathcal{H}_0 - E_n^{(0)}) \Psi_n + \sum_n (a_n \mathcal{H}^{(1)} - i\hbar \dot{a}_n) \Psi_n = 0. \tag{9.45}$$

The first term of Equation 9.45 is trivially zero since Ψ_n was defined to be an eigenfunction of $\mathcal{H}_0$. Multiplying the second term by Ψ_k^* there results

$$\sum_n (\langle k | a_n \mathcal{H}^{(1)} | n \rangle = i\hbar \dot{a}_n \delta_{kn} e^{(i/\hbar)(E_k^{(0)} - E_n^{(0)})t}.$$

Defining

$$\omega_{kn} = \frac{1}{\hbar}(E_k^{(0)} - E_n^{(0)}),$$

this may be written as

$$i\hbar \dot{a}_k = \sum_n a_n \mathcal{H}_{kn}^{(1)} e^{i\omega_{kn}t}. \tag{9.46}$$

Equation 9.46 is impossible to solve exactly, since it relates the derivative of the kth coefficient to all of the others. However, for a small perturbing potential it is not a bad approximation to assume that the rate of change of each coefficient a_n is so small that a_n can be taken as a constant on the right hand side. Suppose that at $t = 0$ the system is in the state Ψ_j so that $a_j(0) = 1$. Then, after a time t we assume that $a_j(t)$ is still nearly 1, and so

$$i\hbar \dot{a}_k(t) \approx \mathcal{H}_{kj}^{(1)} e^{i\omega_{kj}t}. \tag{9.47}$$

From Equation 9.47,

$$a_j(t) = 1 - \frac{i}{\hbar} \int_0^t \mathcal{H}_{jj}^{(1)} \, dt \tag{9.48}$$

and

$$a_k(t) = -\frac{i}{\hbar} \int_0^t \mathcal{H}_{kj} e^{i\omega_{kj}t} \, dt, \quad \text{for} \quad k \neq j. \tag{9.49}$$

For a constant perturbation that is turned on at $t = 0$, Equations 9.48 and 9.49 may be integrated easily to obtain the development of the expansion coefficients with time. Thus,

$$a_j(t) = 1 - \frac{i}{\hbar} \mathcal{H}_{jj}^{(1)} t \tag{9.50}$$

and

$$a_k(t) = \frac{\mathcal{H}_{kj}^{(1)}}{\hbar \omega_{kj}} (1 - e^{i\omega_{kj}t}), \quad \text{for} \quad k \neq j. \tag{9.51}$$

The quantity $|a_k(t)|^2$ is the probability that the state Ψ_k is occupied at time t. Hence, it is a measure of the probability that the system will make a transition from the initial state Ψ_j to Ψ_k in time t. Squaring Equation 9.51, we have

$$P_k = |a_k(t)|^2 = \left| \frac{2 \mathcal{H}_{kj}^{(1)}}{E_k^{(0)} - E_j^{(0)}} \right|^2 \cdot \sin^2 \frac{(E_k^{(0)} - E_j^{(0)})t}{2\hbar}, \quad \text{for} \quad k \neq j. \tag{9.52}$$

Note that the probability oscillates at the angular frequency given by $\frac{1}{2}\omega_{kj}$ but that its effect is appreciable only for values of k near j, where the amplitude reaches its maximum value. Recall that the derivation of Equation 9.52 assumes that the a_k remain small; this implies that $\mathcal{H}_{kj}^{(1)} \ll E_k^{(0)} - E_j^{(0)}$.

For large t the amplitude $a_k(t)$ in Equation 9.52 resembles the delta function, $\delta(E - E_j^{(0)})$, discussed in section 7 of Chapter 4. Assuming a quasi-continuum of closely-spaced states, the area under the graph showing $|a_k(t)|^2$ versus E represents the total transition probability. Then, letting $\rho(E)$ be the density of states,

$$P = \int_{-\infty}^{\infty} |a_k(t)|^2 \, \rho(E) \, dE = |2\,\mathcal{H}_{kj}^{(1)}|^2 \, \rho(E) \cdot \frac{t}{2\hbar} \int_{-\infty}^{\infty} \frac{\sin^2 x}{x^2} \, dx$$

$$= \frac{2\pi t}{\hbar} \cdot |\mathcal{H}_{kj}^{(1)}|^2 \cdot \rho(E), \tag{9.53}$$

which says that the total probability of a transition out of the initial state Ψ_j is directly proportional to the time that the perturbation acts on the system. Both $\rho(E)$ and $\mathcal{H}_{kj}^{(1)}$ are assumed to be slowly varying and are regarded as constants for the integration of Equation 9.53. This is not a bad assumption, since the integrand is a sharply-peaked function centered around $E_j^{(0)}$. The *transition rate* from the initial state j may be defined as,

$$R = \frac{dP}{dt} = \frac{2\pi}{\hbar} |\mathcal{H}_{kj}^{(1)}|^2 \, \rho(E), \tag{9.54}$$

which is often referred to[3] as *Fermi's Golden Rule Number 2*. It will be used later in our discussion of scattering theory.

A few words should be said about energy conservation during the transitions discussed in this section. In Equation 9.52, energy is conserved to within the limits of the uncertainty principle expressed by $\Delta E \cdot \Delta t \sim \hbar$. Although the system can oscillate between states of widely separated energies, the greater this separation, the shorter the lifetime of the excited state. However, as the resonance condition is approached, $\Delta E \to 0$ and $\Delta t \to \infty$, implying that a long-lived transition can occur. In this case the radiation field insures that the total energy is conserved by creating or annihilating a photon. In Equations 9.53 and 9.54 energy conservation is built into the formalism by the assumption that all transitions occur between states which are closely packed around the energy $E_j^{(0)}$.

6. PERTURBATIONS THAT ARE HARMONIC IN TIME

Harmonic perturbations are extremely important in physics, since nearly all of the weak interactions of electromagnetic radiation with matter may be approximated by this model. Such diverse phenomena as magnetic resonance

---

[3] E. Fermi, *Nuclear Physics*, University of Chicago Press, Chicago (1950), p. 142.

in solids and the study of resonance absorption in optical spectroscopy are both easily described in the harmonic approximation.

Since either the electric vector or the magnetic vector (or both) of the incident radiation may interact with an atom, let us omit the details of the interaction for the time being and write the perturbation as

$$2\mathcal{H}^{(1)}\cos\omega t = \mathcal{H}^{(1)}(e^{i\omega t} + e^{-i\omega t}). \tag{9.55}$$

Substituting this into Equation 9.49 we obtain for the expansion coefficient $a_k(t)$,

$$a_k(t) = -\frac{i}{\hbar}\int_0^t \mathcal{H}_{kj}^{(1)}(e^{i(\omega_{kj}+\omega)t} + e^{i(\omega_{kj}-\omega)t})\, dt$$

$$= -\mathcal{H}_{kj}^{(1)}\left\{\frac{\exp[i(\omega_{kj} + \omega)t] - 1}{\hbar(\omega_{kj} + \omega)} + \frac{\exp[i(\omega_{kj} - \omega)t] - 1}{\hbar(\omega_{kj} - \omega)}\right\}. \tag{9.56}$$

It is important to notice that the first term of Equation 9.56 completely dominates the second term when $\omega \sim -\omega_{kj}$, that is, when

$$\hbar\omega \sim E_j^{(0)} - E_k^{(0)}.$$

This corresponds to the physical process of *stimulated emission* discussed in section 4 of Chapter 2. An incident photon of energy $\hbar\omega$ induces a transition from the state $E_j^{(0)}$ to a state $E_k^{(0)}$ of lower energy. This process has a maximum probability when $\hbar\omega = E_j^{(0)} - E_k^{(0)}$. In like manner, the second term of Equation 9.56 dominates the first term when $\omega \sim \omega_{kj}$, that is, when

$$\hbar\omega \sim E_k^{(0)} - E_j^{(0)}.$$

This will be recognized as the process of *resonant absorption*, in which the photon excites the system to a higher energy state $E_k^{(0)} > E_j^{(0)}$. For values of ω which are not close to $\pm\omega_{kj}$ neither term of Equation 9.56 gets large and $a_k(t)$ is negligibly small. In the limit as ω goes to zero, Equation 9.56 reduces to the case of a constant perturbation given by Equation 9.51.

Since only one term of Equation 9.56 is important at any time, we may write the probability of absorption or emission as

$$|a_k(t)|^2 = \left|\frac{2\mathcal{H}_{kj}^{(1)}}{\hbar(\omega_{kj} \mp \omega)}\right|^2 \sin^2\left[\frac{(\omega_{kj} \mp \omega)t}{2}\right], \tag{9.57}$$

where the minus signs correspond to absorption and the plus signs to emission. Using the delta function approximation for large t, the transition rate to the kth state from the jth state may be expressed as

$$R_k = \frac{2\pi}{\hbar}|\mathcal{H}_{kj}^{(1)}|^2\,\delta(E_{kj} \mp E), \tag{9.58}$$

where $E = \hbar\omega$ is the photon energy.

As an example of a harmonic perturbation, let us consider the case of electric dipole radiation. In the dipole approximation we neglect the spatial variation of the electric field over the region of interaction (that is, over the dimensions of the atom) and write the perturbation as

$$\mathscr{H}^{(1)}(\vec{r}, t) = -e\vec{\mathscr{E}} \cdot \vec{r}(e^{i\omega t} + e^{-i\omega t}) = \vec{\mathscr{E}} \cdot \vec{D}(e^{i\omega t} + e^{-i\omega t}). \tag{9.59}$$

Here, $\vec{\mathscr{E}}_0 = \hat{i}\mathscr{E}_x + \hat{j}\mathscr{E}_y + \hat{k}\mathscr{E}_z$ is the amplitude of the electric field and $\vec{D} = -e\vec{r}$ is the electric dipole operator. Then, from Equation 9.58, the rate of electric dipole transitions from the jth to the kth state is given by

$$R_k = \frac{2\pi}{\hbar} |\langle k| \vec{\mathscr{E}} \cdot \vec{D} |j\rangle|^2 \delta(E_k^{(0)} - E_j^{(0)} \mp E). \tag{9.60}$$

If the incident radiation is unpolarized, we may write

$$\mathscr{E}_x = \mathscr{E}_y = \mathscr{E}_z,$$

$$\mathscr{E}_x^2 = \mathscr{E}_y^2 = \mathscr{E}_z^2 = \tfrac{1}{3}\mathscr{E}_0^2,$$

since $\mathscr{E}_0^2 = \mathscr{E}_x^2 + \mathscr{E}_y^2 + \mathscr{E}_z^2$. Furthermore, since the energy density of an electromagnetic wave is given by

$$I(\omega) = \frac{1}{4\pi} \langle \mathscr{E}_0^2 \rangle,$$

when the permeability is unity, the matrix element of Equation 9.60 may be expressed as

$$|\langle k| \vec{\mathscr{E}} \cdot \vec{D} |j\rangle|^2 = \frac{4\pi I(\omega)}{3} |\langle k| \vec{D} |j\rangle|^2 = \frac{4\pi I(\omega)}{3} |\vec{D}_{kj}|^2, \tag{9.61}$$

where

$$|\vec{D}_{kj}|^2 = (D_x)_{kj}^2 + (D_y)_{kj}^2 + (D_z)_{kj}^2.$$

Then

$$R_k = \frac{8\pi^2 I(\omega)}{3\hbar} \cdot |D_{kj}|^2 \delta(E_k^{(0)} - E_j^{(0)} \mp E). \tag{9.62}$$

Recall that B_{kj}, the Einstein coefficient for induced emission or absorption of photons, was defined in terms of the radiation energy density as (see section 4 of Chapter 2),

$$R_k = B_{kj} I(\omega). \tag{9.63}$$

By equating 9.62 and 9.63, we see that the Einstein coefficient may be expressed in terms of the electric dipole matrix elements; that is,

$$B_{kj} = \frac{8\pi^2}{3\hbar} |D_{kj}|^2. \tag{9.64}$$

It is interesting to note that the electric dipole model does not permit a direct calculation of the Einstein coefficient for spontaneous emission, A_{kj}, although the latter can be obtained from the ratio given in section 4 of Chapter 2. In order to treat spontaneous transitions rigorously, it is necessary to use a model in which the electromagnetic field is quantized so that the physical system consists of the quantum states of the field as well as those of the atom. The absorption of a photon is then described by the simultaneous annihilation of a photon state in the field and an upward transition in the atom. Emission is regarded as the creation of a photon state in the field, accompanied by a downward transition in the atom. In the harmonic oscillator approximation, the operators of Chapter 5 may be used. Although this topic belongs in a more advanced course, the interested reader may wish to consult the excellent book by Heitler.[4]

The selection rules for electric dipole transitions have been mentioned previously, but it will be instructive to discuss them in the present context. Note that R_k will be zero whenever Equation 9.61 is zero. Let us write the hydrogenic wave functions in ket form as $\psi_{n\ell m} = |n\rangle\,|\ell\rangle\,|m\rangle$. Then the matrix elements of the components of $\vec{D}$ are:

$$\langle\psi'|\,D_x\,|\psi\rangle = \langle\psi'|\,er\sin\theta\cos\phi\,|\psi\rangle = e\,\langle n'|\,r\,|n\rangle\,\langle\ell'|\sin\theta\,|\ell\rangle\,\langle m'|\cos\phi\,|m\rangle$$

$$= e\,\langle n'|\,r\,|n\rangle\,\delta_{\ell',\ell\pm1}\,\delta_{m',m\pm1}$$

$$\langle\psi'|\,D_y\,|\psi\rangle = e\,\langle n'|\,r\,|n\rangle\,\langle\ell'|\sin\theta\,|\ell\rangle\,\langle m'|\sin\phi\,|m\rangle$$

$$= e\,\langle n'|\,r\,|n\rangle\,\delta_{\ell',\ell\pm1}\,\delta_{m',m\pm1}$$

$$\langle\psi'|\,D_z\,|\psi\rangle = e\,\langle n'|\,r\,|n\rangle\,\langle\ell'|\cos\theta\,|\ell\rangle\,\langle m'\,|\,m\rangle$$

$$= e\,\langle n'|\,r\,|n\rangle\,\delta_{\ell',\ell\pm1}\,\delta_{m',m}\,.$$

The electric dipole selection rules are (neglecting spin):

$$\Delta\ell = \pm1$$
$$\Delta m = 0,\ \pm1. \tag{9.65}$$

If spin is included, they become: $\Delta j = 0,\ \pm1;\ j = 0 \nrightarrow j = 0$

$$\Delta m = 0,\ \pm1;\ m = 0 \nrightarrow m = 0 \text{ when } \Delta j = 0.$$

PROBLEM 9-22

Verify the following matrix elements:
(a) $\langle\ell'|\sin\theta\,|\ell\rangle = \delta_{\ell',\ell\pm1}$
(b) $\langle\ell'|\cos\theta\,|\ell\rangle = \delta_{\ell',\ell\pm1}$
(c) $\langle m'|\sin\phi\,|m\rangle = \delta_{m',m\pm1}$
(d) $\langle m'|\cos\phi\,|m\rangle = \delta_{m',m\pm1}.$

---

[4] W. Heitler, *The Quantum Theory of Radiation*, 3rd ed., Oxford University Press, Oxford, 1958.

The treatment of atomic radiation may be cast in a somewhat different form which has the advantage of showing how multipole radiation arises. Using the interaction term from Equation 9.31, $(e/mc)\vec{p} \cdot \vec{A}$, let us express the incident radiation as the real part of a plane wave,

$$\vec{A} = \vec{A}_0 e^{i(\vec{k}\cdot\vec{r}-\omega t)}. \tag{9.66}$$

The plane of polarization of this wave is perpendicular to its direction of propagation, that is, $\vec{A}_0 \cdot \vec{k} = 0$. Now, if $\vec{k} \cdot \vec{r} \ll 1$, the exponential may be expanded as follows,

$$\vec{A} = \vec{A}_0 e^{-i\omega t}\left[1 + (i\vec{k} \cdot \vec{r}) + \frac{(i\vec{k} \cdot \vec{r})^2}{2} + \cdots\right]. \tag{9.67}$$

Such an expansion is certainly valid for atomic radiation in the visible spectrum where $\lambda/a_0 \sim 10^4$. Keeping only the first term, the perturbation is

$$\frac{e}{mc}\vec{p} \cdot \vec{A}_0 \cos \omega t.$$

Suppose that the wave is polarized in the x-direction. Then

$$\vec{p} \cdot \vec{A}_0 = p_x A_0$$

and

$$\mathcal{H}^{(1)}_{kj} = \frac{eA_0}{2mc} \langle k| p_x |j\rangle. \tag{9.68}$$

But

$$p_x = m\dot{x} = \frac{im}{\hbar}[\mathcal{H}_0, x],$$

from Equation 6.11. Then,

$$\mathcal{H}^{(1)}_{kj} = \frac{ieA_0}{2c\hbar}[\langle k| \mathcal{H}_0 x |j\rangle - \langle k| x \mathcal{H}_0 |j\rangle]$$

$$= \frac{ieA_0}{2c\hbar}[\langle \mathcal{H}_0 k| x |j\rangle - \langle k| x |\mathcal{H}_0 j\rangle]$$

$$= \frac{ieA_0}{2c\hbar}(E^{(0)}_k - E^{(0)}_j)\langle k| x |j\rangle$$

$$= \frac{iA_0\omega}{2c}\langle k| ex |j\rangle$$

$$= \frac{iA_0\omega}{2c}\langle D_x\rangle_{kj},$$

where we have assumed the resonance condition $\hbar\omega = E^{(0)}_k - E^{(0)}_j$. Since the energy density is

$$I(\omega) = \frac{\omega^2 A_0^2}{8\pi c^2},$$

we write Equation 9.60 as

$$R_k = \frac{4\pi^2}{\hbar} I(\omega) \, |\langle D_x \rangle_{kj}|^2. \tag{9.69}$$

For an unpolarized wave, the average of $\cos^2 \theta$ over solid angle yields a factor of $\frac{2}{3}$ which makes Equation 9.69 consistent with Equation 9.62.

Returning to Equation 9.67, note that the next term of the expansion introduces an additional factor of $i\vec{k} \cdot \vec{r}$ which multiplies the quantity $(\vec{A}_0 \cdot \vec{p})$. These products are components of a second-rank tensor, some of which contain orbital angular momentum while others contain charge times quadratic terms in the coordinates.[5] The former correspond to magnetic dipole moments, while the latter are called *electric quadrupole moments*.

Each higher power of $i\vec{k} \cdot \vec{r}$ introduces higher-ordered multipoles and a corresponding new set of selection rules. We will not discuss them further here except to remark that the appearance of a spectral line which is forbidden in the dipole approximation merely means that other kinds of poles are playing a role in the physical process. For example, the selection rules for electric quadrupole radiation are:

Neglecting spin: $\Delta \ell = 0, \pm 2$

$\qquad\qquad\qquad\Delta m = 0, \pm 1, \pm 2$

Including spin: $\Delta j = 0, \pm 1, \pm 2$

$\qquad\qquad\qquad\Delta m = 0, \pm 1, \pm 2$

$\qquad\qquad$ No parity change

For magnetic dipole transitions the selection rules are:

Neglecting spin: $\Delta \ell = 0$

$\qquad\qquad\qquad\Delta m = 0, \pm 1$

Including spin: $\Delta j = 0$

$\qquad\qquad\qquad\Delta m = 0, \pm 1$

A further example of a harmonic perturbation is provided by the phenomenon of magnetic resonance. Here the magnetic moment operator produces resonant absorption by means of *magnetic dipole transitions*. In a typical experiment the magnetic levels are split by a uniform fixed field, and transitions

---

[5] J. D. Jackson, *Classical Electrodynamics*, John Wiley and Sons, N.Y., 1962, section 9.3; W. K. H. Panofsky and M. Phillips, *Classical Electricity and Magnetism*, Addison-Wesley Publishing Co., Reading, Mass., 1955, section 13-7.

between these levels are induced by an oscillating field. The frequency of the oscillating field is generally in the radio frequency band for nuclear spin resonance, in the microwave band for electron spin resonance, and in the optical band for electronic orbital transitions.

The magnetic term in the Hamiltonian is,

$$(\vec{\mathscr{B}}_0 + 2\vec{\mathscr{B}}_1 \cos \omega t) \cdot \frac{g\mu_B}{\hbar} \vec{J} = g\mu_B m_j \mathscr{B}_0 + \frac{2g\mu_B}{\hbar} \vec{\mathscr{B}}_1 \cdot \vec{J} \cos \omega t, \qquad (9.70)$$

where we have taken $\vec{\mathscr{B}}_0$ to be in the z-direction. For a fixed value of $\mathscr{B}_0$, we already know that the solutions for the Hamiltonian

$$\mathscr{H}_0 + g\mu_B m_j \mathscr{B}_0 \qquad (9.71)$$

are the kets $|j, m_j\rangle$. Therefore, we will now regard Equation 9.71 as the unperturbed Hamiltonian and the kets $|j, m_j\rangle$ as the unperturbed wave functions.

It should be pointed out that Equation 9.70 assumes that g is a scalar quantity. This is a good assumption for free or nearly free atoms, but is not generally valid for ions or atoms in a crystalline lattice. In solids it frequently turns out that the g factor is a tensor, and hence the perturbation must be written as

$$\mathscr{H}^{(1)} = \frac{\mu_B}{\hbar} \vec{\mathscr{B}}_1 \cdot \overleftrightarrow{g} \cdot \vec{J}.$$

However, for simplicity we will regard g as a scalar in the present example. Let us further assume that the oscillating field $\vec{\mathscr{B}}_1$ is in the x-direction. Then the perturbation becomes

$$\mathscr{H}^{(1)} = \frac{g\mu_B}{\hbar} \mathscr{B}_1 \mathbf{J}_x = \frac{g\mu_B}{2\hbar} \mathscr{B}(\mathbf{J}_+ + \mathbf{J}_-). \qquad (9.72)$$

The transition rate from the jth to the kth state at resonance is given by Equation 9.58 as

$$R_k = \frac{2\pi}{\hbar} \left(\frac{g\mu_B \mathscr{B}_1}{2\hbar} \right)^2 [\langle k| \mathbf{J}_+ |j\rangle + \langle k| \mathbf{J}_- |j\rangle]^2,$$

where $\mathbf{J}_+$ governs the absorption process and $\mathbf{J}_-$ the emission. The rate of absorption in the state $|j, m\rangle$ is

$$\frac{\pi}{2\hbar} (g\mu_B \mathscr{B}_1)^2 (j - m)(j + m + 1),$$

and the rate of induced emission is

$$\frac{\pi}{2\hbar} (g\mu_B \mathscr{B}_1)^2 (j + m)(j - m + 1).$$

Of course, the *number* of transitions per second must take into account the populations of the states as discussed in section 4 of Chapter 2.

PROBLEM 9-23

> Show that the transition rate from state $|j, m\rangle$ to $|j, m + 1\rangle$ is equal to the rate from $|j, m + 1\rangle$ to $|j, m\rangle$.

7. ADIABATIC AND SUDDEN PERTURBATIONS

If a perturbation is varied slowly with time so that the physical system always has adequate time to respond to minute changes in the applied forces, then the system will remain in its initial state. *During such an adiabatic process, both the wave function and the energy of the system will change gradually from the values associated with the nth state of the original Hamiltonian to those of the nth state of the final Hamiltonian.* This is exemplified during the tuning of a violin when the string tension is slowly changed while a certain harmonic is bowed. The harmonic used corresponds to a given oscillator state, and the change of pitch is related to the change in the energy of the state. Another example is provided by the elastic collisions of low-energy gas molecules. Since the electronic motions are very rapid compared to the molecular motions, the distortions produced during the collision process may be regarded as adiabatic and the electronic wave functions are restored to their original form. On the other hand, a high-speed molecule can induce a transition to an excited state, and the collision process will no longer be elastic.

The analytical form for the adiabatic approximation may be obtained by integrating Equation 9.47. Since $\mathcal{H}_{kj}^{(1)}$ is now a function of the time, we must integrate by parts:

$$a_k = -\frac{\mathcal{H}_{kj}^{(1)}}{\hbar\omega_{kj}} e^{i\omega_{kj}t} + \frac{1}{\hbar\omega_{kj}} \int_0^t \frac{d}{dt}(\mathcal{H}_{kj}^{(1)}) e^{i\omega_{kj}t}\, dt. \tag{9.73}$$

The second term may be made arbitrarily small in the adiabatic approximation, so that the first term gives the required result. Note that this result would be identical with Equation 9.9 for a constant perturbation if the time factors were included in the wave functions used in section 1 of this chapter.

Now suppose that a system undergoes a sudden change at time $t = 0$ such that its Hamiltonian is given by $\mathcal{H}_0$ for $t < 0$ and by $\mathcal{H} = \mathcal{H}_0 + \mathcal{H}^{(1)}$ for $t > 0$, where

$$\mathcal{H}_0 |i\rangle = E_i |i\rangle,$$

and

$$\mathcal{H} |\eta\rangle = E_\eta |\eta\rangle.$$

That is, the orthonormal eigenfunctions of the original Hamiltonian are designated by Roman letters, whereas the orthonormal eigenfunctions of the new Hamiltonian are designated by Greek letters. Then a general wave function may be expressed as,

$$\psi = \sum_i a_i |i\rangle e^{-(i/\hbar)E_i t}, \quad \text{for} \quad t < 0,$$

$$\psi = \sum_\eta b_\eta |\eta\rangle e^{-(i/\hbar)E_\eta t}, \quad \text{for} \quad t > 0.$$

Since the wave function must be continuous at $t = 0$ (although the derivative will not be in this case), we have

$$\sum_i a_i |i\rangle = \sum_\eta b_\eta |\eta\rangle.$$

Multiplying by $\langle\eta|$ we obtain the sudden approximation:

$$b_\eta = \sum_i a_i \langle\eta | i\rangle. \tag{9.74}$$

Equation 9.74 would give the correct expansion coefficients for an instantaneous perturbation. In a physical problem, the system will require a finite time Δt in order to make the transformation from the state characterized by $\mathcal{H}_0$ to that described by $\mathcal{H}$. In such a case,

$$b_\eta = \sum_i a_i \langle\eta | i\rangle e^{-(i/\hbar)(E_i - E_\eta)t}$$

$$\approx \sum_i a_i \langle\eta | i\rangle \left[1 - \frac{i}{\hbar}(E_i - E_\eta)\Delta t\right], \quad \text{for} \quad \Delta t \ll 1, \tag{9.75}$$

where the exponential has been replaced by the first two terms of its power series expansion. Comparing Equations 9.74 and 9.75, it is evident that the error introduced by the sudden approximation is proportional to Δt and to the energy difference between the initial and final states.

As an example, consider the harmonic oscillator whose spring constant is suddenly reduced to one-half its original value. We wish to know whether this oscillator is likely to make a transition or whether it will remain in its original state. For simplicity assume that the oscillator was initially in its ground state, given by

$$\psi_{i=0} = \left(\frac{\alpha}{\pi}\right)^{\frac{1}{4}} e^{-\alpha x^2/2},$$

where $\alpha = \sqrt{km}/\hbar$. The ground state wave function associated with the new spring constant may be written by replacing α with $\alpha/\sqrt{2}$,

$$\psi_{\eta=0} = \left(\frac{\alpha}{\pi\sqrt{2}}\right)^{\frac{1}{4}} e^{-\alpha x^2/2\sqrt{2}}.$$

Then, from Equation 9.74,

$$b_0 = \langle \psi_{\eta=0} \mid \psi_{i=0} \rangle = \left(\frac{1}{2}\right)^{\frac{1}{8}} \left(\frac{2\sqrt{2}}{1 + \sqrt{2}}\right)^{\frac{1}{2}} \sim 0.99,$$

and the probability that the oscillator will remain in its ground state is proportional to $|b_0|^2 \sim 0.98$. Note that all odd excited states are excluded by conservation of parity. Since the most likely transition would be to the state for which $\eta = 2$, let us calculate the coefficient b_2:

$$b_2 = \langle \psi_{\eta=2} \mid \psi_{i=0} \rangle = \left(\frac{1}{4\sqrt{2}}\right)^{\frac{1}{4}} \left(\frac{\alpha}{\pi}\right)^{\frac{1}{2}} \int_{-\infty}^{\infty} (\sqrt{2}\, \alpha x^2 - 1) \exp\left[-\frac{\alpha(2 + \sqrt{2})}{4}\right] x^2\, dx$$

$$\sim -0.12.$$

Since $|b_2|^2$ is of the order of 0.01, the probability for a transition to the second state is roughly 1 per cent of the probability for the oscillator to remain in the ground state.

PROBLEM 9-24

Calculate the probability that the oscillator discussed above will make a transition to the state for $\eta = 4$. Be sure that ψ_4 is normalized.

PROBLEM 9-25

A harmonic oscillator is slowly displaced at constant velocity so that its equilibrium point is moved from $x = 0$ to $x = x_1$. If the oscillator was initially in its ground state, calculate the coefficient pertaining to a transition to the first excited state.

PROBLEM 9-26

For the oscillator of the previous problem, make the same calculation using the sudden approximation.

PROBLEM 9-27

Use the sudden approximation to calculate the probability that the $1s$ electron of tritium will remain in the $1s$ state of He_3^+ when tritium decays by beta emission.

PROBLEM 9-28

A particle of mass m is in the ground state of an infinitely deep, one-dimensional, square well.

(a) If the wall separation is suddenly doubled, what is the probability that the particle will remain in the ground state?

(b) If the wall separation is suddenly halved, what is the probability that the particle will remain in the ground state?

8. THE VARIATIONAL METHOD

It frequently happens that a physical system which cannot be solved directly is also not amenable to the perturbation method. This may be due to the fact that the system either does not resemble a system which can be solved exactly, or that the perturbation term is too large in relation to the exactly soluble part of the Hamiltonian. In such cases it is possible to apply a variational principle to obtain an upper bound for each of the lowest states of the system.

Suppose that the system has a non-degenerate set of eigenvalues, $E_0 < E_1 < E_2 < \ldots$, corresponding to an orthonormal set of eigenfunctions $|i\rangle$ such that

$$\mathcal{H} |i\rangle = E_i |i\rangle.$$

Then an arbitrary function ψ may be expressed by the superposition,

$$\psi = \sum_i a_i |i\rangle,$$

and the expectation value of the energy is

$$\langle \mathcal{H} \rangle = \frac{\langle \psi | \mathcal{H} | \psi \rangle}{\langle \psi | \psi \rangle} = \frac{\sum_i |a_i|^2 E_i}{\sum_i |a_i|^2}. \tag{9.76}$$

If each eigenvalue on the right is replaced by the *lowest* eigenvalue, E_0, we obtain the inequality,

$$\langle \mathcal{H} \rangle \geq \frac{E_0 \sum_i |a_i|^2}{\sum_i |a_i|^2} = E_0. \tag{9.77}$$

Thus, Equation 9.76 provides an *upper bound* on the ground state of the system, although not a very useful one. If Equation 9.76 can be *minimized*, however, then the upper bound so obtained might well be a good approximation to the actual ground state energy. The usefulness of the method lies in the fact that a *trial wave function* ψ can be chosen such that it contains one or more variational parameters for the minimization procedure. The effectiveness of the method is dependent upon a judicious choice of the trial function. After the ground state energy and wave function are obtained, the method can be applied to the first excited state by choosing a trial function that is orthogonal to the ground state function.

For the purpose of illustrating the method, let us apply it to a problem whose solution we already know, that of finding the ground state energy and wave function for the harmonic oscillator. Choosing a trial function of the form

$$\psi = Ce^{-vx^2},$$

we wish to minimize the expectation value of the energy with respect to v. First, normalizing ψ we have:

$$1 = \langle \psi \mid \psi \rangle = |c|^2 \sqrt{\frac{\pi}{2v}}, \quad \text{or} \quad c = \left(\frac{2v}{\pi}\right)^{\frac{1}{4}}.$$

Then, with

$$\mathcal{H} = -\frac{\hbar^2}{2m}\frac{d^2}{dx^2} + \tfrac{1}{2}kx^2,$$

$$\langle \mathcal{H} \rangle = \langle \psi | \mathcal{H} | \psi \rangle = \frac{k}{8v} + \frac{\hbar^2 v}{2m}. \tag{9.78}$$

Letting

$$\frac{\partial}{\partial v}\langle \mathcal{H} \rangle = 0,$$

$$v^2 = \frac{km}{4\hbar^2}$$

or

$$v = \frac{\sqrt{km}}{2\hbar} = \frac{\alpha}{2},$$

where α is defined as in Chapter 5. Thus the ground state wave function is

$$\psi_0 = \left(\frac{\alpha}{\pi}\right)^{\frac{1}{4}} e^{-\alpha x^2/2},$$

and the ground state energy from Equation 9.78 is

$$E_0 = \langle \mathcal{H} \rangle_{\min} = \frac{k}{4\alpha} + \frac{\hbar^2 \alpha}{4m} = \frac{\hbar \omega}{2}.$$

Because of our choice of the trial function, we obtained the exact solution in this case.

PROBLEM 9-29

Calculate an upper bound on the ground state of hydrogen by using a Gaussian trial function of the form

$$\psi = Ce^{-\beta r^2},$$

where β is the variational parameter.

PROBLEM 9-30

Use the variational method to calculate the lowest energy state of the anharmonic oscillator whose potential energy is $V = \frac{1}{2}\omega^2 m x^2 + bx^4$. Take the trial function as $\psi = \psi_0 + \beta\psi_2$, where ψ_0 and ψ_2 are oscillator eigenfunctions. Let both ω and β be variational parameters.

PROBLEM 9-31

Use the variational method to estimate the ground state energy of a particle of mass m in the potential $V = Cx$, where $x > 0$. The particle is confined to the semi-infinite space defined by $0 \le x \le +\infty$. Take $\psi = xe^{-\alpha x}$ as the trial function.
[Ans.: Upper bound for E_0 is $\frac{9}{4}(2\hbar^2 C^2/3m)^{\frac{1}{3}}$.]

9. THE JWKB SEMICLASSICAL APPROXIMATION

It was seen in Chapters 4 and 5 that the motion of a particle in a region having a constant potential can be represented by a superposition of plane waves traveling to the right and to the left. Each such wave has the form

$$\psi \sim \exp(\pm ikx),$$

where

$$\hbar k = \sqrt{2m(E - V)}.$$

A similar solution may often be used to approximate the real solution in physical cases where the potential is not constant, provided that the potential is *slowly varying* in space. This approximation is quite useful in one dimensional problems and in problems with radial symmetry. Although it has been credited to a number of people, the method is usually called the *WKB method* or the *JWKB method*, after Jeffreys, Wentzel, Kramers and Brillouin[6] who first applied the mathematical techniques to physical problems.

By a "slowly varying potential" we mean a potential whose energy change in a de Broglie wavelength is very much smaller than the kinetic energy of the particle. Thus, the fractional change in kinetic energy per wavelength is much less than one:

$$\frac{1}{(E - V)} \left| \frac{\partial V}{\partial x} \right| \cdot \lambda \ll 1 \qquad (9.79)$$

or

$$\left| \frac{2\hbar}{p^2} \frac{\partial p}{\partial x} \right| \ll 1.$$

This is analogous to an optical medium in which the index of refraction varies continuously, but sufficiently slowly so that no reflection occurs.

Let us designate the slowly-varying potential by $S(x)$ and require that $S(x) = px$ when V is constant. We will assume that $S(x)$ may be represented by an asymptotic expansion in powers of $\hbar$ as follows:[7]

$$S(x) = S_0(x) + \hbar S_1(x) + \frac{\hbar^2}{2} S_2(x) + \cdots, \qquad (9.80)$$

where we also assume that only a few terms will be required.

Writing the wave function as,

$$\psi \sim \exp\left(\frac{iS}{\hbar}\right),$$

we find upon substituting ψ into the Schrödinger equation,

$$\frac{1}{2m} (S')^2 + (V - E) - \frac{i\hbar}{2m} S'' = 0, \qquad (9.81)$$

where the primes denote differentiation with respect to x. Substituting Equation 9.80 into 9.81:

$$\frac{1}{2m} (S_0')^2 + \frac{\hbar^2}{2m} (S_1')^2 + \cdots + \frac{\hbar}{2m} (2S_0'S_1' + \hbar S_0'S_2' + \cdots)$$

$$+ (V - E) - \frac{i\hbar}{2m} (S_0'' + \hbar S_1'' + \frac{\hbar^2}{2} S_2'' + \cdots) = 0.$$

---

[6] H. Jeffreys, *Proc. London Math. Soc.* (2) **23**, 428 (1923); G. Wentzel, *Z. Physik* **38**, 518 (1926); H. A. Kramers, *Z. Physik* **39**, 828 (1926); L. Brillouin, *Compt. rend.* **183**, 24 (1926) and *J. phys. et radium* **7**, 353 (1926).

[7] This treatment will follow that of D. Bohm, *Quantum Theory*, Prentice-Hall, Inc., Englewood Cliffs, N.J., 1951, Chapter 12. Also, see A. Messiah, *Quantum Mechanics*, North-Holland Publishing. Co., Amsterdam, 1958, Chapter 6.

Combining like terms and setting the coefficients of each power of $\hbar$ separately equal to zero, we obtain the following equations:

$$\frac{1}{2m}(S_0')^2 + V - E = 0 \tag{9.82}$$

$$S_0'S_1' - \frac{i}{2}S_0'' = 0 \tag{9.83}$$

$$S_0'S_2' + (S_1')^2 - iS_1'' = 0, \tag{9.84}$$

where all terms beyond the quadratic terms in $\hbar$ have been dropped. Rewriting Equation 9.82 as

$$dS_0 = \sqrt{2m(E - V)}\,dx,$$

its solution may be written as

$$S_0 = \pm \int_{x_0}^{x} \sqrt{2m(E - V)}\,dx. \tag{9.85}$$

In a similar fashion,

$$S_1 = \frac{i}{2}\ln(S_0')$$

or

$$\exp(iS_1) = [2m(E - V)]^{-\frac{1}{4}},$$

and

$$S_2 = \frac{m}{2}\frac{\partial V}{\partial x}\cdot[2m(E - V)]^{-\frac{3}{2}} - \frac{m^2}{2}\int[2m(E - V)]^{-\frac{5}{2}}\left(\frac{\partial V}{\partial x}\right)^2 dx.$$

If V is a slowly-varying function of x so that Equation 9.79 is satisfied and all higher derivatives of V are also small, then it usually is possible to retain only the first two terms of the expansion given by Equation 9.80. Then the wave function may be written as

$$\psi = [E - V]^{-\frac{1}{4}}(Ae^{iS_0} + Be^{-iS_0}), \tag{9.86}$$

where S_0 is expressed by Equation 9.85.

The reader will note that ψ, as given by Equation 9.86, will not be valid as $(E - V)$ approaches zero. This creates serious difficulties since one of the principal applications of the JWKB method is to problems involving barriers, where it is important to match wave functions in two regions. Thus, for the potential shown in Figure 9-1, a classical particle with energy E would have a turning point at $x = a$, where $E - V = 0$. In the quantum mechanical description, however, this particle will have an oscillating wave function for $x < a$ and an exponentially decaying function for $x > a$. The total wave function requires the matching of these two functions at the very point where the JWKB method fails!

The way out of this dilemma is to exclude the two functions from the immediate vicinity of $x = a$ and to construct a *connection formula* to join $\psi_{E>V}$

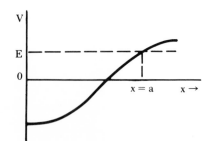

Figure 9-1 Schematic one-dimensional potential. E is the total energy of a particle.

and $\psi_{E<V}$. If the region where the JWKB solutions break down is small enough, the potential can be represented by a straight line within the region. Taking its slope to be that of the potential curve at $x = 0$, the equation of the line is

$$V - E = \left(\frac{\partial V}{\partial x}\right)_{x=a} (x - a).$$

Then the Schrödinger equation in this region can be solved with the aid of Bessel's functions of order $\frac{1}{3}$. The details of deriving the connection formulas will not be given here,[8] but for a simple barrier they may be summarized as follows.

(1) BARRIER TO THE RIGHT.

$$\overbrace{\frac{2}{\sqrt{k_1}} \cos\left(\int_x^a k_1\, dx - \frac{\pi}{4}\right)}^{\psi_{E>V}} \leftrightarrows \overbrace{\frac{1}{\sqrt{k_2}} \exp\left(-\int_a^x k_2\, dx\right)}^{\psi_{E<V}} \qquad (9.87)$$

$$\frac{1}{\sqrt{k_1}} \sin\left(\int_x^a k_1\, dx - \frac{\pi}{4}\right) \leftrightarrows -\frac{1}{\sqrt{k_2}} \exp\left(\int_a^x k_2\, dx\right) \qquad (9.88)$$

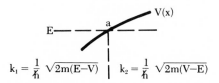

$$k_1 = \frac{1}{\hbar}\sqrt{2m(E-V)} \quad \Big| \quad k_2 = \frac{1}{\hbar}\sqrt{2m(V-E)}$$

(2) BARRIER TO THE LEFT.

$$\overbrace{\frac{1}{\sqrt{k_2}} \exp\left(-\int_x^b k_2\, dx\right)}^{\psi_{E<V}} \rightleftarrows \overbrace{\frac{2}{\sqrt{k_1}} \cos\left(\int_b^x k_1\, dx - \frac{\pi}{4}\right)}^{\psi_{E>V}} \qquad (9.89)$$

$$\frac{1}{\sqrt{k_2}} \exp\left(\int_x^b k_2\, dx\right) \leftrightarrows -\frac{1}{\sqrt{k_1}} \sin\left(\int_b^x k_1\, dx - \frac{\pi}{4}\right) \qquad (9.90)$$

---

[8] See E. C. Kemble, *The Fundamental Principles of Quantum Mechanics*, McGraw-Hill Book Co., New York, 1937, section 21; N. Fröman and P. O. Fröman, *JWKB Approximation, Contributions to the Theory*, North-Holland Publishing Co., Amsterdam, 1965; P. M. Morse and H. Feshbach, *Methods of Theoretical Physics*, McGraw-Hill Book Co., New York, 1953, pp. 1092 ff.

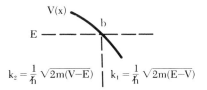

We will illustrate the application of the connection formulas by considering the one-dimensional barrier shown in Figure 9-2. A particle of energy E will have the classical turning points indicated by a and b. For simplicity we will assume that the barrier is symmetric so that the momentum of the particle is given by $k_1 = (1/\hbar)\sqrt{2m(E - V)}$ in both regions I and III. In order for the connection formulas to be valid, the barrier must be high enough and thick enough so that $\int_a^b k_2\, dx = (1/\hbar)\int_a^b \sqrt{2m(V - E)} \gg 1$. If particles are incident only from the left, the solution in region III will consist simply of a wave traveling to the right, having the form

$$\psi_{III} \sim e^{i\int_b^x k_1 dx},$$

where

$$k_1 = \frac{1}{\hbar}\sqrt{2m(E - V)}.$$

For convenience later in applying the connection formulas, we will add a phase factor to ψ_{III}; thus,

$$\psi_{III} = \frac{A}{\sqrt{k_1}}\, e^{i\left(\int_b^x k_1 dx - (\pi/4)\right)}. \tag{9.91}$$

Since A is complex, it can readily absorb the factor $e^{i\pi/4}$. Equation 9.91 may be expanded in the form

$$\psi_{III} = \frac{A}{\sqrt{k_1}}\left[\cos\left(\int_b^x k_1\, dx - \frac{\pi}{4}\right) + i \sin\left(\int_b^x k_1\, dx - \frac{\pi}{4}\right)\right].$$

Using connection formulas 9.89 and 9.90, we obtain for ψ_{II},

$$\psi_{II} = \frac{A}{\sqrt{k_2}}\left[\tfrac{1}{2}\exp\left(-\int_x^b k_2\, dx\right) - i \exp\left(\int_x^b k_2\, dx\right)\right]. \tag{9.92}$$

In order to find ψ_I we must use the connection formulas for the barrier to the

Figure 9-2 A symmetric one-dimensional barrier.

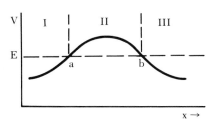

right. First, let us rewrite Equation 9.92 in the following convenient form.

$$\psi_{II} = \frac{A}{\sqrt{k_2}} \left[\tfrac{1}{2} \exp\left(-\int_x^a k_2\,dx - \int_a^b k_2\,dx\right) - i \exp\left(\int_x^a k_2\,dx + \int_a^b k_2\,dx\right) \right]$$

$$= \frac{A}{\sqrt{k_2}} \left[\tfrac{1}{2} \exp\left(-\int_a^b k_2\,dx\right) \cdot \exp\left(\int_a^x k_2\,dx\right) \right.$$

$$\left. - i \exp\left(\int_a^b k_2\,dx\right) \cdot \exp\left(-\int_a^x k_2\,dx\right) \right].$$

Applying connection formulas 9.87 and 9.88, this becomes:

$$\psi_I = -\frac{A}{\sqrt{k_1}} \left[\tfrac{1}{2} \exp\left(-\int_a^b k_2\,dx\right) \cdot \sin\left(\int_x^a k_1\,dx - \frac{\pi}{4}\right) \right.$$

$$\left. + i2 \exp\left(\int_a^b k_2\,dx\right) \cdot \cos\left(\int_x^a k_1\,dx - \frac{\pi}{4}\right) \right]. \tag{9.93}$$

It will be easier to interpret Equation 9.93 if we simplify the notation as follows: Let

$$\alpha = \int_a^b k_2\,dx$$

and

$$\beta = \int_x^a k_2\,dx - \frac{\pi}{4}.$$

Then Equation 9.93 becomes:

$$\psi_I = -\frac{A}{\sqrt{k_1}} \left(\tfrac{1}{2} e^{-\alpha} \sin\beta + i2 e^{\alpha} \cos\beta \right)$$

$$= -\frac{A}{\sqrt{k_1}} \left[\tfrac{1}{2} e^{-\alpha} \cdot \frac{1}{2i} (e^{i\beta} - e^{-i\beta}) + 2i e^{\alpha} \cdot \tfrac{1}{2}(e^{i\beta} + e^{-i\beta}) \right]$$

$$= -\frac{iA}{\sqrt{k_1}} \left[(e^{\alpha} + \tfrac{1}{4} e^{-\alpha}) e^{-i\beta} + (e^{\alpha} - \tfrac{1}{4} e^{-\alpha}) e^{i\beta} \right].$$

Since propagation to the right is represented by negative values of β, the amplitude of the incident wave in region I is

$$\left| \frac{A}{\sqrt{k_1}} (e^{\alpha} + \tfrac{1}{4} e^{-\alpha}) \right|.$$

Then the transmission coefficient may be written:

$$T = \left[\frac{\dfrac{A}{\sqrt{k_1}}}{\dfrac{A}{\sqrt{k_1}}(e^\alpha + \tfrac{1}{4}e^{-\alpha})} \right]^2 = \left[\exp\left(\int_a^b k_2\, dx \right) + \tfrac{1}{4}\exp\left(-\int_a^b k_2\, dx \right) \right]^{-2}.$$

(9.94)

However, since we require $\int_a^b k_2\, dx \gg 1$, the negative exponential can be neglected with respect to the first term in Equation 9.94. Then T is approximately given by

$$T \sim e^{-2\int_a^b k_2\, dx}.$$

For a rectangular barrier of width $2a$ this is approximately $e^{-4k_2 a}$, which is in order-of-magnitude agreement with the approximate result obtained in section 2 of Chapter 5 for the example of tunneling through the nuclear barrier.

The JWKB method can be applied to bound states as well as to scattering problems in unbound systems. Although it has applications in many areas of physics, it is probably fair to say that this method is used only as a last resort; that is, if other methods fail. For this reason, we will not pursue it further here, but will refer the interested reader to the references cited earlier.

SUMMARY

In spite of the fact that very few physical problems in the real world can be solved exactly, methods for approximating complex systems have been developed which are responsible for the widespread application and success of quantum mechanics. The most important and simplest of these approximate methods is perturbation theory, which can be used whenever the actual system can be regarded as a slight modification of a system whose solutions are known. Special techniques are required for the cases of degeneracy and time-dependence of various sorts. These methods are illustrated by numerous examples. When the physical system is not amenable to perturbation theory, a variational method can often be used to obtain an upper bound on the energies of the lowest states. The success of this method hinges upon judicious choices of the trial wave function and the number and kind of variation parameters. A semi-classical method is also described and is applied to a problem in barrier penetration.

SUGGESTED REFERENCES

David Bohm, *Quantum Theory*. Prentice-Hall, Inc., New York, 1951.
Sidney Borowitz, *Fundamentals of Quantum Mechanics*. W. A. Benjamin, Inc., New York, 1967.

R. H. Dicke and J. P. Wittke, *Introduction to Quantum Mechanics*. Addison-Wesley Publishing Co., Inc., Reading, Mass., 1960.

P. A. M. Dirac, *The Principles of Quantum Mechanics*, 3rd ed. Oxford University Press, London, 1947.

R. B. Leighton, *Principles of Modern Physics*. McGraw-Hill Book Co., Inc. New York, 1959.

E. Merzbacher, *Quantum Mechanics*. John Wiley and Sons, Inc., New York, 1961.

Albert Messiah, *Quantum Mechanics*. North-Holland Publishing Co., Amsterdam, 1958.

D. Park, *Introduction to the Quantum Theory*. McGraw-Hill Book Co., Inc., New York, 1964.

L. Pauling and E. B. Wilson, *Introduction to Quantum Mechanics*. McGraw-Hill Book Co., Inc., New York, 1935.

J. L. Powell and B. Crasemann, *Quantum Mechanics*. Addison-Wesley Publishing Co., Inc., Reading, Mass., 1961.

D. S. Saxon, *Elementary Quantum Mechanics*. Holden-Day, Inc., San Francisco, 1968.

Leonard I. Schiff, *Quantum Mechanics*, 3rd. ed. McGraw-Hill Book Co., New York, 1969.

CHAPTER 10

ADDITIONAL APPLICATIONS

I. THE GROUND STATE OF THE HELIUM ATOM

As we have discovered in the previous chapters, quantum mechanics has been very successful in accounting for the ground state and the excited states of the hydrogen atom. This success may be attributed, in the first place, to the fact that the two-body problem is relatively easy to solve exactly. Secondly, refinements to the theory are readily incorporated by means of simple perturbation techniques, as we have seen for the spin-orbit correction, the relativistic effect, the hyperfine interaction, the Stark shift in an electric field, and the Zeeman splittings in a magnetic field.

However, element number two, helium, is not so simple. Since it has two electrons, we are now faced with a three-body problem. Furthermore, the helium nucleus contains four nucleons in contrast with the single proton of the hydrogen nucleus. Because of this additional nuclear structure, the nucleus no longer looks like a point charge and the Coulomb law does not hold rigorously (although it is still a good approximation). There is an additional complication due to the fact that each electron tends to screen some of the nuclear charge "seen" by the other. We must also take into account the strong electrostatic interaction between the two electrons, and since they are indistinguishable particles we must include the quantum mechanical phenomenon of exchange degeneracy discussed in Chapter 8. The electrostatic and exchange effects are large effects, not just refinements like the spin-orbit and spin-spin interactions.

As our point of departure, let us approximate the ground state energy of the helium atom by ignoring all interactions between the electrons, as well as their screening effects upon each other. The Hamiltonian for this system is

$$\mathcal{H} = \left(-\frac{\hbar^2}{2\mu}\nabla_1^2 - \frac{2e^2}{r_1}\right) + \left(-\frac{\hbar^2}{2\mu}\nabla_2^2 - \frac{2e^2}{r_2}\right) = 2\mathcal{H}_0, \qquad (10.1)$$

where $\mathcal{H}_0$ is simply a one-electron Hamiltonian for a hydrogenic atom with $Z = 2$. The ground state energy associated with Equation 10.1 is thus

$$E_1^{(0)} = -2Z^2w_0 = -8w_0 = -108.8 \text{ eV}, \qquad (10.2)$$

where $w_0 = 13.6$ eV, the ionization energy of hydrogen. Since the experimental value for the ground state of helium is -78.62 eV, the reader can readily see how large an error is produced by the effects which have been neglected in Equation 10.1. In spite of the magnitude of the corrections needed, it is an attractive idea to try a perturbation calculation, since the interaction between the electrons is an obvious choice for the perturbing potential. Pursuing this approach, the Hamiltonian becomes,

$$\mathcal{H} = 2\mathcal{H}_0 + \frac{e^2}{r_{12}}, \tag{10.3}$$

where $r_{12} = |\vec{r}_1 - \vec{r}_2|$ is the distance between the two electrons. The unperturbed wave functions are eigenfunctions of Equation 10.1, namely, products of the two one-electron wave functions. Let us designate the unperturbed ground state by *

$$|0\rangle = \psi_{1s}(1) \cdot \psi_{1s}(2) \cdot |s, m_s\rangle = \frac{1}{\pi}\left(\frac{Z}{a_0}\right)^3 e^{-Z(r_1 + r_2)/a_0} |s, m_s\rangle. \tag{10.4}$$

The nuclear charge Z will be carried for greater generality in spite of the fact that $Z = 2$ for this example. The ket $|s, m_s\rangle$ stands for the normalized antisymmetric two-electron spin function given in Equation 8.22. This is the only spin function that can be used for the ground state of helium, since the spatial wave function must be symmetric. Therefore, the ground state of helium is a non-degenerate singlet state. Using the wave function given in Equation 10.4, the expectation value of the energy is

$$\langle 0| \mathcal{H} |0\rangle = -2Z^2 w_0 + \langle 0| \frac{e^2}{r_{12}} |0\rangle, \tag{10.5}$$

where

$$\langle 0| \frac{e^2}{r_{12}} |0\rangle = \frac{e^2}{\pi^2}\left(\frac{Z}{a_0}\right)^6 \left\langle \exp\left[-\frac{Z}{a_0}(r_1 + r_2)\right] \left| \frac{1}{r_{12}} \right| \exp\left[-\frac{Z}{a_0}(r_1 + r_2)\right] \right\rangle$$

$$\langle s, m_s \,|\, s, m_s \rangle$$

$$= \frac{e^2}{\pi^2}\left(\frac{Z}{a_0}\right)^6 \int_{\Omega_1} d\Omega_1 \int_0^\infty dr_1 r_1^2 \exp\left(-\frac{2Zr_1}{a_0}\right) \int_{\Omega_2} d\Omega_2 \int_0^\infty dr_2 \frac{r_2^2}{r_{12}} \exp\left(-\frac{2Zr_2}{a_0}\right). \tag{10.6}$$

For spherically symmetric wave functions of the type used here, this integration may be easily performed by regarding it as the electrostatic energy of two overlapping spherical charge distributions. Thus, we first calculate the potential at a point $\vec{r}_1$ due to the infinite charge distribution defined by $\vec{r}_2$ and having charge density given by $\exp[-(2Zr_2/a_0)]$. Since this potential is constant for $r_2 < r_{12}$ but varies as $1/r_2$ for $r_2 > r_{12}$, we have the two integrals

---

*Although the simple product function is satisfactory for the present discussion, the reader should note that the proper form is the symmetrized spatial function given on page 306.

(except for the constant factor):

$$V_1 = \int_0^{r_1} e^{-2Zr_2/a_0} \frac{1}{r_1} \cdot 4\pi r_2^2 \, dr_2 + \int_{r_1}^{\infty} e^{-2Zr_2/a_0} \frac{1}{r_2} \cdot 4\pi r_2^2 \, dr_2$$

$$V_1 = 4\pi \left(\frac{a_0}{2Z}\right)^2 \left\{\frac{1}{x_1}[2 - e^{-x_1}(x_1^2 + 2x_1 + 2)] + (x_1 + 1)e^{-x_1}\right\},$$

where

$$x_1 = \frac{2Zr_1}{a_0}.$$

Now we calculate the energy of the system by integrating the charge density e^{-x_1} over all of space,

$$\langle 0|\frac{e^2}{r_{12}}|0\rangle = \frac{e^2}{\pi^2}\left(\frac{Z}{a_0}\right)^3 \cdot \int_0^{\infty} e^{-x_1} V_1 4\pi x_1^2 \, dx_1 = \frac{5e^2 Z}{8a_0} = \tfrac{5}{4}Zw_0. \qquad (10.7)$$

PROBLEM 10-1

Verify the integrations leading to Equation 10.7.

A method of integrating Equation 10.6 which can be extended to non-spherical charge distributions is to expand $1/r_{12}$ in spherical harmonics as follows:[1]

$$\frac{1}{r_{12}} = \frac{1}{r_1}\sum_{\ell=0}^{\infty}\left(\frac{r_2}{r_1}\right)^{\ell} P_\ell(\cos\theta), \quad \text{for} \quad r_1 > r_2$$

$$\frac{1}{r_{12}} = \frac{1}{r_2}\sum_{\ell=0}^{\infty}\left(\frac{r_1}{r_2}\right)^{\ell} P_\ell(\cos\theta), \quad \text{for} \quad r_1 < r_2,$$

where θ is the angle between $\vec{r}_1$ and $\vec{r}_2$. Because of the orthonormality of the spherical harmonics, performing the integration over Ω_1 and Ω_2 before the radial integrations will eliminate all of the terms containing P_ℓ except for $\ell = 0$. Then Equation 10.6 becomes:

$$\langle 0|\frac{e^2}{r_{12}}|0\rangle = \frac{e^2}{2}\left(\frac{Z}{a_0}\right)\int_0^{\infty} dx_1 x_1^2 e^{-x_1}\left\{\left[\frac{1}{x_1}\int_0^{x_1} dx_2 x_2^2 e^{-x_2} + \int_{x_1}^{\infty} dx_2 x_2 e^{-x_2}\right]\right\}$$

$$= \frac{5}{8}\frac{e^2 Z}{a_0} = \tfrac{5}{4}Zw_0, \qquad (10.8)$$

in agreement with Equation 10.7.

[1] L. I. Schiff, *Quantum Mechanics*, 3rd ed. McGraw-Hill Book Co., N.Y., 1969, p. 258.

Verify Equation 10.8.

The corrected ground state energy, Equation 10.5, is now

$$E_{\text{ground}} = -2Z^2 w_0 + \frac{5}{4} Z w_0 = -\frac{11}{2} w_0 = -74.8 \text{ eV}, \qquad (10.9)$$

for $Z = 2$. This is in surprisingly good agreement with the experimental value of -78.62 eV, particularly since perturbation theory would not be expected to be valid when the perturbation term and the initial term are of the same order of magnitude.

Let us now attempt to take into account the effect of the partial screening of the nucleus by the second electron. We will proceed by treating the nuclear charge Z as a variational parameter in the minimization of the ground state energy. Using Equation 10.4 as the normalized trial wave function, the expectation value of the ground state energy is

$$\langle E_{\text{ground}} \rangle = \langle 0 | \, 2 \mathcal{H}_0 + \frac{e^2}{r_{12}} | 0 \rangle$$

$$= 2 \langle 0 | \, \mathcal{H}_0 | 0 \rangle + \langle 0 | \frac{e^2}{r_{12}} | 0 \rangle.$$

The second integral was just calculated above, and the first will be left as a problem. The result is

$$\langle E_{\text{ground}} \rangle = 2(Z^2 - 4Z) w_0 + \tfrac{5}{4} Z w_0. \qquad (10.10)$$

Minimizing Equation 10.10 by setting $(d/dZ)\langle E_{\text{ground}} \rangle = 0$, we obtain an effective nuclear charge of $Z = \tfrac{27}{16}$ for the screened helium nucleus. Using this value in Equation 10.10 the upper bound on E_{ground} becomes $-5.7 w_0 = -77.5$ eV, which is in good agreement with the experimental value of -78.62 eV.

It is also possible to make an estimate of the first ionization energy of helium from the difference in the energies of the $1s^2$ and the $1s$ configurations. This amounts to

$$5.7 w_0 - 4 w_0 = 1.7 w_0,$$

whereas the experimental value is $1.81 w_0$.

Show that

$$\langle 0 | \, \mathcal{H}_0 | 0 \rangle = (Z^2 - 4Z) \cdot \frac{e^2}{2 a_0}, \qquad (10.11)$$

where $|0\rangle$ is given by Equation 10.4 and $\mathcal{H}_0$ is given by Equation 10.1.

PROBLEM 10-4

(a) Apply the model used for helium to singly ionized lithium in the $1s^2$ configuration and set up the integrals. Note whether any are the same as in the helium problem.

(b) Show that $\langle 0| \mathcal{H}_0 |0\rangle = (Z^2 - 6Z)w_0$ in this case.

(c) Find an upper bound on the ground state of Li^+. (Ans.: $-14.4w_0$.)

PROBLEM 10-5

The first ionization energy of lithium can be calculated by finding the difference in energy for the configurations $1s^2$ and $1s^2 2s$. Choose a product wave function for the $1s^2 2s$ configuration, set up the integrals, and explain the procedure to be followed. Do not evaluate the integrals.

PROBLEM 10-6

Find the first ionization energy for Li^+. (This is the energy difference between the ground states of Li^+ and Li^{++}.) The experimental value is $5.56w_0$.

PROBLEM 10-7

Find the first ionization energy for Be^{++}. The experimental value is $11.31w_0$.

2. THE LOWEST EXCITED STATES OF HELIUM

The lowest excited states of helium are those for which one electron remains in the $1s$ state and the other electron is raised to a $2s$ or $2p$ state. Since the electrons are indistinguishable, for each pair of quantum states there will be four possible spin wave functions as given in Equation 8.22. There are also four two-electron spatial quantum states represented by n and n', namely, $\psi_{100}\psi_{200}$, $\psi_{100}\psi_{210}$, $\psi_{100}\psi_{211}$, and $\psi_{100}\psi_{21\bar{1}}$. Neglecting all interactions, then, there are 16 degenerate wave functions representing the first excited state of helium. In reality, of course, we cannot turn off the Coulomb interactions, so that the 16-fold degeneracy mentioned above does not physically exist. It turns out that the four states associated with $\psi_{1s}\psi_{2s}$ are not raised as much as

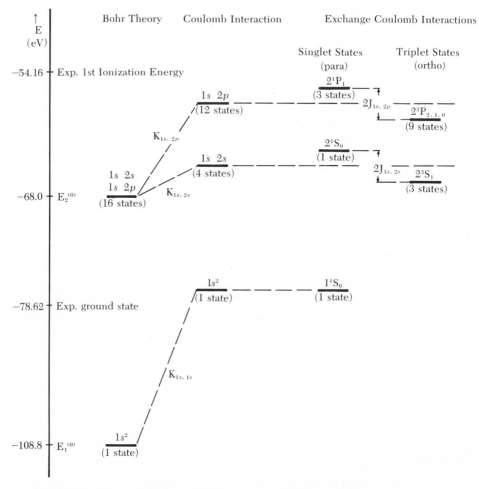

Figure 10-1 Schematic diagram of the lowest states of helium. The spin-orbit interaction has been neglected since it is only $\sim 10^{-4}$ eV. Degeneracies of levels are given in parentheses.

the twelve states associated with the $\psi_{1s}\psi_{2p}$ wave functions. This is shown in Figure 10-1. The remaining degeneracies are split further by spin-orbit interactions and the correlation effects arising jointly from symmetry and the Coulomb repulsion as discussed in sections 5 and 6 of Chapter 8. In order to see this more clearly, let us look in detail at a perturbation treatment of the four degenerate wave functions formed from $\psi_{1s}\psi_{2s}$. We write ψ as,

$$\psi = \frac{1}{\sqrt{2}}\left[\psi_{1s}(1)\psi_{2s}(2) \pm \psi_{2s}(1)\psi_{1s}(2)\right] \cdot |s, m_s\rangle, \qquad (10.12)$$

where the plus sign refers to the singlet state and the minus sign to the triplet state. The ket $|s, m_s\rangle$ can be any one of the three symmetric spin functions when the minus sign is used; for the plus sign, the ket $|s, m_s\rangle$ must be the one antisymmetric spin function. It will not be necessary to express the spin functions exactly because the Coulomb perturbation is independent of the spin coordinates. Then the first order correction to the energy of the first excited state of helium is

$$E_2^{(1)} = \langle \psi | \frac{e^2}{r_{12}} | \psi \rangle$$

$$= \frac{e^2}{2} \langle \psi_{1s}(1)\psi_{2s}(2) \pm \psi_{2s}(1)\psi_{1s}(2) | \frac{1}{r_{12}} | \psi_{1s}(1)\psi_{2s}(2) \pm \psi_{2s}(1)\psi_{1s}(2) \rangle \langle s, m_s | s, m_s \rangle$$

$$= K \pm J, \tag{10.13}$$

where

$$K = e^2 \langle \psi_{1s}(1)\psi_{2s}(2) | \frac{1}{r_{12}} | \psi_{1s}(1)\psi_{2s}(2) \rangle,$$

$$\tag{10.14}$$

$$J = e^2 \langle \psi_{1s}(1)\psi_{2s}(2) | \frac{1}{r_{12}} | \psi_{1s}(2)\psi_{2s}(1) \rangle.$$

The integral K is the *direct or ordinary Coulomb integral* and the integral J is called the *exchange Coulomb integral*. Notice that when K is evaluated, the integration over the coordinates of one of the electrons involves only *one* of the functions ψ_{1s} or ψ_{2s}. On the other hand, when J is evaluated each configuration integral involves *both* ψ_{1s} and ψ_{2s} and, in fact, is a measure of the degree of overlap of these functions.

The energy of the first excited state of helium is, then,

$$E_2 = E_2^{(0)} + K \pm J,$$

where the plus sign refers to the singlet state and the minus sign to the triplet state. The calculation of the following energies will be left to the problems:

$$K_{1s,2p} = e^2 \langle 1s(1) \cdot 2p(2) | \frac{1}{r_{12}} | 1s(1) \cdot 2p(2) \rangle \sim 10.0 \text{ eV}$$

$$K_{1s,2s} = e^2 \langle 1s(1) \cdot 2s(2) | \frac{1}{r_{12}} | 1s(1) \cdot 2s(2) \rangle \sim 9.1 \text{ eV}$$

$$\tag{10.15}$$

$$J_{1s,2s} = e^2 \langle 1s(1) \cdot 2s(2) | \frac{1}{r_{12}} | 1s(2) \cdot 2s(1) \rangle \sim 0.4 \text{ eV}$$

$$J_{1s,2p} = e^2 \langle 1s(1) \cdot 2p(2) | \frac{1}{r_{12}} | 1s(2) \cdot 2p(1) \rangle \sim 0.1 \text{ eV}.$$

Note that the energies of both the singlet and triplet states are raised by the direct Coulomb interaction, but that the exchange interaction raises the singlet and lowers the triplet (see Figure 10-1).

Calculate the integrals K and J for the $1s2s$ configuration of helium.

3. THE HEISENBERG EXCHANGE INTERACTION AND MAGNETISM

The *exchange splitting* of Equation 10.13 is given by the quantity,

$$\Delta E_{\text{exch}} = 2J. \tag{10.16}$$

Heisenberg and Dirac have shown[2] that the phenomenological vector coupling model provides a very simple spin Hamiltonian which gives the correct energy splitting for this case. The Hamiltonian is

$$\mathscr{H}_{\text{exch}}^{(1)} = -2J\vec{S}_1 \cdot \vec{S}_2, \tag{10.17}$$

where $\vec{S}_1$ and $\vec{S}_2$ are the spin vectors for the two electrons. Since the total spin is $\vec{S} = \vec{S}_1 + \vec{S}_2$,

$$\vec{S}^2 = (\vec{S}_1 + \vec{S}_2)^2 = S_1^2 + S_2^2 + 2\vec{S}_1 \cdot \vec{S}_2,$$

and

$$2\vec{S}_1 \cdot \vec{S}_2 = s(s+1) - s_1(s_1+1) - s_2(s_2+1) = s(s+1) - \tfrac{3}{2}.$$

Now, for the singlet state, where the spins are antiparallel,

$$(\vec{S}_1 \cdot \vec{S}_2)_{\text{singlet}} = -\tfrac{3}{4}$$

and for the triplet state where the spins are parallel,

$$(\vec{S}_1 \cdot \vec{S}_2)_{\text{triplet}} = \tfrac{1}{2}(2 - \tfrac{3}{2}) = \tfrac{1}{4}.$$

Then,

$$\Delta E_{\text{exch}} = -2J[(\vec{S}_1 \cdot \vec{S}_2)_{\text{singlet}} - (\vec{S}_1 \cdot \vec{S}_2)_{\text{triplet}}] = -2J \cdot (-\tfrac{3}{4} - \tfrac{1}{4}) = 2J,$$

in agreement with Equation 10.16. Thus the vector model gives the correct energy difference between the singlet and triplet states, and says that the lower energy state is achieved by parallel spins when J is positive. It should be evident that this phenomenon, applied here to a simple two-electron atom, is the clue to the mystery surrounding the Hund rule for parallel spins in many-electron atoms (see section 8 of Chapter 8). This, in turn, accounts for the presence of permanent magnetic moments due to uncompensated spins

---

[2] W. Heisenberg, *Z. f. Physik* **49**, 619 (1928); P. A. Dirac, *Proc. Roy. Soc.* A **123**, 714 (1929). See J. H. Van Vleck, *The Theory of Electric and Magnetic Susceptibilities*, Oxford University Press, Oxford, 1932, Chapter 12.

in certain atoms of the transition metal and rare earth series of the periodic table. Moreover, it is now generally believed that this same exchange mechanism is the origin of the intense internal field postulated by Weiss to account for the parallel ordering of the spins in ferromagnets and the antiparallel ordering in antiferromagnets (where the exchange integral J is negative).

A serious defect of Equation 10.17 is that its form is apt to lead one to believe that the exchange energy has its origin in a spin-spin interaction (that is, a dipole-dipole interaction). This certainly is not the case, and it was known even at the time of Weiss that the dipolar forces were three or four orders of magnitude too weak to account for spontaneous magnetic ordering. The reader should bear in mind that the energy difference is due to an *electrostatic interaction* which arises from the symmetry requirements of fermions. The fact that this energy difference can be accounted for by means of a scalar product of spin vectors should be regarded as fortuitous. In spite of this minor shortcoming, the Heisenberg-Dirac-Van Vleck Hamiltonian, Equation 10.17, is of such great importance that the overwhelming majority of papers in the field of magnetism use it as a starting point.

4. THE HARTREE SELF-CONSISTENT METHOD FOR MULTI-ELECTRON ATOMS

In the discussion of perturbation theory it was pointed out that the smaller the perturbation, the better the approximation. Therefore, since the Coulomb repulsion between electrons is so large, we cannot hope to apply perturbation theory alone to atoms with a large number of electrons—in spite of its fortuitous success in the case of helium. On the other hand, the inclusion of the electron interactions in the unperturbed part of the Hamiltonian would make it an insoluble many-body problem.

A way out of this dilemma was suggested by Hartree,[3] who showed that the spherically symmetric part of the electron interaction energy can be included in the unperturbed Hamiltonian by means of a potential derived from the average charge density of all electrons but one. That is, a single electron is visualized as moving in an effective potential due to all of the other electrons as well as the nucleus. The contribution to the potential seen by the kth electron due to all the other electrons is calculated from the spatial average of each unperturbed one-electron wave function $\psi_i^{(0)}$, or

$$V_k = \sum_{i \neq k} \int_{\tau_i} \frac{e \rho_i}{r_{ik}} \, d\tau_i = \sum_{i \neq k} \langle \psi_i^{(0)} | \frac{e}{r_{ik}} | \psi_i^{(0)} \rangle. \tag{10.18}$$

Then this potential is used along with the nuclear potential to solve Schrödinger's equation self-consistently. That is, the equation

$$\left[-\frac{\hbar^2}{2m} \nabla_k^2 - \frac{Ze^2}{r_k} + V_k \right] \psi_k^{(1)} = E_k \psi_k^{(1)}, \tag{10.19}$$

---

[3] D. R. Hartree, *Proc. Cambridge Phil. Soc.* **24,** 89, 111, 426 (1928).

is solved for each electron. Then the new wave functions $\psi_i^{(1)}$ are used in Equation 10.18 to calculate an improved potential V_k'. The iterations are continued until the potential ceases to change appreciably; that is, until it becomes self-consistent.

In recent years the self-consistent method has been used extensively because it is readily programmed for digital computation.[4] Some of the short-comings of the model are that it neglects correlations and the Pauli principle, and it retains only the spherically symmetric part of the Coulomb interactions. An improvement due to Fock[5] consists of including exchange by using properly symmetrized product functions for the initial wave function.

5. THE THOMAS-FERMI STATISTICAL MODEL OF THE ATOM

Another central-field approximation that has been applied to atoms of large Z is due to Thomas and Fermi.[6] The model is more crude than the Hartree model, but it has the advantage of being much easier to solve. It assumes that the central potential varies so slowly that it can be regarded as essentially constant over several de Broglie wavelengths. Thus, several electrons can be localized in a region of quasi-constant potential. Under such conditions the electrons may be treated as an ideal gas of non-interacting particles known as a Fermi gas, which fills the potential.

The density of states of a free electron gas is obtained by applying periodic boundary conditions and box normalization to a cell that is small with respect to the volume of the gas, but large with respect to a de Broglie wavelength. This has been done in Problem 7-2, with the result that

$$n(E) = \frac{L^3}{2\pi^2\hbar^3}(2m)^{\frac{3}{2}}E^{\frac{1}{2}}, \tag{10.20}$$

where an additional factor of 2 has been included here because there are two spin states associated with each spatial state. Notice that the energy of the gas increases as the number of electrons increases. The energy of a collection of Z electrons is obtained as follows:

$$Z = \int_0^{E_F} n(E)\,dE = \frac{L^3(2m)^{\frac{3}{2}}}{2\pi^2\hbar^3}\int_0^{E_F} E^{\frac{1}{2}}\,dE = \frac{L^3(2m)^{\frac{3}{2}}}{3\pi^2\hbar^3}E_F^{\frac{3}{2}},$$

or

$$E_F = \frac{\hbar^2}{2m}(3\pi^2\rho)^{\frac{2}{3}}. \tag{10.21}$$

In Equation 10.21 the quantity $\rho = Z/L^3$ is the number of electrons per unit volume, and E_F is called the Fermi energy. Note that the Fermi energy

---

[4] See D. R. Hartree, *The Calculation of Atomic Structures*, John Wiley & Sons, New York, 1957; J. C. Slater, *Quantum Theory of Matter*, 2nd ed., McGraw-Hill Book Co., N.Y., 1968.

[5] V. Fock, *Z. Physik* **61**, 126 (1930).

[6] L. H. Thomas, *Proc. Cambridge Phil. Soc.* **23**, 542 (1927); E. Fermi, *Z. Physik* **48**, 73 (1928).

is simply the energy of the highest filled state. In the many-electron atom which we are treating, we have $E_F = -V(r)$, and from Equation 10.21,

$$\rho = \frac{Z}{L^3} = \frac{(2m)^{\frac{3}{2}}}{3\pi^2\hbar^3}[-V(r)]^{\frac{3}{2}}. \tag{10.22}$$

Now, the self-consistency condition is that the potentials due to the electron density in Equation 10.22, as well as that due to the nuclear charge, properly reproduce the potential energy, $-V(r)$. Consequently, the charge density, $-e\rho$, and the electrostatic potential, $-[V(r)/e]$, must satisfy Poisson's equation,

$$-\frac{1}{e}\nabla^2 V = -4\pi(-e\rho),$$

or

$$\nabla^2 V = \frac{1}{r^2}\frac{d}{dr}\left(r^2\frac{dV}{dr}\right) = 4\pi e^2\rho. \tag{10.23}$$

Equations 10.22 and 10.23 may be solved simultaneously[7] for ρ and V, imposing the boundary condition $V(r) = -(Ze^2/r)$ as $r \to 0$, and $rV(r) \to 0$ as $r \to \infty$. The result is that the "size" of a heavy atom is approximately

$$a = \frac{\hbar^2}{me^2Z^{\frac{1}{3}}} = \frac{a_0}{Z^{\frac{1}{3}}}$$

and, accordingly, the screened Coulomb potential seen by an outer electron may be represented by

$$V(r) = -\frac{Ze^2}{r}e^{-r/a}. \tag{10.24}$$

The validity of the Thomas-Fermi model improves as the number of electrons increases. However, at best, its description of electron densities is very crude, since all of the structure of the radial probability density curves is smoothed out in the statistical averaging process. For light atoms it overestimates the potential and the charge density at small radial distances.

PROBLEM 10-9

Using Equation 10.20, show that the average energy of a particle in a Fermi gas is $\frac{3}{5}E_F$. What is the total energy of a Fermi gas of N particles?

---

[7] L. I. Schiff, *Quantum Mechanics*, 3rd ed., McGraw-Hill Book Co., New York, 1968, p. 430.

Taking the density of the gas of conduction electrons in copper to be $\sim 10^{23}$ cm^{-3}, estimate the Fermi energy. How does this energy compare with the thermal energy of an electron at room temperature? Discuss the consequences of this with respect to the participation of electrons in thermal processes in metals.

6. THE HYDROGEN MOLECULE: THE COVALENT BOND

We have seen in the foregoing sections how, in principle, the eigenfunctions for the electronic states of a many-electron atom can be constructed from one-electron eigenfunctions. A similar approach is often used for the electronic states of molecules, although with much less success since even the simplest molecules deviate drastically from spherical symmetry. As an illustrative example of a possible approach to this problem, let us look at the method used for the first successful calculation of the dissociation energy of the hydrogen molecule.[8]

The Hamiltonian for the molecule, which is shown schematically in Figure 10-2, is

$$\mathcal{H} = \mathcal{H}_{CM} + \mathcal{H}_N + \mathcal{H}_1 + \mathcal{H}_2 + \mathcal{H}_{12} \qquad (10.25)$$

where

$$\mathcal{H}_{CM} = -\frac{\hbar^2}{2m_{\text{total}}} \nabla_{CM}^2$$

$$\mathcal{H}_N = -\frac{\hbar^2}{m_p} \nabla_N^2 + \frac{e^2}{R}$$

$$\mathcal{H}_1 = -\frac{\hbar^2}{2m_e} \nabla_1^2 - \frac{e^2}{r_1} - \frac{e^2}{r_{1b}}$$

$$\mathcal{H}_2 = -\frac{\hbar^2}{2m_e} \nabla_2^2 - \frac{e^2}{r^2} - \frac{e^2}{r_{2a}}$$

$$\mathcal{H}_{12} = \frac{e^2}{r_{12}}$$

Here $m_p/2$ is the reduced mass of the two protons and m_e is the reduced mass of the electron relative to a stationary proton. For simplicity we will neglect the translational and rotational energy of the molecule and will regard the distance between the protons, R, to be fixed. (We are also ignoring the exchange symmetry of the protons.) We expect both electrons to be in $1s$

---

[8] W. Heitler and F. London, *Z. Physik* **44**, 455 (1927).

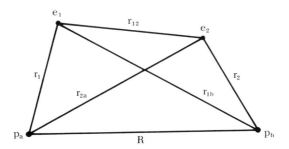

Figure 10-2 Schematic diagram of the hydrogen molecule.

states in the ground state of the molecule. Then, taking into account the exchange of the electrons, there are two possible approximate electronic wave functions which are products of hydrogenic functions, namely,

$$\psi_{ab} = \psi_a(1) \cdot \psi_b(2)$$
$$\psi_{ba} = \psi_b(1) \cdot \psi_a(2),$$
(10.26)

where ψ_a and ψ_b are both $1s$ wave functions. Recall that in the helium atom, where there is only one nucleus, ψ_{ab} and ψ_{ba} are degenerate eigenfunctions of the *same* Hamiltonian. Here, however, ψ_{ab} and ψ_{ba} are *not* degenerate, since one is regarded as an eigenfunction of $\mathscr{H}_1$ and the other is an eigenfunction of $\mathscr{H}_2$.

A variation method may be used to solve the problem by writing a trial function that is a linear combination of the functions in Equation 10.26. Thus we may write

$$\psi = \psi_{ab} + A\psi_{ba},$$

where A is the variation parameter. Then,[9]

$$E_{\text{ground}} \le \frac{(1 + A^2)B + 2AD}{1 + A^2 + 2AC},$$
(10.27)

where

$$B = \langle\psi_{ab}| \mathscr{H} |\psi_{ab}\rangle = \langle\psi_{ba}| \mathscr{H} |\psi_{ba}\rangle$$
$$C = \langle\psi_{ab} | \psi_{ba}\rangle$$
$$D = \langle\psi_{ab}| \mathscr{H} |\psi_{ba}\rangle = \langle\psi_{ba}| \mathscr{H} |\psi_{ab}\rangle.$$

For fixed R, the derivative of Equation 10.27 with respect to A is

$$\frac{2(1 - A^2)(D - BC)}{(1 + A^2 + 2AC)^2} = 0,$$

which yields a minimum of E_{ground} for $A = +1$ and a maximum for $A = -1$.

[9] See L. I. Schiff, *op. cit.*, page 450.

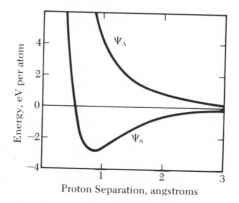

Figure 10-3 Energy per atom for the symmetric and antisymmetric electron spatial wave functions for the hydrogen molecule.

Writing $\psi = \psi_{ab} + \psi_{ba}$,

$$E_{\text{ground}} \leq \frac{B + D}{1 + C}. \tag{10.28}$$

Upon evaluation of the integrals B, C, and D, the difference between the energy obtained from Equation 10.28 and the electronic energy of two separated hydrogen atoms (about 27 eV) should be the dissociation energy of the hydrogen molecule. The experimental value for the latter is 4.72 eV.

Physically, one could have guessed at the proper linear combination of the wave functions in Equation 10.26, since an antisymmetric spin function would require a symmetric spatial function in order that the total wave function be antisymmetric. That the symmetric spatial function has a lower energy than its antisymmetric counterpart can be seen by reviewing the example of the double wells in Chapter 5 (see section 4). The energies of these functions versus nuclear separation are shown schematically in Figure 10-3. Since the antisymmetric function can have no bound state, it is often called the *antibonding orbital*, and the symmetric function is called the *bonding orbital*.

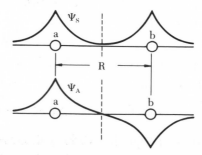

Figure 10-4 Schematic cross sections of the spatial wave functions for the hydrogen molecule.

7. MAGNETIC SUSCEPTIBILITY OF AN ATOM

The behavior of an atom in a magnetic field may be determined by means of the Hamiltonian given in Equation 9.33. Calling the direction of $\vec{\mathscr{B}}$ the

z-direction, the Hamiltonian becomes,

$$\mathscr{H} = \mathscr{H}_0 - g\mu_B m_j \mathscr{B} + \frac{e^2}{8mc^2} \sum_i (x_i^2 + y_i^2)\mathscr{B}^2$$

$$= \mathscr{H}_0 + \mathscr{H}_1 \mathscr{B} + \mathscr{H}_2 \mathscr{B}^2, \tag{10.29}$$

where the summation is over all of the i electrons in the atom and where $\mathscr{H}_0$ includes all of the contributions to the total energy which persist when the magnetic field $\vec{\mathscr{B}}$ is turned off. Classically, the magnetic moment along the field direction may be defined as

$$\mu_z = -\frac{\partial E}{\partial \mathscr{B}}, \tag{10.30}$$

where the energy E is

$$E = -\vec{\mu} \cdot \vec{\mathscr{B}}.$$

The quantum mechanical equivalent of Equation 10.30, which is

$$\langle \mu_z \rangle_{nn} = -\frac{\partial}{\partial \mathscr{B}} \langle E \rangle_{nn}, \tag{10.31}$$

has been verified by Van Vleck.[10] The quantity $\langle E \rangle_{nn}$ is calculated from perturbation theory as follows:

$$\langle E \rangle_{nn} = \langle n| \mathscr{H}_0 |n\rangle + \mathscr{B} \langle n| \mathscr{H}_1 |n\rangle + \mathscr{B}^2 \langle n| \mathscr{H}_2 |n\rangle + \mathscr{B}^2 \sum_{n \neq n'} \frac{|\langle n| \mathscr{H}_1 |n'\rangle|^2}{E_{n'}^{(0)} - E_n^{(0)}}$$

$$= E_n^{(0)} + \mathscr{B} E_n^{(1)} + \mathscr{B}^2 E_n^{(2)} + \cdots \tag{10.32}$$

The perturbation $\mathscr{H}_1$ has been carried to second order, since all quadratic terms in $\mathscr{B}$ should be considered if any are retained. Then Equation 10.31 becomes:

$$\langle \mu_z \rangle_{nn} = -E_n^{(1)} - 2\mathscr{B} E_n^{(2)} + \cdots$$

$$= -\langle n| \mathscr{H}_1 |n\rangle - 2\mathscr{B} \langle n| \mathscr{H}_2 |n\rangle - 2\mathscr{B} \sum_{n \neq n'} \frac{|\langle n| \mathscr{H}_1 |n'\rangle|^2}{E_{n'}^{(0)} - E_n^{(0)}}$$

$$= \langle n| g\mu_B m_j |n\rangle - \mathscr{B} \sum_i \frac{e^2}{4mc^2} \langle n| x_i^2 + y_i^2 |n\rangle + 2\mathscr{B} \sum_{n \neq n'} \frac{|\langle n| g\mu_B m_j |n'\rangle|^2}{E_n^{(0)} - E_{n'}^{(0)}}$$

$$= \langle \mu_0 \rangle_{nn} - \frac{e^2 \mathscr{B}}{4mc} \langle n| x_i^2 + y_i^2 |n\rangle + 2\mathscr{B} \sum_{n \neq n'} \frac{|\langle n| \mu_0 |n'\rangle|^2}{E_n^{(0)} - E_{n'}^{(0)}}. \tag{10.33}$$

The middle term in Equation 10.33 is always negative and is the source of the *diamagnetic* contribution to the magnetic moment; this term will be neglected in what follows. Let us now obtain the paramagnetic susceptibility associated with the induced moment of the last term of Equation 10.33. Assuming a

---

[10] J. H. Van Vleck, *op. cit.*, section 36 of Chapter VI.

Boltzmann distribution* over the states n', the atomic susceptibility may be expressed as

$$\chi = \frac{\langle \mu_z \rangle_{nn}}{\mathscr{B}} = \frac{\sum_n \mu_z e^{-E_n/kT}}{\mathscr{B} \sum_n e^{-E_n/kT}}.$$ (10.34)

But, from Equation 10.32,

$$e^{-E_n/kT} = e^{-E_n^{(0)}/kT} \cdot e^{-\mathscr{B}E_n^{(1)}/kT} \cdots \sim e^{-E_n^{(0)}/kT}\left(1 - \frac{\mathscr{B}E_n^{(1)}}{kT}\right),$$

where second order terms have been dropped in the expansion of the exponential. Then, for the numerator of Equation 10.34, we write

$$\sum_n \mu_z e^{-E_n/kT} \sim \sum_n (-E_n^{(1)} - 2\mathscr{B}E_n^{(2)})\left(1 - \frac{\mathscr{B}E_n^{(1)}}{kT}\right)e^{-E_n^{(0)}/kT}$$

$$\sim \sum_n \left[-E_n^{(1)} + \mathscr{B}\frac{(E_n^{(1)})^2}{kT} - 2\mathscr{B}E_n^{(2)}\right]e^{-E_n^{(0)}/kT},$$

where only first order terms in $\mathscr{B}$ are retained. Now the first term in the bracket represents the net moment in zero field, which is zero. Then,

$$\sum_n \mu_z e^{-E_n/kT} \sim \mathscr{B}\sum_n \left[\frac{(E_n^{(1)})^2}{kT} - 2E_n^{(2)}\right]e^{-E_n^{(0)}/kT},$$

and

$$\chi = \frac{\sum_n \left[\frac{(E_n^{(1)})^2}{kT} - 2E_n^{(2)}\right]e^{-E_n^{(0)}/kT}}{\sum_n e^{-E_n^{(0)}/kT}},$$ (10.35)

where only the field-independent term is retained in the denominator. Substituting the values for $E_n^{(1)}$ and $E_n^{(2)}$ into Equation 10.35,

$$\chi = \frac{\sum_n \left[\frac{|\langle \mu_0 \rangle_{nn}|^2}{kT} + 2\sum_{n' \neq n} \frac{|\langle n| \mu_0 |n'\rangle|^2}{E_n^{(0)} - E_{n'}^{(0)}}\right]e^{-E_n^{(0)}/kT}}{\sum_n e^{-E_n^{(0)}/kT}}.$$ (10.36)

If the separation of the states is much greater than kT, only the ground state contributes to the first term of Equation 10.36, and it becomes

$$\frac{\sum_{m_j} |\langle m_j| \mu_0 |m_j\rangle|^2}{kT} = \frac{g^2\mu_B^2}{kT}\frac{\sum_{-J}^{J} m_j^2}{2J+1} = \frac{g^2\mu_B^2 J(J+1)}{3kT},$$ (10.37)

which represents the susceptibility due to the permanent dipole moment.

_____

* See Appendix B of Chapter 2.

The second term in Equation 10.36 gives rise to the so-called *Van Vleck temperature-independent susceptibility.* Here there are two contributions, one due to widely separated values of n and n', and the other due to widely separated J values.[11] Although these contributions are usually neglected, it is worth noting that they arise when the Zeeman effect is carried to second order in perturbation theory.

SUGGESTED REFERENCES

David Bohm, *Quantum Theory.* Prentice-Hall, Inc., New York, 1951.

R. H. Dicke and J. P. Wittke, *Introduction to Quantum Mechanics.* Addison-Wesley Publishing Co., Inc., Reading, Mass., 1960.

H. Eyring, J. Walter, and G. E. Kimball, *Quantum Chemistry.* John Wiley and Sons, Inc., New York, 1944.

E. Merzbacher, *Quantum Mechanics.* John Wiley and Sons, Inc., New York, 1961.

Albert Messiah, *Quantum Mechanics.* North-Holland Publishing Co., Amsterdam, 1958.

L. Pauling and E. B. Wilson, *Introduction to Quantum Mechanics.* McGraw-Hill Book Co., New York, 1935.

D. Park, *Introduction to the Quantum Theory.* McGraw-Hill Book Co., New York, 1964.

J. L. Powell and B. Crasemann, *Quantum Mechanics.* Addison-Wesley Publishing Co., Inc., Reading, Mass., 1961.

D. S. Saxon, *Elementary Quantum Mechanics.* Holden-Day, Inc., San Francisco, 1968.

Leonard I. Schiff, *Quantum Mechanics.* .3rd ed. McGraw-Hill Book Co. New York, 1969.

---

[11] J. H. Van Vleck, *op. cit.*, section 55, Chapter IX.

CHAPTER II

SCATTERING THEORY

I. THE PARTIAL WAVE TREATMENT OF SCATTERING

An introduction to the classical study of unbound systems of particles by means of scattering experiments was given in sections 2 and 3 of Chapter 3. We will now discuss a useful quantum mechanical approximation for the case of elastic scattering. In analogy with the phenomenon of diffraction in optics and acoustics, one may visualize the incident particle flux as a superposition of plane waves. Furthermore, the scattering center itself may be regarded as the source of a weak spherical wave similar to a Huygens wavelet. The effect of the scattering event near the scatterer may then be determined by the superposition of the incident plane wave and the newly generated spherical wave. Far from the scatterer, in the forward direction, this solution asymptotically approaches a plane wave again. In the case of elastic scattering, the propagation constant for this asymptotic plane wave has the same magnitude as that for the incident plane wave, but its direction is altered.

A spherical wave may be represented by $(1/r) \exp (ikr)$, where the factor r in the denominator insures that the intensity obeys the inverse square law. In order to provide for some angular asymmetry, it is common practice to include an angular factor, $f(\theta)$, in the amplitude. Then the outgoing spherical wave has the form,

$$\psi_{\text{sph}} = f(\theta) \cdot \frac{e^{ikr}}{r} .$$

Letting the particle beam be directed along the z-axis, the incident plane wave is given by

$$\psi_{\text{in}} = Ae^{i\vec{k}\cdot\vec{r}} = Ae^{ikz}.$$

Hence the scattered wave is the superposition

$$\psi_{\text{sc}} = Ae^{ikz} + f(\theta) \cdot \frac{e^{ikr}}{r} . \tag{11.1}$$

Since the interaction region is assumed to be restricted to the immediate vicinity of the scattering center, the beam fluxes far from the scatterer do not interfere and may be considered separately. The intensity of the incident beam is

$$I = |\psi_{in}|^2 \cdot \frac{\hbar k}{m} = \frac{\hbar k}{m},$$

for ψ_{in} normalized, while the intensity of the beam scattered at an angle θ is

$$N(\theta) = \left| \frac{f(\theta) \cdot e^{ikr}}{r} \right|^2 \cdot \frac{\hbar k}{m} = |f(\theta)|^2 \cdot \frac{\hbar k}{mr^2}.$$

Then the number of particles scattered through the angle θ into an element of area da, normal to the beam, is

$$dN(\theta) = |f(\theta)|^2 \cdot \frac{\hbar k}{m} \frac{da}{r^2} = |f(\theta)|^2 \cdot \frac{\hbar k}{m} d\Omega.$$

From Equation 3.7, the differential scattering cross section is then

$$\frac{d\sigma}{d\Omega} = |f(\theta)|^2. \tag{11.2}$$

The factor $f(\theta)$ is called the *scattering amplitude*, since its square gives the differential scattering cross-section.

Equation 11.1 is a solution of the free particle Schrödinger equation. It is convenient to write it in a different form, however, by expanding the plane wave in terms of the Legendre polynomials. Thus,

$$e^{ikz} = \sum_{\ell=0}^{\infty} i^\ell (2\ell + 1) j_\ell(kr) \cdot P_\ell(\cos \theta), \tag{11.3}$$

where the $j_\ell(kr)$ are the spherical Bessel functions.[1]

A general formula for these functions is

$$j_\ell(kr) = (-1)^\ell \left(\frac{r}{k}\right)^\ell \left(\frac{1}{r}\frac{d}{dr}\right)^\ell \frac{\sin kr}{kr}, \tag{11.4}$$

and the first two are:

$$j_0(kr) = \frac{\sin kr}{kr}$$

$$j_1(kr) = \frac{\sin kr}{(kr)^2} - \frac{\cos kr}{kr}.$$

The asymptotic form of these functions for $kr \gg \ell$ is

$$j_\ell(kr) \rightarrow \frac{\sin \left(kr - \frac{\ell\pi}{2}\right)}{kr} \tag{11.5}$$

---

[1] G. N. Watson, *Theory of Bessel Functions*, Macmillan, N.Y., 1944, rev. ed., p. 128.

Thus, at *large* distances from the scattering center, Equation 11.3 may be written as

$$e^{ikz} = \sum_{\ell=0}^{\infty} i^{\ell}(2\ell + 1) \cdot P_{\ell}(\cos \theta) \cdot \frac{\sin\left(kr - \dfrac{\ell\pi}{2}\right)}{kr} \tag{11.6}$$

Since Equation 11.6 is the free particle solution, it must also be the solution for a non-zero potential *at large distances from the scatterer*. We assume that in the interaction region its form cannot change radically, but that the potential must only alter the phase of the sinusoidal function. For the asymptotic form of the scattered wave we then assume

$$\psi_{\rm sc} \sim \frac{\sin\left(kr - \dfrac{\ell\pi}{2} + \delta_{\ell}\right)}{kr} \; ,$$

where δ_{ℓ} represents a phase shift of the ℓth partial wave. Rewriting Equation 11.1,

$$\sum_{\ell=0}^{\infty} b_{\ell}P_{\ell}(\cos \theta) \cdot \frac{\sin\left(kr - \dfrac{\ell\pi}{2} + \delta_{\ell}\right)}{kr}$$

$$= \sum_{\ell=0}^{\infty} i^{\ell}(2\ell + 1) \cdot P_{\ell}(\cos \theta) \cdot \frac{\sin\left(kr - \dfrac{\ell\pi}{2}\right)}{kr} + f(\theta) \cdot \frac{e^{ikr}}{r} \tag{11.7}$$

we find that

$$b_{\ell} = i^{\ell}(2\ell + 1)e^{i\delta_{\ell}} \tag{11.8}$$

and

$$f(\theta) = \frac{1}{k}\sum_{\ell=0}^{\infty} (2\ell + 1)e^{i\delta_{\ell}} \sin \delta_{\ell} \cdot P_{\ell}(\cos \theta). \tag{11.9}$$

Note that the scattering amplitude is zero if all of the phase shifts vanish. The total scattering cross-section calculated from Equation 11.8 takes the simple form

$$\sigma = \int_{\Omega} |f(\theta)|^2 \, d\Omega = \frac{4\pi}{k^2}\sum_{\ell=0}^{\infty} (2\ell + 1) \sin^2 \delta_{\ell}. \tag{11.10}$$

PROBLEM 11-1

Show that Equation 11.5 is a solution of the Schrödinger equation for a free particle.

PROBLEM 11-2

Verify Equations 11.8 and 11.9.

PROBLEM 11-3

Derive Equation 11.10.

From the notion of the conservation of probability, it is evident that when scattering occurs, the removal of particles from the beam should affect the forward scattering amplitude, $f(0)$. We therefore expect to find a relationship between the total cross section and $f(0)$. Using Equation 11.9, we write

$$f(0) = \frac{1}{k} \sum_{\ell=0}^{\infty} (2\ell + 1) \cos \delta_\ell \sin \delta_\ell + \frac{i}{k} \sum_{\ell=0}^{\infty} (2\ell + 1) \sin^2 \delta_\ell.$$

Comparing the second term with Equation 11.10, we observe that

$$\sigma = \frac{4\pi}{k} \operatorname{Im} [f(0)]. \tag{11.11}$$

This relationship is known as the *optical theorem;* its validity is far more general than the present derivation implies.

In order to apply the partial-wave method of analysis to a scattering experiment it is necessary to be able to calculate the phase shifts, δ_ℓ. In general, the phase shifts will be positive for an attractive potential, which means that the phase is advanced and the wave is pulled in toward the scatterer in the interaction region. The peak of the outgoing wave is thus spatially behind the free particle wave (see Figure 11-1). The reverse is true for a repulsive potential. The scattering cross-section, however, does not depend upon the sign of δ_ℓ. The δ_ℓ are obtained by matching the radial solutions of Equation 7.70 for $V = 0$ to the radial functions which include the effect of the scattering potential, Equation 11.7. An explicit expression relating the phase shifts to the scattering potential and the radial eigenfunctions can be obtained by means

Figure 11-1 Phase shifts produced by scattering from attractive and repulsive potentials. $V = 0$ for the top figure, V is attractive for the middle figure, and V is repulsive for the bottom figure.

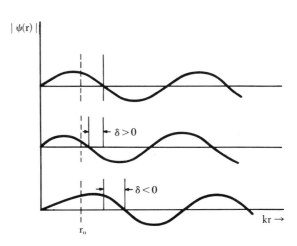

of the Green's function treatment of the scattering problem.[2] However, the partial-wave method is frequently applied to situations where only the first one or two phase shifts are required, as the following argument will show. Each partial wave corresponds to a definite value of angular momentum which may be related to the impact parameter and the linear momentum of the incoming particle by the expression, $\ell = \hbar k s$.* Since there will be essentially no scattering for values of s greater than the range of the potential, r_0, it is evident that the relationship

$$\ell \leq \hbar k r_0 \tag{11.12}$$

provides a simple guide for determining how many values of ℓ will be required to represent the scattering. Thus, for low energy particles (small k) or a short-range potential (small r_0) only the lowest ℓ values are required. In fact, it often turns out that $kr_0 < 1$ so that $\ell = 0$ is the only angular momentum state that need be considered. Such an event is called S-wave scattering, in analogy with the terminology of spectroscopy. S-wave scattering is isotropic (in the C-frame), since from Equation 11.9 we note that

$$f(\theta) = \frac{1}{k} e^{i\delta_0} \sin \delta_0,$$

when all phase shifts vanish except for that associated with $\ell = 0$. The total scattering cross section for S-wave scattering is

$$\sigma = \frac{4\pi}{k^2} \sin^2 \delta_0. \tag{11.14}$$

The isotropy of low energy scattering may easily be predicted from the wave picture. Since the de Broglie wavelength of the incident particle is much greater than the range of the scattering, the detector must be able to resolve phase shifts of the order of $2\pi \cdot (r_0/\lambda) = kr_0$ if anisotropies are to be observed. Hence, if k and r_0 are both small, no phase shifts will be detected.

Applying the S-wave approximation to the classical collision of two hard spheres, the maximum phase shift is $\delta_0 = kR$, where R is the sum of the radii of the spheres. For $kR \ll 1$, $\sin \delta_0 \sim \delta_0$, and the total scattering cross section is $4\pi R^2$, which is four times the classical value obtained for the geometrical cross section in Problem 3-1. The additional cross section in the quantum mechanical case is due to interference and diffraction effects.

A more important example is that of square-well scattering. Let the potential be $V(r) = -V_0$ for $r < a$ and $V(r) = 0$ for $r > a$. Then a suitable wave function inside the well for the case of $\ell = 0$ is

$$\psi_{\text{in}} = A \sin k_1 r, \quad \text{for} \quad r < a,$$

---

[2] See, for example, E. Merzbacher, *Quantum Mechanics*, John Wiley and Sons, N.Y., 1961, section 12.7. For a different method of evaluating the phase shifts see L. I. Schiff, *Quantum Mechanics*, McGraw-Hill Book Co., N.Y., 3rd ed., 1968, p. 121.

* See section 2 of Chapter 3.

where

$$\hbar k_1 = \sqrt{2m(E + V_0)}.$$

In a similar manner, the S-wave solution outside the well is

$$\psi_{\text{out}} = B \sin(k_2 r + \delta_0), \quad \text{for} \quad r > a,$$

where $\hbar k_2 = \sqrt{2mE}$. The boundary condition requires that $(1/\psi)(\partial\psi/\partial r)$ be continuous at $r = a$. Thus,

$$k_1 \cot k_1 a = k_2 \cot (k_2 a + \delta_0)$$

or

$$\tan(k_2 a + \delta_0) = \frac{k_2}{k_1} \tan k_1 a.$$

(11.15)

At low energies k_2 and δ_0 are both small so that

$$\tan(k_2 a + \delta_0) \sim k_2 a + \delta_0,$$

and then

$$\delta_0 \simeq \frac{k_2}{k_1} \tan k_1 a - k_2 a$$

or

$$\sin \delta_0 \simeq k_2 a \left(\frac{\tan k_1 a}{k_1 a} - 1\right).$$

From Equation 11.15, the total scattering cross section for S-waves is

$$\sigma = 4\pi a^2 \left(\frac{\tan k_1 a}{k_1 a} - 1\right)^2 . \cdot$$

(11.16)

The scattering cross section becomes infinite here for $k_1 a = \pi/2, 3\pi/2, \ldots$ where the tangent function becomes infinite. Since Equation 11.16 was derived from the assumption that $\delta_0 \ll k_1 a$, the infinite cross section is inadmissible. Returning to the exact expression, Equation 11.15, we see that $\sin \delta_0 \sim 1$ when $k_1 a = \pi/2$ and $k_2 a \ll 1$, and so the maximum cross section is

$$\sigma_{\text{res}} = \frac{4\pi}{k_2^2} .$$

(11.17)

Such maxima are called *S-wave resonances*.

If the energy of the incident particles is high enough to overcome the centrifugal barrier, a similar procedure may be carried out for higher ℓ values and one may find *P-wave resonances*, and so on. Such resonances are often called *virtual states*, since the incident particle appears to be temporarily bound in spite of the fact that its total energy is positive.

> By means of the optical theorem, Equation 11.11, show that the upper bound for σ for S-wave scattering is Equation 11.17, regardless of the potential.

The *Ramsauer-Townsend effect* is also predicted by Equation 11.16, since $\sigma = 0$ when $k_1a = \tan k_1a$, and $\delta_0 = \pi,\ 2\pi, \ldots$. Nearly complete transmission of electrons is observed for scattering by rare gas atoms at energies around 0.7 eV. For energies higher and lower than this value a scattering cross section appreciably greater than zero is measured. This phenomenon is similar to the transparency of barriers (for wave functions that match up properly) discussed in section 2 of Chapter 5.

2. SCATTERING AS A PERTURBATION: THE BORN APPROXIMATION

Time-dependent perturbation theory may be applied to the scattering problem by treating the perturbation potential as being "turned on" when the incident particle gets within its range, and "turned off" again when the scattered particle exceeds its range. The scattering process is thus regarded as a time-dependent transition from an initial plane wave state $|i\rangle$ to a final plane wave state $|f\rangle$. The transition rate per incident particle is given by Fermi's Golden Rule, Equation 9.54, as

$$R = \frac{2\pi}{\hbar}\,|\langle f|\,V\,|i\rangle|^2\,\rho(E_f),\qquad(11.18)$$

where V is the scattering potential and $\rho(E_f)$ is the density of final states. Since the final states here are free particle states, we may use the electron gas density of states, Equation 10.20, omitting the factor of two for spin states,

$$\rho(E_f) = \frac{mL^3}{2\pi^2\hbar^2}\cdot\left(\frac{2mE_f}{\hbar^2}\right)^{\frac{1}{2}} = \frac{mL^3k_f}{2\pi^2\hbar^2}\qquad(11.19)$$

We are interested in those states for which the vector $\vec{k}_f$ lies within a small solid angle centered around $\vec{k}_f$. Therefore, the density of states, Equation 11.19, must be multiplied by the factor $d\Omega/4\pi$. Then the transition rate per unit solid angle centered around $\vec{k}_f$ due to a particle flux of intensity

$$I = \frac{v}{L^3} = \frac{\hbar k_0}{L^3 m}$$

is

$$\frac{R}{d\Omega} = I\frac{d\sigma}{d\Omega}.$$

Thus, the differential scattering cross section is

$$\frac{d\sigma}{d\Omega} = \frac{R}{I \, d\Omega} = \left(\frac{mL^3}{2\pi\hbar^2}\right)^2 \cdot \frac{k_f}{k_0} \cdot |\langle f | \, V \, |i\rangle|^2. \tag{11.20}$$

For elastic scattering of plane waves, $k_f = k_0$ and Equation 11.20 becomes

$$\frac{d\sigma}{d\Omega} = \left(\frac{mL^3}{2\pi\hbar^2}\right)^2 \left[\int_\tau L^{-3} e^{-i\vec{k}_f \cdot \vec{r}} V(r) e^{i\vec{k}_0 \cdot \vec{r}} \, d\tau\right]^2,$$

and

$$f(\theta) = -\frac{m}{2\pi\hbar^2} \cdot \int_\tau V(r) e^{i(\vec{k}_0 - \vec{k}_f) \cdot \vec{r}} \, d\tau. \tag{11.21}$$

This integral is known as the *Born approximation*. Since the volume normalization factor $L^{-\frac{3}{2}}$ drops out, the integration is independent of the size of the box. Hence, contributions to the scattering cross section arise only where $V(r)$ is not small, regardless of the limits of integration on r. This approximation is equivalent to assuming that only single scattering events are important, and thus the first-order term in $V(r)$ is all that appears in Equation 11.21. We will see in the next section how multiple scattering events are treated.

Equation 11.21 may be simplified by means of the substitution

$$\vec{K} = \vec{k}_0 - \vec{k}_f \quad .$$

Since $\vec{k}_0$ and $\vec{k}_f$ are the same length, the magnitude of $\vec{K}$ is given by $K = 2k_0 \cdot \sin \theta/2$, where θ is the angle of scattering between $\vec{k}_0$ and $\vec{k}_f$ (see Figure 11-2). By defining an auxiliary set of polar coordinates with the polar axis directed along $\vec{K}$,

$$\int_\tau V(r) e^{i\vec{K}\cdot\vec{r}} \, d\tau = 2\pi \int_0^\infty V(r) r^2 \, dr \int_0^\pi e^{iKr\cos\theta} \sin\theta \, d\theta$$

$$= \frac{4\pi}{K} \int_0^\infty V(r) \sin Kr \cdot r dr.$$

Then the scattering amplitude is

$$f(\theta) = -\frac{2m}{\hbar^2 K} \int_0^\infty V(r) \cdot \sin Kr \cdot dr. \tag{11.22}$$

Equation 11.22 is sometimes referred to as the *standard form* of the Born approximation.

A discussion of the range of validity of the Born approximation is a complex

Figure 11-2 The wave vectors used for the Born approximation.

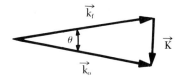

subject, but we may make a few general remarks in the context of the assumptions used here. First, we have used perturbation theory, which presupposes that the amplitude of the scattered wave is small compared to that of the incident wave and that $|V(r)| \ll E$, the energy of the incident particles. Thus, the energy of the incident particles must be large enough so that they may be still represented by plane waves after the scattering occurs.

As an example, let us apply the Born approximation to the scattering of electrons by a screened Coulomb potential,

$$V(r) = -\frac{Ze^2}{r} e^{-r/a},$$

where a is the screening radius. From Equation 11.22,

$$f(\theta) = -\frac{2mZe^2}{\hbar^2 K} \int_0^\infty \sin (Kr) \, e^{-r/a} \, dr.$$

The integral has the value,

$$\frac{1}{K^2 + \left(\frac{1}{a}\right)^2} \left[e^{-r/a} \left(-\frac{1}{a} \sin Kr - K \cos Kr \right) \right]_0^\infty = \frac{K}{K^2 + \left(\frac{1}{a}\right)^2}.$$

Thus, the scattering amplitude may be written in the form

$$f(\theta) = \frac{2mZe^2}{\hbar^2 K^2} \left[1 - \frac{1}{1 + (Ka)^2} \right]. \tag{11.23}$$

PROBLEM 11-5

(a) Apply a limiting process to Equation 11.23 to obtain $f(\theta)$ for the unscreened Coulomb potential. How does this compare with Equation 3.15?

(b) Sketch the curves of $(d\sigma/d\Omega)$ vs θ for these two cases on the same graph.

PROBLEM 11-6

Calculate the differential scattering cross section for the attractive square well potential,

$$V(r) = -V_0, \quad \text{for} \quad 0 < r < r_0,$$
$$V(r) = 0, \quad \text{for} \quad r > r_0,$$

using the Born approximation. Sketch $(d\sigma/d\Omega)$ vs Kr_0.

$$\left[\text{Ans.:} \quad \frac{d\sigma}{d\Omega} = \left(\frac{2mV_0}{\hbar^2 K^3} \right)^2 \cdot [\sin (Kr_0) - Kr_0 \cos (Kr_0)]^2 \right]$$

PROBLEM 11-7

Find the total scattering cross section in the limits of high and low energies for the square well of the previous problem.

PROBLEM 11-8

Calculate the differential scattering cross section for the attractive Gaussian potential,

$$V(r) = -V_0 \exp\left[-\left(\frac{r}{r_0}\right)^2\right],$$

using the Born approximation.

$$\left[\text{Ans.:}\quad \frac{d\sigma}{d\Omega} = \pi \left(\frac{mV_0 r_0^3}{\hbar^2}\right)^2 \exp\left[-\tfrac{1}{2}(Kr_0)^2\right]\right]$$

PROBLEM 11-9

A particle is scattered by the repulsive potential,

$$V(r) = \tfrac{1}{2}br^2 \exp\left[-\left(\frac{r}{r_0}\right)^2\right],$$

where b is positive.
(a) Use the Born approximation to obtain the differential scattering cross section.
(b) Estimate the energy for which phase shifts associated with $\ell > 0$ will become appreciable.

PROBLEM 11-10

Using the Born approximation, find the scattering amplitude for the spherical, triangular potential shown in the figure.

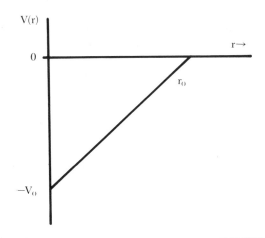

3. THE GREEN'S FUNCTION METHOD FOR SCATTERING

A solution of the equation

$$(\nabla^2 + k^2)\psi(\vec{r}) = -4\pi\rho(\vec{r}) \tag{11.24}$$

can be constructed from a superposition of the solutions for point sources of unit strength, $\rho(\vec{r}')$, if we define $\rho(\vec{r})$ by

$$\rho(\vec{r}) = \int \delta(\vec{r} - \vec{r}')\rho(\vec{r}')\, d\vec{r}', \tag{11.25}$$

where $\int \delta(\vec{r} - \vec{r}')\, d\vec{r}' = 1$. Suppose that a solution $G(\vec{r}, \vec{r}')$ can be found for the equation

$$(\nabla^2 + k^2)G(\vec{r}, \vec{r}') = -4\pi\delta(\vec{r} - \vec{r}'). \tag{11.26}$$

Then $\psi(\vec{r})$, the desired solution of Equation 11.24, is readily shown to be

$$\psi(\vec{r}) = \int G(\vec{r}, \vec{r}')\rho(\vec{r}')\, d\vec{r}'. \tag{11.27}$$

To verify that Equation 11.27 is a solution of 11.24, we utilize Equations 11.25 and 11.26 as follows:

$$(\nabla^2 + k^2)\int G(\vec{r}, \vec{r}')\rho(\vec{r}')\, d\vec{r}' = \int (\nabla^2 + k^2)G(\vec{r}, \vec{r}')\rho(\vec{r}')\, d\vec{r}'$$

$$= \int -4\pi\delta(\vec{r} - \vec{r}')\rho(\vec{r}')\, d\vec{r}'$$

$$= -4\pi\rho(\vec{r}).$$

In order to apply the above technique to the scattering problem, we will write the Schrödinger equation in the form,

$$(\nabla^2 + k^2)\psi(\vec{r}) = \frac{2m}{\hbar^2} V(\vec{r})\psi(\vec{r}), \tag{11.28}$$

where $\hbar^2 k^2 = 2mE$. Then a solution of Equation 11.28 is

$$\psi(\vec{r}) = -\frac{2m}{\hbar^2}\int G(\vec{r}, \vec{r}')V(\vec{r}')\psi(\vec{r}')\, d\vec{r}, \tag{11.29}$$

where $G(\vec{r}, \vec{r}')$ is the solution of

$$(\nabla^2 + k^2)G(\vec{r}, \vec{r}') = -\delta(\vec{r} - \vec{r}'), \tag{11.30}$$

which must now be determined. Using the solutions of Poisson's equation in electrostatics as a model, we write the solution as,

$$G(\vec{r}, \vec{r}') = \frac{1}{4\pi} \frac{e^{ik|\vec{r}-\vec{r}'|}}{|\vec{r} - \vec{r}'|}.$$ (11.31)

PROBLEM 11-11

Show that Equation 11.31 satisfies 11.30.

Using Equations 11.29 and 11.31, Equation 11.1 may be written as

$$\psi_{sc} = e^{ik_0 z} - \frac{m}{2\pi\hbar^2} \int \frac{e^{ik|\vec{r}-\vec{r}'|}}{|\vec{r} - \vec{r}'|} V(\vec{r}')\psi(\vec{r}')\, d\vec{r}'.$$ (11.32)

In the limit as $\vec{r}$ gets very large, the directions of $\vec{k}$ and $\vec{r}$ tend to coincide and

$$k|\vec{r} - \vec{r}'| \sim k|r - \hat{r}\cdot\vec{r}'| = kr - \vec{k}'\cdot\vec{r}',$$

where $\vec{k}' = k\hat{r}$ (see Figure 11-3). That is, $\vec{k}'$ has the same magnitude as the initial wave vector $\vec{k}$, but a different direction. Then, for large r,

$$\frac{e^{ik|\vec{r}-\vec{r}'|}}{|\vec{r} - \vec{r}'|} \sim \frac{e^{i(kr-\vec{k}'\cdot\vec{r}')}}{r},$$

and

$$\psi_{sc} = e^{ik_0 z} - \frac{m}{2\pi\hbar^2} \cdot \frac{e^{ikr}}{r} \int e^{-ik'\cdot\vec{r}'} V(\vec{r}')\psi(\vec{r}')\, d\vec{r}'.$$ (11.33)

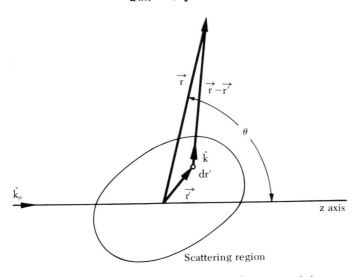

Figure 11-3 Schematic diagram of the scattering process for an extended scatterer.

Comparing this with Equation 11.1, we see that the scattering amplitude is given by

$$f(\theta) = -\frac{m}{2\pi\hbar^2}\int e^{-i\vec{k}'\cdot\vec{r}'}V(\vec{r}')\psi(\vec{r}')\,d\vec{r}', \tag{11.34}$$

where the scattering angle, θ, is the angle between the incident beam and $\vec{r}$.

A calculational difficulty still exists in that the total scattered wave function appears in the integrand of Equation 11.34. Note that if this total wave function is replaced by only the *first term* of Equation 11.32, namely, the incident plane wave, then Equation 11.34 is identical with the result for the Born approximation, Equation 11.21. An analytical justification for this substitution may be sketched as follows. Assuming the convergence of iterative solutions, the solution for $\psi(\vec{r}')$ may be written as

$$\psi(\vec{r}') = e^{i\vec{k}\cdot\vec{r}'} - \frac{m}{2\pi\hbar^2}\int G(\vec{r}',\vec{r}'')V(\vec{r}'')\psi(\vec{r}'')\,d\vec{r}'',$$

which, upon substitution into Equation 11.33, gives:

$$\psi_{sc} = e^{i\vec{k}_0\cdot\vec{r}} - \frac{m}{2\pi\hbar^2}\int G(\vec{r},\vec{r}')V(r')e^{i\vec{k}\cdot\vec{r}'}\,d\vec{r}'$$

$$+ \left(\frac{m}{2\pi\hbar^2}\right)^2\int G(\vec{r},\vec{r}')V(\vec{r}')G(\vec{r}',\vec{r}'')V(\vec{r}'')\psi(\vec{r}'')\,d\vec{r}'', \tag{11.35}$$

and so on. Each successive iteration replaces the wave function in the integrand by the solution in the next lowest approximation. Thus, the first integral contains the wave function in the zero-order approximation, which is the incident plane wave. This has already been described as the Born approximation. The second integral in Equation 11.35 may then be evaluated by replacing $\psi(\vec{r}'')$ by the wave function obtained from the Born approximation. This is called the *second Born* approximation. In general, the nth Born approximation consists of replacing the wave function in the nth integrand of the iterative expansion by the solution obtained in the $(n-1)^{th}$ approximation.

The physical interpretation of the iterative expansion, the first few terms of which are given in Equation 11.35, is that the higher order terms represent multiple scatterings.[3] An integrand containing one Green's function is an integration over single scattering events; two Green's functions characterize double scatterings; and so forth. The Green's functions may be regarded as "propagators" which carry the wave to the next scattering point. Since one would not expect many multiple scattering events for a weakly interacting potential, it is easy to see why the first Born approximation has such a wide range of applicability.

---

[3] R. P. Feynman, *Phys. Rev.* **76**, 749 (1949).

4. SCATTERING OF ELECTRONS BY ATOMS

The atomic potential seen by an incoming electron may be represented by the function

$$V(r) = - Ze^2 \int \frac{\rho_T(\vec{r}') \, d\vec{r}'}{|\vec{r} - \vec{r}'|} \, ,$$

where $\rho_T(\vec{r}') = \delta(\vec{r}') + \rho(\vec{r}')$, is the total charge density, that is, nuclear plus electronic, at the point $\vec{r}'$. Then Equation 11.21 becomes

$$f(\theta) = \frac{mZe^2}{2\pi\hbar^2} \iint \frac{\rho_T(\vec{r}')e^{i\vec{K}\cdot\vec{r}}}{|\vec{r} - \vec{r}'|} \, d\vec{r}' \, d\vec{r}$$

$$= \frac{mZe^2}{2\pi\hbar^2} \iint \rho_T(\vec{r}') \frac{e^{i\vec{K}\cdot\vec{r}}e^{i\vec{K}\cdot\vec{r}'}e^{-i\vec{K}\cdot\vec{r}'}}{|\vec{r} - \vec{r}'|} \, d\vec{r}' \, d\vec{r}$$

$$= \frac{mZe^2}{2\pi\hbar^2} \int \rho_T(\vec{r}')e^{i\vec{K}\cdot\vec{r}'} \, d\vec{r}' \int \frac{e^{i\vec{K}\cdot(\vec{r}-\vec{r}')}}{|\vec{r} - \vec{r}'|} \, d\vec{r}.$$

The integral on the right is evaluated by taking $\vec{r}' = 0$ as the origin for the integration over $\vec{r}$ and by including the integrating factor $e^{-\alpha r}$. Then,

$$\int \frac{e^{i\vec{K}\cdot\vec{r}}}{r} \, d\vec{r} = \lim_{\alpha \to 0} \int_0^{2\pi}\int_0^\pi\int_0^\infty e^{-\alpha r}e^{i\vec{K}\cdot\vec{r}} r \, dr \, d\theta \, d\phi$$

$$= 2\pi \lim_{\alpha \to 0} \int_0^\infty e^{-\alpha r}r \, dr \int_\pi^0 e^{i\vec{K}\cdot\vec{r} \cos\theta} \, d(\cos\theta)$$

$$= 2\pi \lim_{\alpha \to 0} \int_0^\infty e^{-\alpha r}r \, dr \left[\frac{e^{iKr} - e^{-iKr}}{iKr}\right]$$

$$= \frac{4\pi}{K} \lim_{\alpha \to 0} \int_0^\infty e^{-\alpha r} \sin Kr \, dr$$

$$= \frac{4\pi}{K} \lim_{\alpha \to 0} \frac{K}{\alpha^2 + K^2} = \frac{4\pi}{K^2}$$

It follows that

$$f(\theta) = \frac{2mZe^2}{\hbar^2 K^2} \int \rho_T(\vec{r}')e^{i\vec{K}\cdot\vec{r}'} \, d\vec{r}'$$

$$= \frac{2mZe^2}{\hbar^2 K^2} \int [\delta(\vec{r}') - \rho(\vec{r}')]e^{i\vec{K}\cdot\vec{r}'} \, d\vec{r}'$$

$$= \frac{2mZe^2}{\hbar^2 K^2} [1 - \int \rho(\vec{r}')e^{i\vec{K}\cdot\vec{r}'} \, d\vec{r}']$$

$$= \frac{2mZe^2}{\hbar^2 K^2} [1 - F(K)]. \qquad (11.36)$$

Here $F(K)$ is the *atomic scattering factor*, which is a measure of the amount of shielding of the nuclear charge by the electrons. It is given by,

$$F(K) = \int \rho(\vec{r}')e^{i\vec{K}\cdot\vec{r}'}d\vec{r}'.$$ (11.37)

Note that the atomic scattering factor is the Fourier transform of the electronic density. For a spherically symmetric charge distribution, Equation 11.37 may be integrated over the angle variables and simplified to

$$F(K) = \frac{4\pi}{K} \int \rho(r') \sin (Kr') \ r' \ dr'.$$ (11.38)

PROBLEM 11-12

Integrate Equation 11.37 over solid angle to obtain Equation 11.38.

PROBLEM 11-13

Assuming an electronic charge density of the Gaussian form,

$$\rho(r') = \left(\frac{1}{\sqrt{\pi}\ r_0}\right)^3 e^{-(r/r_0)^2},$$

obtain the atomic scattering factor $F(K)$. Here r_0 is the atomic radius.

$$\left[\text{Ans.:}\ F(K) = \exp\left[-\left(\frac{Kr_0}{2}\right)^2\right].\right]$$

The quantity $\hbar^2K^2$ in Equation 11.36 can be written in terms of the energy of the incident electrons by means of the relationship derived from Figure 11-2,

$$\hbar^2K^2 = 4\hbar^2k^2 \sin^2 \frac{\theta}{2} = 8mE \sin^2 \frac{\theta}{2}.$$

Then,

$$f(\theta) = \frac{Ze^2}{4E \sin^2 \dfrac{\theta}{2}} [1 - F(K)].$$ (11.39)

This gives the Rutherford result for large angle scattering, Equation 3.15, when the energy E is much greater than the ionization energy so that the shielding is ineffective. However, for very small scattering angles the shielding effect is great enough to prevent the singularity in $|f(\theta)|^2$ at $\theta = 0$, which is predicted by the Rutherford formula.

5. THE EFFECTS OF EXCHANGE SYMMETRY AND SPIN ON SCATTERING CROSS SECTIONS

In the case of a two-particle scattering event in the C-frame, the two particles will go off in opposite directions after they interact (see Figure 3-5b). That is, one particle appears at the angle θ and the other at the angle $(\pi - \theta)$. Now, if the particles are indistinguishable, it is impossible to tell which is the scatterer and which is the scattered particle. Thus, it is necessary to combine the scattering amplitudes, $f(\theta)$ and $f(\pi - \theta)$, before calculating the scattering cross sections. For spinless particles, such as the α-particle, the wave function must be symmetric (section 5 of Chapter 8) and the scattering amplitudes add with the same sign. Hence, in the C-frame,

$$(symmetric\ spatial\ function) \quad d\sigma/d\Omega = |f(\theta) + f(\pi - \theta)|^2, \quad (11.40)$$

for two-particle scattering events involving identical, spinless bosons. The reader should note that the classical result for this case would be

$$\frac{d\sigma}{d\Omega} = |f(\theta)|^2 + |f(\pi - \theta)|^2,$$

where no interference terms are present.

On the other hand, suppose we consider the elastic scattering of identical spin 1 particles by particles of the same kind. For the symmetric spin case we must also use the symmetric spatial function (since the total wave function must be symmetric), and the differential cross section is given by Equation 11.40. However, we can also obtain a total wave function that is symmetric by using both antisymmetric spatial and spin functions. For this case the differential scattering cross section is given by

$$(antisymmetric\ spatial\ function) \quad \frac{d\sigma}{d\Omega} = |f(\theta) - f(\pi - \theta)|^2. \quad (11.41)$$

Since the symmetric and antisymmetric scattering events are independent events, we expect the differential scattering cross sections to be the sum of Equations 11.40 and 11.41, with proper consideration being given to their statistical weights. From the table in section 10 of Chapter 8 (see also Problem 8-29), we note that the ratio of symmetric to antisymmetric spin states is $2:1$. Therefore, the differential scattering cross section for the elastic scattering of unpolarized, indistinguishable, spin 1 particles is

$$\frac{d\sigma}{d\Omega} = \tfrac{2}{3}|f(\theta) + f(\pi - \theta)|^2 + \tfrac{1}{3}|f(\theta) - f(\pi - \theta)|^2. \quad (11.42)$$

A similar procedure is followed for fermions. For example, for either $e - e$, $p - p$, or $n - n$ scattering, where the spin is $\frac{1}{2}$, we have already seen that the symmetric spatial function can be combined with only one antisymmetric spin function, whereas the antisymmetric spatial function can combine with three different symmetric spin functions. Therefore, the differential scattering cross section for the elastic scattering of identical, unpolarized, spin $\frac{1}{2}$ particles is

$$\frac{d\sigma}{d\Omega} = \tfrac{1}{4} \, | \, f(\theta) \, + f(\pi - \theta)|^2 + \tfrac{3}{4} \, | \, f(\theta) \, - f(\pi - \theta)|^2. \qquad (11.43)$$

SUMMARY

Three approaches to scattering problems have been introduced in Chapter 11. In the partial wave method, the incident particle is represented by a plane wave which is expanded in the Legendre polynomials and the spherical Bessel functions. The scattering event is then treated as the superposition of this incident wave and a spherical wave generated by the scattering center. By comparing the forms of the scattered and incident waves at large distances from the scatterer (the asymptotic limit), the effect of the scatterer can be described in terms of phase shifts in the arguments of sine and cosine functions. The calculation of these phase shifts constitutes a solution to the problem. In the perturbation method, a scattering event is regarded as a transition from one state in the continuum of positive energy states to another. Hence, the mathematics used in calculating transition probabilities, including Fermi's Golden Rule, can be applied directly. The Green's function method is equivalent to using an infinite perturbation expansion. It can be applied to both elastic and inelastic scattering, and multiple scattering events to any order can be included. The Green's function is regarded as a propagator which carries the particle from one scattering center to another. The Born approximation; which is valid when the particle energy is large with respect to the interaction energy, is derived and applied to simple potentials.

SUGGESTED REFERENCES

David Bohm, *Quantum Theory*. Prentice-Hall, Inc., New York, 1951.

R. H. Dicke and J. P. Wittke, *Introduction to Quantum Mechanics*. Addison-Wesley Publishing Co., Inc., Reading, Mass., 1960.

E. Merzbacher, *Quantum Mechanics*. John Wiley and Sons, Inc., New York, 1961.

Albert Messiah, *Quantum Mechanics*. North-Holland Publishing Co., Amsterdam, 1958.

D. Park, *Introduction to the Quantum Theory* McGraw-Hill Book Co., Inc., New York, 1964.

J. L. Powell and B. Crasemann, *Quantum Mechanics*. Addison-Wesley Publishing Co., Inc., Reading, Mass., 1961.

D. S. Saxon, *Elementary Quantum Mechanics*. Holden-Day, Inc., San Francisco, 1968.

Leonard I. Schiff, *Quantum Mechanics*, 3rd ed. McGraw-Hill Book Co., New York, 1969.

EPILOGUE

It is difficult to terminate a book at this point because there are so many additional topics that one would like to include. There is no natural end to the subject, since quantum mechanics is continually being applied either to new phenomena or to old phenomena in new ways.

My hope is that those readers who have come this far will have gained new insight into the physical behavior of our universe on the microscopic level, as well as some degree of mastery of the mathematical techniques necessary for the solution of problems hitherto undreamed of. It is my further hope that the reader will proceed either to apply quantum mechanics to specialized problems in solids, nuclei, biological systems, and so forth, or to move forward into the realms of relativistic quantum mechanics and quantum field theory.

AUTHOR INDEX

SUBJECT INDEX ━━━━

(Italics refer to illustration page numbers.)